POPULATION AND COMMUNITY ECOLOGY FOR INSECT MANAGEMENT AND CONSERVATION

PROCEEDINGS OF THE ECOLOGY AND POPULATION DYNAMICS SECTION OF THE 20TH INTERNATIONAL CONGRESS OF ENTOMOLOGY FLORENCE/ITALY/25-31 AUGUST 1996

Population and Community Ecology for Insect Management and Conservation

Edited by

Johann Baumgärtner
International Centre of Insect Physiology and Ecology, Nairobi, Kenya

Pietro Brandmayr
Università della Calabria, Cosenza, Italy

Bryan F.J. Manly
University of Otago, Dunedin, New Zealand

A.A. BALKEMA/ROTTERDAM/BROOKFIELD/1998

Photo cover supplied by Dr Colleen Jamieson.
The magpie moth *Nyctemera annulata* a daytime flying New Zealand native.

The texts of the various papers in this volume were set individually by typists under the supervision of each of the authors concerned.

Published by
A.A.Balkema, P.O.Box 1675, 3000 BR Rotterdam, Netherlands
(Fax: +31.10.413.5947)
A.A.Balkema Publishers, Old Post Road, Brookfield, VT 05036-9704, USA
(Fax: 802.276.3837)

ISBN 90 5410 930 0

Printed in the Netherlands

Population and Community Ecology for Insect Management and Conservation, Baumgärtner et al. (eds)
© 1998 Balkema, Rotterdam, ISBN 90 5410 930 0

Table of contents

Population and Community Ecology for Insect Management and Conservation, Baumgärtner et al. (eds)
© 1998 Balkema, Rotterdam, ISBN 90 5410 930 0

Preface

One of the major themes of the 20th International Congress of Entomology which was held in Florence, Italy, from 25-31 August 1996 was Ecology and Population Dynamics, with six sessions on this topic. During the Congress it was agreed that a selection of the papers would be published, and we are pleased to present them in this book.

The papers are grouped into three categories: Single species dynamics, Population interactions, and Community ecology. Within these categories a great variety of topics are discussed, and we believe that taken as a whole these proceedings provide a valuable guide to the latest research in these areas, with some emphasis on quantitative methods. As such these papers will be of interest to mathematicians and statisticians working in population modelling and estimation as well as population ecologists in general and entomologists.

To ensure the quality of the papers, authors were asked to act as referees for each other. We thank them for the efforts made and believe that it has helped to improve the final product. We also thank the individual authors who persevered with us when we requested formatting changes so as to ensure consistency as far as possible. Finally, special thanks are due to Dr Colleen Jamieson of Dunedin who agreed at short notice to review one of the papers, although she is not herself an author, and to Irene Goodwin who made a number of final format changes to the manuscript before it was sent to the publisher.

1 Single species dynamics

Population and Community Ecology for Insect Management and Conservation, Baumgärtner et al. (eds)

Sampling and modelling of insect populations

Bryan F.J. Manly
University of Otago, Dunedin, New Zealand

ABSTRACT: This paper gives a general review of various topics that are discussed in some of the papers that follow, under the general theme of the sampling and modelling of insect populations. The topics covered includes models for the development of age-structured or stage-structured populations, various questions related to studying the abundance of a population or several populations over a number of generations, the dispersion of a population in space, and the measurement of the correlation of spatial patterns, and testing this for significance. An attempt is made to put recent developments into an historic framework.

1 GENERAL THEMES

There are four general themes that are discussed by the Invited Speakers for the Symposium on Sampling and Modelling:

- models for the development of age-structured or stage-structured populations (Di Cola and Baumgärtner, Worner);
- sampling and modelling questions related to studying the abundance of a population or several populations over a number of generations (McArdle);
- modelling of the dispersion of a population in space (Sharov); and
- the measurement of the correlation of spatial patterns, and testing this for significance (Liebhold and Sharov, Perry).

Alternatively, these can be thought of as two topics concerning populations in time, and two topics concerning populations in space. All the contributions are characterised as having a quantitative emphasis, and being quite mathematical in their approaches.

In this introductory paper I have chosen to look at the themes one by one, with something of an historical perspective, and to fit the papers as best I can within the framework that this produces. Several of these topics are covered in more detail in a longer review that I have produced recently (Manly, 1994a).

2 MODELLING AGE-STRUCTURED AND STAGE-STRUCTURED POPULATIONS

There are three different analytical approaches that are used to model reproducing populations. The first of these approaches is based on the integral equation model of Sharpe and Lotka (1911) and Lotka (1939) which relates the number of births at time t to the number of individuals born at earlier times, the survival rates to different ages, and the reproduction rates at different ages. The second approach uses continuous time differential equations for either age-structured or stage-structured populations, as exemplified for example by the work of Metz et al. (1988) and Wood and Nisbet (1991), building on the equations of McKendrick (1926) and von Foerster (1959). The third approach uses discrete time matrix models for age-structured or stage-structured populations, as described for example by Caswell (1989) and Manly (1990, Chapter 6), building on the work of Bernardelli (1941), Lewis (1942), and Leslie (1945, 1948).

As noted by Caswell (1989, p. 24) various arguments have been used in favour of using one of these approaches for modelling instead of the others but the choice is really a matter of personal taste. However, matrix models are relatively easy to adapt to changed assumptions, are well suited for numerical calculations, and they reflect the true nature of many insect life cycles with discrete stages. They will therefore often be the most appropriate of the classical analytical methods in entomological applications.

The basis of the Bernadelli-Lewis-Leslie model and its extension for stage-structured populations is as follows. Let $n(x,t)$ denote the number of females in the age group x at time t, $p(x)$ denote the probability that a female in the age group x at time t will survive to be in the age group $x+1$ at time $t+1$, and $B(x)$ denote the average number of female offspring born to females aged from x to $x+1$ in a unit period of time that survive to the end of that period. The number of females in age group x at time $t+1$ will then be the sum of the offspring from females of different ages, so that

$$n(0,t+1) = B(0)n(0,t) + B(1)n(1,t) + \ldots + B(k)n(k,t),$$

where $k+1$ is the maximum possible age. It also follows from the definitions that

$$n(x+1,t+1) = p(x)n(x,t),$$

for $x = 0, 1, \ldots, k-1$. These equations can be written together as the matrix equation

$$\begin{bmatrix} \text{n(0,t+1)} \\ \text{n(1,t+1)} \\ \vdots \\ \text{n(k,t+1)} \end{bmatrix} = \begin{bmatrix} \text{B(0)} & \text{B(1)} & \ldots & \text{B(k-1)} & \text{B(k)} \\ \text{p(0)} & 0 & \ldots & & 0 \\ \vdots & & & & \\ 0 & 0 & \ldots & \text{p(k-1)} & 0 \end{bmatrix} \begin{bmatrix} \text{n(0,t)} \\ \text{n(1,t)} \\ \vdots \\ \text{n(k,t)} \end{bmatrix}$$

or

$$\boldsymbol{N}_{t+1} = \boldsymbol{M}\boldsymbol{N}_t.$$

It then follows that

$$\boldsymbol{N}_t = \boldsymbol{M}^t\boldsymbol{N}_0. \tag{1}$$

The matrix $\mathbf{M}$, whose elements are the fecundity rates $B(x)$ and the survival probabilities $p(x)$, is often called the Leslie matrix. The last equation shows that the numbers in different age groups at an arbitrary time t are determined by the numbers in the age groups at time zero ($\mathbf{N}_0$) and the Leslie matrix raised to the power t. Subject to certain mild assumptions it is possible to show that a population following this model will eventually reach a stable distribution for the relative numbers of individuals with different ages, and be growing or declining at a constant rate, with the long term behaviour of the population determined by the dominant eigenvalue of the Leslie matrix.

Lefkovitch (1963,1964a,1964b,1965) modified the Bernardelli-Leslie-Lewis model to allow a population to be grouped by life stages rather than by age by allowing the number in stage j at time $t+1$ to depend on the numbers in all previous stages at time t. Thus if $f_j(t)$ is the number of individuals in stage j at time t then for q stages his model is, in matrix notation,

$$\begin{bmatrix} f_1(t+1) \\ f_2(t+1) \\ \vdots \\ f_q(t+1) \end{bmatrix} = \begin{bmatrix} m_{11} & m_{12} & \cdots & m_{1q} \\ m_{21} & m_{22} & & m_{2q} \\ & & & \\ m_{q1} & m_{q2} & \cdots & m_{qq} \end{bmatrix} \begin{bmatrix} f_1(t) \\ f_2(t) \\ \vdots \\ f_q(t) \end{bmatrix}$$

or

$$\boldsymbol{F}_{t+1} = \boldsymbol{M}\boldsymbol{F}_t,$$

so that

$$\boldsymbol{F}_t = \boldsymbol{M}^t\boldsymbol{F}_0. \tag{2}$$

The typical entry in the matrix $\boldsymbol{M}$ in equation (2), m_{ij}, reflects how the number in stage i at time $t+1$ depends on the number in stage j at time t. Equation (2) is similar to equation (1) but the matrix M of the latter equation does not have the simple structure of a Leslie matrix, with its many necessarily zero elements. There is an implicit assumption with Lefkovitch's model that the age distribution within stages is constant enough to make any variation in the m_{ij} values with time unimportant.

Lefkovitch's model is not as straightforward as the Bernardelli-Leslie-Lewis model to study from a theoretical point of view. Nevertheless the long term behaviour of the population will be determined by the eigenvalue of $\mathbf{M}$ with maximum modulus, with its corresponding eigenvector. To be precise, let T_1 be the eigenvalue of the matrix $\mathbf{M}$ with the largest absolute value, with corresponding eigenvector v_1. Then the proportions in different stages in the population should tend towards the proportions in the vector v_1, and the numbers in each stage should increase by the factor T_1 per generation. If the dominant eigenvalue of the matrix $\mathbf{M}$ is complex then cyclic population changes are implied. In this case it may be that the real situation is that the m_{ij} values were not constant over the sampling period. A negative dominant eigenvalue is biologically meaningless.

The main problem with using Lefkovitch's model for data analysis is the large number of coefficients to be estimated in the transition matrix. However, this can be overcome to some extent if the time between samples is small so that the possibility of an individual developing through more than one stage in this time can be discounted. A model based upon this assumption was developed by Usher (1966, 1969) in the context of the management of a forest. He assumed that equation (2) can be simplified to

$$\begin{bmatrix} f_1(t+1) \\ f_2(t+1) \\ f_3(t+1) \\ \vdots \\ f_{q-1}(t+1) \\ f_q(t+1) \end{bmatrix} = \begin{bmatrix} B_1 & B_2 & B_3 & \ldots & B_{q-1} & B_q \\ b_1 & a_2 & 0 & \ldots & 0 & 0 \\ 0 & b_2 & a_3 & \ldots & 0 & 0 \\ \vdots & & & & & \\ 0 & 0 & 0 & \ldots & a_{q-1} & 0 \\ 0 & 0 & 0 & \ldots & b_{q-1} & a_q \end{bmatrix} \begin{bmatrix} f_1(t) \\ f_2(t) \\ f_3(t) \\ \vdots \\ f_{q-1}(t) \\ f_q(t) \end{bmatrix}$$

Here B_j is the contribution to the number in stage 1 at time $t+1$ that comes from those in stage j at time t. For $j = 1$ this contribution comes from those that remain alive but do not develop to stage 2, and also possibly from the reproduction of stage 1 individuals. For $j > 1$ the contribution is from reproduction only. Also, a_j is the probability that an individual in stage i at time t is still in stage j at time $t+1$, while b_j is the probability that an individual in stage j at time t moves to stage $j+1$ by time $t+1$. The sum $a_j + b_j$ gives the survival rate between two sample times for an individual in stage j at the first of these times.

As for Lefkovitch's model, the dominant eigenvalue of the transition matrix indicates the long term behaviour of the population and the corresponding eigenvector indicates the stable proportions in different stages. If the dominant eigenvalue is T, which is real, then the population size will be multiplied by T each time unit once these stable proportions have been reached.

Caswell (1989) and Manly (1990, Chapter 6) discuss further modifications and generalizations of matrix models for population dynamics, including the introduction of stochastic variation in the development process. They also review methods for the estimation of the elements of transition matrices, and the use of bootstrap and simulation methods for assessing the accuracy of estimates. Recent developments of these and other modelling methods for stage-structured populations include Söndgerath and Richter's (1990) extension of the Bernardelli-Leslie-Lewis model, Wood's (1993, 1994) use of spline functions,

Escos et al.'s (1994) model with density dependent fecundity, the papers of Costantino et al. (1995) and Shimada and Tuda (1996) on comparisons between the behaviour of models and laboratory populations of beetles as a function of demographic parameters, and Caswell's (1996) explanation of the uses for second derivatives of the population growth rate.

In addition to these various approaches for modelling the dynamics of reproducing populations there are a host of methods available for the development of one or several cohorts of a population through a series of development stages until the eventual death of all the individuals (Manly, 1990, Chapters 4 and 5). The models are relatively simple because reproduction does not have to be taken into account.

The modelling of the population dynamics of age-structured populations using differential equation and matrix models is the subject of Di Cola and Baumgärtner's contribution to these *Proceedings.* They find that in many cases the quality of model predictions are improved if physiological age is used in place of chronological age, to reflect the effects of temperature dependent factors.

Worner's contribution to the *Proceedings* is also concerned with the effects of temperature dependent factors, but more specifically with the problem of modelling data in order to predict the timing of important events in the life cycle of an insect population for control purposes. She notes the well known difficulties associated with constructing models that allow for the influence of varying temperature on development rates (Worner, 1992; Liu et al., 1995) and discusses an alternative approach that involves the use of long term insect, climate and host plant data.

3 TIME SERIES OF POPULATION ABUNDANCE

The modelling of time series of the abundance of insect and other populations over successive generations has been of continuing interest for many years. Three questions that have been of particularly interest are:

- To what extent are populations regulated by density dependent processes?
- To what extent is variation due to deterministic chaos as distinct from stochastic variation?
- How are extinction rates of populations related to their temporal variability?

I will briefly comment on each of these in turn.

3.1 *Population Regulation by Density Dependence*

The extent to which natural populations are regulated by density dependent processes has been controversial. Nicholson (1933) and others (e.g., Hassell et al., 1989; Berryman, 1991; Murdoch, 1994; Agrell et al., 1995) have argued that populations must exist in states of balance that are regulated by density dependent processes because otherwise they would not remain in existence for as long as they do. But Andrewartha and Birch (1954) and others (e.g., Den Boer, 1991; Wolda, 1989, 1991) have argued that this idea is not necessarily true, and is not supported by the available data when analyzed in an appropriate manner. More recently, Wolda (1995) has suggested that the whole debate has been fruitless and should be discontinued in favour of studies of the relative importance of regulatory and disruptive processes on populations.

The statistical problem at the heart of this argument is about how to determine whether the changes in a time series are a function of earlier values in the series. To be precise, suppose that $N_1, N_2, \ldots, N_G$ are the sizes for G generations of a population. Then one matter of interest is whether the changes $N_{i+1} - N_i$ are negatively related to N_i, so that a large population size tends to be followed by a negative change and a small population size tends to be followed by positive change. Because overlapping generations confuse the situation, many studies have concentrated on insect populations where the individuals emerge at the start of one generation, produce the eggs for the next generation, and then die. A complicating factor is that in practice population sizes can only be estimated for natural populations and sampling errors in the estimates may lead to apparent density dependent changes.

A variety of tests for density dependence have been suggested in this situation (Bulmer, 1975; Pollard et al., 1987; Reddingius and Den Boer, 1989; Den Boer and Reddingius, 1989; Den Boer, 1990; Crowley, 1992; Crowley and Johnson, 1992; Dennis and Taper, 1994) and questions about the relative merits of the different tests have also caused a debate (Holyoak, 1993a,b; Holyoak and Crowley, 1993; Holyoak and Lawton, 1993; Hanski et al., 1993; Wolda and Dennis, 1993; Wolda et al., 1994; Fox and Ridsdill-Smith, 1995). Several of the proposed tests have been based on the computer intensive methods of randomization and bootstrapping. These will now be briefly described, although it should be noted that Fox and Ridsdill-Smith (1995) have argued that the benefits of such tests do not justify the heavy computations that are needed, when they are compared with Bulmer's (1975) first test, which is much easier to carry out.

Pollard et al.'s (1987) test involves three potential models. For the first model it is assumed that $N_{i+1} = N_i \exp(e_i)$, where e_i is a random value from a normal distribution with mean zero. Taking logarithms then gives the linear equation

$$x_{i+1} = x_i + e_i, \tag{3}$$

where $x_i = \log_e(N_i)$. This is a random walk model for which the variance of x_i increases indefinitely with i, although the mean is constant. For the second model it is assumed that $N_{i+1} = \exp(r) N_i \exp(e_i)$, so that

$$x_{i+1} = r + x_i + e_i. \tag{4}$$

This is an unregulated model where the variance of x_i increases with i and a trend in the mean is also allowed. For the third model $N_{i+1} = \exp(r) N_i^{\beta} exp(e_i)$, so that

$$x_{i+1} = r + \beta x_i + e_i. \tag{5}$$

This is a first order autoregressive series which is stationary providing that $|\beta| < 1$, so that the population is regulated.

To compare models (3) and (4), Pollard et al. proposed that the statistic $(x_n - x_1)/\sqrt{\{\sum(x_{i+1} - x_i)^2\}}$ should be tested against the t-distribution with $n - 1$ degrees of freedom, where the summation is from 2 to n. This was criticized by Link and Hoover (1989) who show that it is better to compare a modified statistic with the t-distribution with $n - 2$ degrees of freedom. However, the more interesting test in the context of the arguments for and against density dependent regulation of a population is between models (4) and (5). For this comparison, Pollard et al. propose comparing the observed correlation between x_i and $d_i = x_{i+1} - x_i$ with the distribution obtained for this statistic when new time series are constructed on a computer by starting from x_1 and then adding on the observed values of d_i in a random order.

If $\beta = 0$ in equation (5) then $x_{i+1} = r + e_i$. Thus the population size is a random value from a constant distribution with mean r. Such a population is regulated. In fact, it is regulated to the maximum possible extent, but this regulation is not density dependent. This has been discussed to a surprising extent lately (Wolda and Dennis, 1993; Holyoak and Lawton, 1993; Hanski et al., 1993; Wolda et al. 1994).

The Pollard et al. test has the advantage of not being affected by a trend in population sizes. However, it is worrying that sampling errors in populations estimates will have the same effect as density dependence. This is obvious when it is considered that a large over-estimate in x_i will almost certainly be followed by an under-estimate of $d_i = x_{i+1} - x_i$. Therefore if sampling errors are not taken into account it is not possible to know whether a significant negative correlation between d_i and x_i is due to density dependent population changes rather than sampling errors.

To overcome the effects of sampling errors in population estimates Crowley (1992) developed what he called the random walk attraction test. This compares the extent to which a time series moves towards a central range of attraction with what is obtained from generating a large number of random walks with observation errors. He also devised another similar test based on randomizing the changes in a time series, which he called the randomized attraction test. These tests are interesting because they are not based on any particular parametric model. However they have been found to have relatively low power when compared to the alternatives (Holyoak and Crowley, 1993).

In order to overcome the limitations that they perceived with other tests, Dennis and Taper (1994) proposed the use of a discrete time stochastic logistic model which they suggested is suitable for a wide variety of situations. With this model

$$N_{i+1} = N_i \exp(a + bN_i + cZ_i), \tag{6}$$

where a and b are constants, c is a positive constant, and Z_i is a random value from the standard normal distribution. They consider in particular three special cases of the model. First, setting $a = b = 0$ gives

$$N_{i+1} = N_i \exp(cZ_i),$$

or, taking logarithms to base e,

$$x_{i+1} = x_i + cZ_i,, \tag{7}$$

which is a random walk (model 0). Second, setting $b = 0$ gives

$$x_{i+1} = x_i + a + cZ_i,, \tag{8}$$

which is a random walk with trend (model 1). Finally, if a and b are non-zero then

$$x_{i+1} = x_i + a + b\{\exp(x_i)\} + cZ_i,, \tag{9}$$

so that density dependence is present (model 2).

Estimates of the parameters for the three models can be obtained by linear regression. However, the usual assumptions for linear regression do not hold for model 2. Dennis and Taper therefore suggest obtaining confidence limits for the parameters of this model using jackknifing or bootstrapping. They also propose a parametric bootstrap likelihood ratio (PBLR) test for comparing the goodness of fit of models 1 and 2 for a set of data. The use of the test on adult rangeland grasshopper densities is described by Kemp and Dennis (1993).

The tests described so far here have been for direct density dependence of the population size in one generation on the size in the previous generation. It should be noted that testing for delayed density dependence has also been attracting increased interest lately (e.g. see Holyoak, 1994a,b; Agrell et al., 1995; Williams and Liebhold, 1995).

3.2 *Deterministic Chaos and Stochastic Variation*

While the interest in modelling and detecting density dependence has been continuing among some population ecologists there has been a parallel interest in the concept that much of the apparently random variation in population sizes from generation to generation may in fact be chaos that is produced by deterministic processes. This is an interesting idea. Certainly the underlying relationships in biological systems are likely to be complicated and non-linear. Therefore chaotic dynamics seems plausible although it also seems likely that the impact of random disturbances may be quite effective in hiding these relationships.

The important property of chaotic systems is that they are highly sensitive to initial conditions. Therefore, if a deterministic population has this property then a slight change to the composition at one point in time will develop into large changes quite quickly. In effect, an important difference between a deterministic system that is chaotic and a deterministic system that is not chaotic but is effected by stochastic variation is that the behaviour of the chaotic system can be predicted quite well for a short period in the future but is not predictable in the long term. The predictability of stochastic systems always depends on the size of the random effects, but this may not change much with time.

No attempt will be made here to discuss the large and increasing literature on the theory of chaos as it relates to population sizes. Reviews are provided by Hastings et al. (1993), Ellner and Turchin (1995), Guckenheimer (1995), and Stone and Ezrati (1996). The second of these is particularly interesting because it starts with the assumption, which is surely correct, that population sizes are determined by a mixture of deterministic and stochastic effects. For recent developments and examples see Falck et al. (1995) and Turchin (1995).

3.3 *Extinction Rates and Temporal Variability*

Schoener and Spiller (1992) report the results of an investigation that they made into the relationship between extinction rates and temporal variability in population size using data for 174 populations of seven species of orb spiders on 106 subtropical islands. They found that the conclusions from their data analysis depended very much on whether or not they included zero population sizes (which might be real or due to sampling errors), and how they transformed the data. Their final comment was 'Clearly, a substantial amount of research, both statistical and biological, is needed to develop this inchobate but very important area of population biology'.

The problems involved in measuring the temporal variation in populations for these and other purposes have been reviewed in detail by Gaston and McArdle (1994). Another recent contribution to the debate is Stewart-Oaten et al.'s (1995) proposal for a new method for adjusting estimates of temporal variability to allow for the spatial variation in a population. In his paper in these Proceedings, McArdle returns again to the topic of the study of the relationship between temporal variation and extinction, and the difficulties that this entails, but in terms of modelling the time trends in population abundance at several sites simultaneously.

4 DISPERSING POPULATIONS

Early studies of the dispersion of animal populations took the random walk as a basic model, where this implies that if an organism starts at the point (x_0, y_0) at time 0, then the position (X_t, Y_t) at some time $t > 0$ will have a bivariate normal distribution with mean (x_0, y_0),$\mathrm{Var}(X_t) = \mathrm{Var}(Y_t) = \sum^2 t$, and $\mathrm{Cov}(X_t, Y_t) = 0$ (Dobzhansky and Wright, 1943; Skellam, 1951, 1973; Okuba, 1980; Kareiva, 1983; Manly and Chatterjee, 1992; Matsuda and Akamine, 1994). In reality, however, the comparison of field data with this model makes it clear that the random walk is often not altogether realistic for various reasons, including the need to allow for geographical constraints and sampling intensities in different areas (Barraclough, 1978; Smith et al., 1990; Porter and Dooley, 1993), and the fact that organisms often tend to move in one direction and are sometimes attracted or repelled from a starting point, objects in the environment, or each other (Manly, 1977; Taylor, 1981a,b; Kareiva and Shigesada, 1983; Thórarinsson, 1986; Bovett and Benhamou, 1988; Marsh and Jones, 1988; Othmer et al., 1988; Routledge, 1990; Holmes, 1993; Brookes and Butlin, 1994; Kleiber and Hampton, 1994; Crist and Wiens, 1995; Preisler and Akers, 1995; Worton 1995).

The study of the dispersion of animal populations takes on a particular interest when the animal is a pest that is reproducing and expanding in an area that was previously unoccupied. In this case the modelling of the rate of spread involves taking into account the reproduction and survival of the members of the population as well as their mode of movement (Skellam, 1951, 1973; Hegeveld, 1989; Van den Bosch et al., 1992; Holmes, 1993; Shaw, 1995). Sharov's contribution to the *Proceedings* is in this area. He considers the gypsy moth (*Lymantria dispar*), which was introduced into North America near Boston in 1869 and has subsequently spread west and south. Sharov models the spread of this pest in terms of the establishment of isolated colonies beyond a moving population front and uses this model to estimate the rate of spread of the population, and to evaluate the effectiveness of barrier zones set up by the United States Department of Agriculture. His method for the estimation of the extent of the gypsy moth population at any point in time may be of particular interest here.

5 CORRELATION IN SPACE

The final topic that I will discuss is the assessment of the correlation between two spatial distributions. In this situation there are a number of related problems that can be considered, as exemplified by the following three examples:

(a) Table 1 shows the counts of newly emergent and one year old bramble canes (*Rubus fruticosus*) in 0.5 by 0.5m quadrats covering a 4.5 by 9m area of ground as derived from a study conducted by Hutchinson (1979). Here an interesting question concerns how to measure the association between the two distributions of quadrat counts for the two categories of bramble, and how to test whether this association is significantly different from what would be expected by chance alone if in fact the distributions are unrelated. In doing this it must be accepted that one or both of the two distributions may be non-random.

(b) Figure 1 shows the locations of 45 nests of the ant *Messor wasmani* and 15 nests of the ant *Cataglyphis bicolor* in a 240 by 250 foot area (Harkness and Isham, 1983; Särkkä, 1993, Figure 5.8). *Messor* (species 1) collects seeds for food while *Cataglyphis* (species 2) eats dead insects, mostly *Messor* ants. We may ask whether there is any evidence of a relationship between the two distributions, given that one or both of the two species may be non-randomly distributed. This is related to the situation for example (a) in the sense that it would be possible to divide the ant study area into quadrats and compare the two species by quadrat counts in the same way as for the brambles.

Table 1: Counts of two types of bramble in 0.5 by 0.5 metre quadrats covering an area of 4.5 by 9 metres. The position of the counts in the table matches the position in the study area for both types of bramble so that the counts match up. For example, the quadrat at the bottom on the right-hand side had a count of three for newly emergent brambles and a count of two for one year old brambles.

(a) Newly emergent brambles

0	0	0	1	1	0	0	1	2	1	0	0	0	0	0	0	0	0
0	0	1	4	1	0	1	2	3	1	1	0	0	0	0	0	0	0
1	0	2	1	0	0	0	0	4	2	4	3	0	0	0	0	1	4
0	2	2	0	0	0	1	0	1	3	3	3	0	0	0	1	1	1
2	2	0	1	1	0	0	0	1	1	5	2	1	0	0	1	1	1
0	0	2	1	2	4	0	0	0	2	1	2	1	0	0	2	0	4
1	2	2	3	2	0	2	0	0	0	0	6	0	0	0	0	3	2
1	3	4	1	5	3	5	1	0	0	0	0	1	0	0	0	1	2
0	0	0	2	5	1	2	3	1	2	0	0	0	0	1	2	1	3

(b) One year old brambles

0	1	1	1	3	0	0	0	1	1	0	0	0	0	0	0	0	0
0	0	2	5	1	0	1	1	1	4	3	0	0	0	0	0	1	3
1	1	2	0	0	0	0	0	3	0	3	3	0	0	0	0	7	2
0	2	2	0	0	0	0	0	1	2	1	4	0	0	0	0	3	1
1	1	3	3	1	0	0	0	0	1	4	2	3	0	0	1	1	1
0	1	1	2	0	1	1	0	0	4	0	1	1	0	0	2	5	2
1	4	2	4	1	0	1	1	0	3	0	9	2	0	0	0	2	4
2	4	3	0	2	3	1	1	0	0	0	1	2	0	0	3	2	2
0	0	1	1	4	2	0	2	1	1	0	0	1	1	1	6	1	2

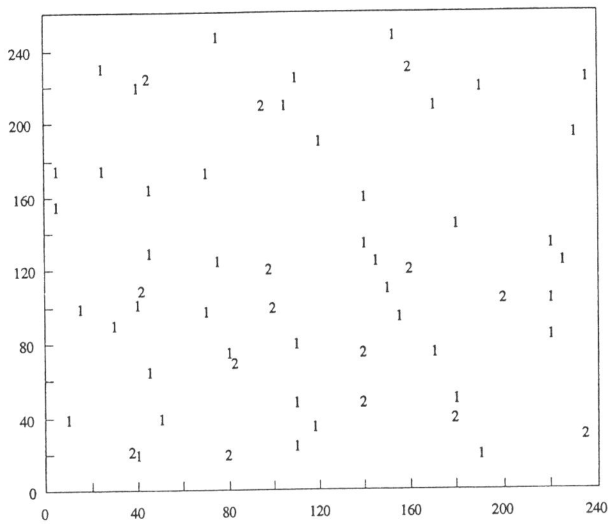

Figure 1: Location of 45 nests of *Messor wasmanni* (species 1) and 15 nests of *Cataglyphis bicolor* (species 2) in a 240ft by 250ft study area.

(c) A Norwegian research programme was started in 1972 in response to widespread concern in Scandinavian countries about the effects of acid precipitation (Overrein et al., 1980; Mohn and Volden, 1985). As part of this study, regional surveys of small lakes in Norway were carried out from 1974 to 1978, with some extra sampling done in 1981. Data were recorded for pH, sulphate (SO_4) concentration and certain other variables at each sampled lake. Figure 2 shows the the pH values and the SO_4 values in 1976 for 46 lakes, plotted against the locations of the lakes that they relate to. Here a question that might be asked is whether there is a significant relationship between the pH and sulphate measurements, taking into account any patterns that exist in the spatial distributions for each of the measurements considered individually. Note that only some of the lakes in the study region were sampled.

All of these examples are concerned with spatial association. However, the conditions are a little different in each case. For (a) the question is whether the quadrat counts of two types of bramble are correlated, for (b) the question is whether the positions of two types of ants nests are related in some way that is undefined as yet, and for (c) the question is whether the values of two continuous variables that are sampled over space are correlated.

An appropriate analysis for assessing the correlation between two sets of quadrat counts is not easy to find. Besag and Diggle (1977) and Besag (1978) proposed a randomization test that is a generalization of Mead's (1974) test for randomness in the spatial distribution of a single species that can be used when the number of quadrats is a multiple of 16. For each set of 16 contiguous quadrats the Spearman correlation coefficient is calculated for the counts of the two classes of object being considered. The mean of these correlation coefficients is then the test statistic, and this is compared with the distribution of such statistics that is obtained by a process of randomly permuting the quadrat counts for one of the types of object within contiguous sets of four quadrats. This procedure is supposed to maintain the spatial pattern for the individual types of object, but it can be argued that this is not the case, and that the use of the test is questionable unless the blocks of four quadrats that are randomized are separated far enough to make them effectively independent (Manly, 1991, p. 150).

In his contribution to the Symposium Perry suggests a different approach to handling the problem of testing for association between quadrat counts. This is based on an algorithm for producing permutations of quadrat counts with a fixed level of aggregation and a fixed

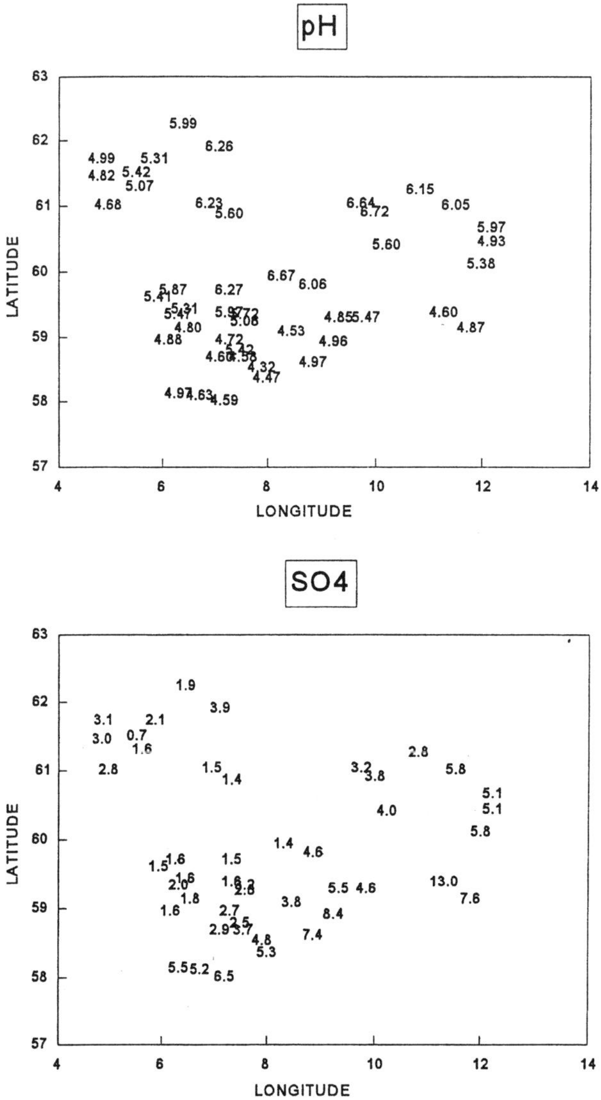

Figure 2: Values for pH and SO_4 concentrations (milligrams per litre) of lakes in Norway plotted against the latitude and longitude of the lakes.

centroid over the study region (Perry, 1996a), combined with a test statistic based on the minimum amount of movement between quadrats that is required in order to produce counts that are all equal Perry (1996b). The idea is to keep the counts of species 1 fixed for the quadrats and to produce a large number of alternative sets of data for which the quadrat counts for species 2 are permutations of the original ones. A randomization test can then be carried out to see whether the test statistic for the observed data is unusual when compared to the distribution obtained if the species have independent distributions.

With the comparison of the positions of the nests for two species of ants (Figure 1) it is possible to devise a simpler type of randomization test. In this context Lotwick and Silverman (1982) suggested that a point pattern in a rectangular region can be converted to a pattern over a larger area by simply copying the pattern from the original region to similar sized regions above, below, to the left and to the right, and then copying the copies as far away from the original region as required. A test for independence between two patterns then involves comparing the test statistic observed for the points over the original region with the distribution of this statistic that is obtained when the rectangular 'window' for the species 1 positions is randomly shifted over an enlarged region for the species 2 positions. Harkness and Isham (1983) used this type of analysis with the ant data and concluded that there is evidence of a relationship between the positions of the two types of nest. See also Andersen's (1992) review of these types of analysis in ecology.

As Lotwick and Silverman note, the need to reproduce one of the point patterns over the edge of the region studied in an artificial way is an unfortunate aspect of this procedure. It can be avoided by taking the rectangular window for the type 1 points to be smaller than the total area covered and calculating a test statistic over this smaller area. The distribution of the test statistic can then be determined by randomly placing this small window within the larger area a large number of times. In this case the positions of any type 1 points outside the small window are ignored and the choice of the positioning of the small window within the larger region is arbitrary.

Another idea involves considering a circular region and arguing that if two point patterns within the region are independent then this means that they have a random orientation with respect to each other. Therefore a distribution that can be used to assess a test statistic is the one that is obtained from randomly rotating one of the sets of points about the centre point of the region. A considerable merit with this idea is that the distribution can be determined as accurately as desired by rotating one of the sets of points about the centre of the study area from zero to 360 degrees in suitable small increments (Manly, 1991, Section 8.7). Reflections as well as rotations are also possible (Perry, 1996c).

The third example described above concerned the correlation between two water chemistry variables that have been measured on 46 lakes in Norway. There are several differences between this example and the previous ones. First, there were many more lakes in Norway that could have been studied. Therefore the data are a sample from the population of lakes in the study region. By contrast, the brambles data and the ant nest data represent the whole of the populations concerned. Second, the problem is not to compare two spatial distributions. Rather, the problem is to assess the correlation between two variables taking into account the fact that the values of these variables may tend to be similar for lakes that are close in space, for a variety of reasons. Thus the need is to study the correlation between the two variables with an allowance for the possible lack of independence between the results from different lakes.

One approach to analyzing data of this type has been through the use of matrix randomization tests (Mantel, 1967; Manly, 1994b, Section 5.6). To this end, two distance matrices can be set up. For the lake data the first matrix would have 46 rows and 46 columns, and the entry in row i and column j would be a measure of the distance between the pH values for lake i and lake j. For example, this could be the absolute value of the pH difference. The second matrix would be of the same size, and the ith row and j^{th} column would contain a measure of the SO_4 difference between lakes i and j, where again this might just be the absolute difference. A randomization test can then be used to see whether the corresponding elements in the two matrices show any association, e.g., whether lakes with small pH distances also tend to have small SO_4 differences. Here the randomization involves a permutation of the lake labels for one of the variables, which ensures that the spatial structure in the data is maintained for both of the individual variables. It was thought at one time that using matrix randomization tests allows completely for the spatial correlation in the variables being analysed if the distance between sample locations is included as a covariate in the analysis. However, there are complications involved that still need to be resolved (Manly, 1993).

Geostatistical type methods (Rossi et al., 1992) offer an interesting alternative to ones that are based on distance matrices. One such approach for assessing the correlation between two variables is the topic of Liebhold and Sharov's contribution to the Symposium. Sample variograms are first estimated for each of the variables using standard methods (Cressie, 1991). These variograms then quantify the amount of correlation that exists for sample points that are separated by different distances. Data are then simulated in such a way that the frequency distribution and the variogram are the same as those observed for the two variables being considered, but these two variables are uncorrelated. The significance of the correlation between the variables in the original data is then determined by comparison with the values from simulated data, which is a type of Monte Carlo test. The calculations that are required for this procedure are considerable, but Liebhold and Sharov raise the interesting possibility that tables can be constructed for critical values of the correlation coefficient.

At first sight the Liebhold and Sharov test for correlation between two continuous variables measured at different points over a region seems quite different from Perry's test for correlation between quadrat counts. However, in reality the principle being used is the same in both cases: a test statistic is calculated for the observed set of data, and this is compared with the distribution for the same statistic that is obtained by generating new sets of data that match the observed data in certain ways that are judged to be important. The key need with this type of procedure is to choose the right characteristics of the observed set of data to match with the generated sets of data whilst still allowing enough to depend on chance so that there is good power to detect effects of interest.

6 CONCLUDING REMARKS

The papers delivered at the Symposium demonstrate the considerable activity that is taking place in the development of the theory and methods for sampling and modelling of insect populations. Comparing the papers with those presented at a conference on the same topic that took place eight years ago (McDonald et al., 1989) it can be seen that the interest in the modelling of population size is continuing, as seems inevitable, but there is increased attention to movement and the spatial aspect of data. It seems likely that interest in spatial data will continue to grow in the future, presumably with more work that considers the modelling of variation in space and time simultaneously.

Another point worth noting is the increased use of the computer intensive techniques of bootstrapping, randomization and Monte Carlo simulation for determining confidence limits and carrying out tests of significance. To some extent this may reflect the reviewer's bias. But there is no doubt that these methods are receiving increased use in biology in general (Manly, 1991; Crowley, 1992).

REFERENCES

Agrell, J., S. Erlinge, J. Nelson, C. Nilsson, and I. Persson, 1995. Delayed density dependence in a small rodent population. *Proceedings of the Royal Society of London* B262: 65-70.

Andersen, M. 1992. Spatial analysis of two-species interactions. *Oecologia* 91: 134-40.

Andrewartha, H.G. and L.C. Birch, 1954. *The Distribution and Abundance of Animals.* Chicago: University of Chicago Press.

Barraclough, G.F. 1978. Sampling bias in dispersal studies based on finite area. *Bird Banding* 49: 333-41.

Bernardelli, H. 1941. Population waves. *Journal of the Burma Research Society* 31: 1-18.

Berryman, A.A. 1991. Stabilization or regulation: what it all means! *Oecologia* 86: 140-3.

Besag, J. 1978. Some methods of statistical analysis for spatial pattern. *Bulletin of the International Statistical Institute* 47: 77-92.

Besag, J. and P.J. Diggle, 1977. Simple Monte Carlo tests for spatial pattern. *Applied Statistics* 26: 327-33.

Bovet, P. and S. Benhamou, 1988. Spatial analysis of animals' movements using a correlated random walk model. *Journal of Theoretical Biology* 131: 419-33.

Brookes, M.I. and R.K. Butlin, 1994. Estimates of male dispersal in *Yponomeuta padellus* (Lepidoptera: Yponomeutidea), the small ermine moth, by means of simulation. *Heredity* 73: 207-14.

Bulmer, M.G. 1975. The statistical analysis of density dependence. *Biometrics* 31: 901-11.

Caswell, H. (1989). *Matrix Population Models: Construction, Analysis, and Interpretation.* Sunderland, Massachusetts: Sinauer Associates.

Caswell, H. 1996. Second derivatives of population growth rate: calculation and applications. *Ecology* 77: 870-9.

Constantino, R.F., J.M. Cushing, B. Dennis, and R.A. Desharnais, 1995. Experimentally induced transitions in the dynamic behaviour of insect populations. *Nature* 375: 227-30.

Cressie, N. 1991. *Statistics for Spatial Data.* New York: Wiley.

Crist, T.O. and J.A. Wiens, 1995. Individual movements and estimation of population size in darkling beetles (Coleopterra: Tenebrionidae). *Journal of Animal Ecology* 64: 733-46.

Crowley, P.H. 1992. Resampling methods for computation-intensive data analysis in ecology and evolution. *Annual Review of Ecology and Systematics* 23: 405-47.

Crowley, P.H. 1992. Density dependence, boundedness, and attraction: detecting stability in stochastic systems. *Oecologia* 90: 246-54.

Crowley, P.H. and D.M. Johnson, 1992. Variability and stability of a dragonfly assemblage. *Oecologia* 90: 260-9.

Den Boer, P.J. 1990. On the stabilization of animal numbers. Problems of testing. 3. What do we conclude from significance test results? *Oecologia* 83: 38-46.

Den Boer, P.J. 1991. Seeing the trees for the wood: random walk or bounded fluctuations of population size? *Oecologia* 86: 484-91.

Den Boer, P.J. and J. Reddingius, 1989. On the stabilization of animal numbers. Problems of testing. 2. Confrontation with data from the field. *Oecologia* 79: 143-9.

Dennis, B. and M.L. Taper, 1994. Density dependence in time series observations of natural populations: estimation and testing. *Ecological Monographs* 64: 205-24.

Dobzhansky, T. and S. Wright, 1943. Genetics of natural populations. X. Dispersion rates in *Drosophila pseudoobscura*. *Genetics* 28: 304-40.

Ellner, S. and P. Turchin, 1995. Chaos in a noisy world: new methods and evidence from time series analysis. *American Naturalist* 145: 343-75.

Escos, J., C.L. Alados, and J.M. Emlen, 1994. Application of the stage-projection model with density-dependent fecundity to the population dynamics of Spanish ibex. *Canadian Journal of Zoology* 72: 731-7.

Falck, W., O.N. Bjørnstad, and, N.C. Stenseth 1995. Bootstrap estimated uncertainty of the dominant Lyapunov exponent of Holarctic microtine rodents. *Proceedings of the Royal Society of London* B261: 159-65.

Fox, D.R. and J. Ridsdill-Smith, 1995. Tests for density dependence revisited. *Oecologia* 103: 435-43.

Gaston, K.J. and B.H. McArdle, 1994. The temporal variability of animal abundance: measures, methods and patterns. *Philosophical Transactions of the Royal Society of London* B345: 335-58.

Guckenheimer, J. 1995. Dynamical systems theory for ecologists: a brief overview. In *Ecological Time Series* (eds. T.M. Powell and J.H. Steele), pp. 54-69. New York: Chapman and Hall.

Hanski, I., I. Woiwod, and J. Perry, 1993. Density dependence, population persistence, and largely futile arguments. *Oecologia* 95: 595-8.

Harkness, R.D. and V. Isham, 1983. A bivariate spatial point pattern of ants' nests. *Applied Statistics* 32: 293-303.

Hassell, M.P., J. Latto, and R.M. May, 1989. Seeing the wood for the trees: detecting density dependence from existing life table studies. *Journal of Animal Ecology* 58: 883-92.

Hastings, A., C.L. Hom, S. Ellner, P. Turchin, and H.C.J. Godfray, 1993. Chaos in ecology: is mother nature a strange attractor? *Annual Review of Ecology and Systematics* 24: 1-33.

Hegeveld, R. 1989. *Dynamics of Biological Invasions.* London: Chapman and Hall.

Holmes, E.E. 1993. Are diffusion models too simple? A comparison with telegraph models of invasion. *American Naturalist* 142: 779-95.
Holyoak, M. 1993a. New insights into testing for density dependence. *Oecologia* 93: 435-44.
Holyoak, M. 1993b. The frequency of detection of density dependence in insect orders. *Ecological Entomology* 18: 339-47.
Holyoak, M. 1994a. Appropriate time scales for identifying lags in density dependent processes. *Journal of Animal Ecology* 63: 479-83.
Holyoak, M. 1994b. Identifying delayed density dependence in time series data. *Oikos* 70: 296-304.
Holyoak, M. and P.H. Crowley, 1993. Avoiding erroneously high levels of detection in combinations of semi-independent tests. *Oecologia* 95: 103-14.
Holyoak, M. and J.H. Lawton, 1993. Comments arising from a paper by Wolda and Dennis: using and interpreting the results of tests for density dependence. *Oecologia* 95: 592-4.
Hutchings, M.J. 1979. Standing crop and pattern in pure stands of *Merculialis perennis* and *Rubus fruticus* in mixed deciduous woodland. *Oikos* 31: 351-7.
Kareiva, P.M. 1983. Local movement in herbivorous insects: applying a passive diffusion model to mark-recapture field experiments. *Oecologia* 57: 322-7.
Kareiva, P.M. and N. Shigesada, 1983. Analyzing insect movement as a correlated random walk. *Oecologia* 56: 234-8.
Kemp, W.P. and B. Dennis, 1993. Density dependence in rangeland grasshoppers (Orthoptera: acrididae). *Oecologia* 96: 1-8.
Kleiber, P. and J. Hampton, 1994. Modelling effects of FADs and islands on movement of skipjack tuna (*Katsuwonus pelamis*): estimating parameters from tagging data. *Canadian Journal of Fisheries and Aquatic Sciences* 51: 2642-53.
Lefkovitch, L.P. 1963. Census studies on unrestricted populations of *Lasioderma serricorne* (F.) (Coleoptera: Anobiidae). *Journal of Animal Ecology* 32: 221-31.
Lefkovitch, L.P. 1964a. The growth of restricted populations of *Lasioderma serricorne* (F.) (Coleoptera: Anobiidae). *Bulletin of Entomological Research* 55: 87-96.
Lefkovitch, L.P. 1964b. Estimating the Malthusian parameter from census data. *Nature* 204: 810.
Lefkovitch, L.P. 1965. The study of population growth in organisms grouped by stages. *Biometrics* 21: 1-18.
Leslie, P.H. 1945. On the use of matrices in certain population mathematics. *Biometrika* 33: 182-212.
Leslie, P.H. 1948. Some further notes on the use of matrices in population mathematics. *Biometrika* 35: 213-45.
Lewis, E.G. 1942. On the generation and growth of a population. *Sankhya* 6: 93-6.
Link, W.A. and B.A. Hoover, 1991. The detection of density dependence from a series of annual censuses - a comment. *Ecology* 72: 1514-5.
Liu, S., G. Zhang, and J. Zhu, 1995. Influence of temperature variations on rate of development in insects: analysis of case studies from entomological literature. *Annals of the Entomological Society of America* 88: 107-19.
Lotka, A.J. 1939. A contribution to the theory of self-renewing aggregates, with special reference to industrial replacement. *Annals of Mathematical Statistics* 10: 1-25.
Lotwick, H.W. and B.W. Silverman, 1982. Methods for analysing spatial processes of several types of point. *Journal of the Royal Statistical Society* B44: 406-13.
Manly, B.F.J. 1977. A model for dispersion experiments. *Oecologia* 31: 119-30.
Manly, B.F.J. 1991. *Randomization and Monte Carlo Methods in Biology.* London: Chapman and Hall.
Manly, B.F.J. 1993. A review of computer intensive multivariate methods in ecology. In *Multivariate Environmental Statistics* (eds. G.P. Patil and C.R. Rao), 307-46. Amsterdam: Elsevier Science Publishers.
Manly, B.F.J. 1994a. Ecological statistics. In *Handbook of Statistics 12: Environmental*

Statistics (eds. G.P. Patil and C.R. Rao), 307-76. Amsterdam: Elsevier.
Manly, B.F.J. 1994b. *Multivariate Statistical Methods: A Primer*, 2nd Edit. London: Chapman and Hall.
Manly, B.F.J. and C. Chatterjee, 1992. A model for mark-recapture data allowing for animal movement. In *Marked Individuals in the Study of Bird Populations* (eds. J.D. Lebreton and P.M. North), 309-22. Basel: Birkhäuser Verlag.
Marsh, L.M. and R.E. Jones, 1988. The form and consequences of random walk models. *Journal of Theoretical Biology* 133: 113-31.
Matsuda, H. and T. Akamine, 1994. Simultaneous estimation of mortality and dispersal rates of an artificially released population. *Researches on Population Ecology* 36:73-8.
McDonald, L.L., B.F.J. Manly, J.A. Lockwood, and J.A. Logan, eds. 1989. *Estimation and Analysis of Insect Population.* Berlin: Springer-Verlag.
McKendrick, A.G. 1926. Applications of mathematics to medical problems. *Proceedings of the Edinburgh Mathematics Society* 44: 98-130.
Mead, R. 1974. A test for spatial pattern at several scales using data from a grid of contiguous quadrats. *Biometrics* 30: 295-307.
Metz, J.A.J. A.M., de Roos and F. van den Bosch, 1988. Population models incorporating physiological structure: a quick survey of the basic concepts and an application to size-structured population dynamics in waterfleas. In *Size-Structured Populations* (eds. B. Ebenman and L. Persson), 106-26. Berlin: Springer-Verlag.
Mohn, E. and R. Volden, 1985. Acid precipitation: effects on small lake chemistry. In *Data Analysis in Real Life Environment: Ins and Outs of Solving Problems* (eds. J.F. Marcotorchino, J.M. Proth and J. Janssen), 191-6. Amsterdam: Elsevier.
Murdoch, W.W. 1994. Population regulation in theory and practice. *Ecology* 75: 271-87.
Nicholson, A.J. 1933. The balance of animal populations. *Journal of Animal Ecology 2* (Supplement): 132-588.
Okubo, A. 1980. *Diffusion and Ecological Problems: Mathematical Models.* Berlin: Springer-Verlag.
Othmer, H.G., S.R. Dunbar, and W. Alt, 1988. Models of dispersal in biological systems. *Journal of Mathematical Biology* 26: 263-98.
Overrein, L.N., H.M. Seip, and A. Tollan, 1980. *Acid Precipitation - Effects on Forest and Fish: Final Report.* Norwegian Institute for Water Research, Oslo.
Perry, J.N. 1996a. Simulating spatial patterns of counts in agriculture and ecology. *Computers and Electronics in Agriculture* (in press).
Perry, J.N. 1996b. Measuring the spatial pattern of animal counts with indices of crowding and regularity. Submitted for publication.
Perry, J.N. 1996c. Spatial association for counts of two species. Submitted for publication.
Pollard, E., K.H. Lakhani, and P. Rothery, 1987. The detection of density dependence from a series of annual censuses. *Ecology* 68: 2046-55.
Porter, J.H. and J.L. Dooley, 1993. Animal dispersal patterns: a reassessment of simple mathematical models. *Ecology* 74: 2436-43.
Preisler, H.K. and R.P. Akers, 1995. Autoregressive-type models for the analysis of bark beetle tracks. *Biometrics* 51: 259-67.
Reddingius, J. and P.J. Den Boer, 1989. On the stabilization of animal numbers. Problems of testing. 1. Power estimates and estimation errors. *Oecologia* 78: 1-8.
Rossi, R.E., D.J. Mulla, A.G. Journel, and E.H. Franz, 1992. Geostatistical tools for modelling and interpreting ecological spatial data. *Ecological Monographs* 62: 277-314.
Routledge, R.D. 1990. Spatial patterns arising from plant dispersal as modelled by a correlated random walk. *Journal of Applied Probability* 27: 1-13.
Särkkä, A. 1993. Pseudo-likelihood approach for pair potential estimation of Gibbs processes. Jyväskylä *Studies in Computer Science, Economics and Statistics* 22. University of Jyväskylä, Finland.
Sharpe, F.R. and A.J. Lotka, 1911. A problem in age distribution. *Philosophical Magazine* 21: 435-8.

Schoener, T.W. and D.A. Spiller, 1992. Is extinction rate related to temporal variability in population size? An empirical answer for orb spiders. *American Naturalist* 139: 1176-207.

Shaw, M.W. 1995. Simulation of population expansion and spatial pattern when individual dispersal distributions do not decline exponentially with distance. *Proceedings of the Royal Society of London* B259: 243-8.

Shimada, M. and M. Tuda, 1996. Delayed density dependence and oscillatory population dynamics in overlapping-generation systems of a seed beetle *Callosobruchus chinensis*: matrix population model. *Oecologia* 105: 116-25.

Skellam, J.G. 1951. Random dispersal in theoretical populations. *Biometrika* 38: 196-218.

Skellam, J.G. 1973. The formulation and interpretation of mathematical models of diffusionary processes in population biology. In *The Mathematical Theory of the Dynamics of Biological Populations* (M.S. Bartlett and R.W. Hiorns, eds), 63-85. London: Academic Press.

Smith, B.D., G.A. McFarlane, and A.J. Cass, 1990. Movement and mortality of tagged male and female Lingcod in the Strait of Georgia, British Columbia. *Transactions of the American Fisheries Society* 119: 813-24.

Sändgerath, D. and O. Richter, 1990. An extension of the Leslie matrix model for describing population dynamics of species with several development stages. *Biometrics* 46: 595-607.

Stewart-Oaten, A., W.W. Murdoch, and S.J. Walde, 1995. Estimation of temporal variability in populations. *American Naturalist* 146: 519-35.

Stone, L. and S. Ezrati, 1996. Chaos, cycles and spatiotemporal dynamics in plant ecology. *Journal of Ecology* 84: 279-91.

Taylor, R.A.J. 1981a. The behavioral basis of redistribution. I. The Δ-model concept. *Journal of Animal Ecoogy* 50: 573-56.

Taylor, R.A.J. 1981b. The behavioral basis of redistribution. II. Simulations of the Δ-model. *Journal of Animal Ecology* 50: 587-604.

Thórarinsson, K. 1986. Population density and movement: a critique of Δ-models. *Oikos* 46: 70-81.

Turchin, P. 1995. Chaos in microtine populations. *Proceedings of the Royal Society of London* B262: 357-61.

Usher, M.B. 1966. A matrix approach to the management of renewable resources, with special reference to selection forests. *Journal of Applied Ecology* 3: 355-67.

Usher, M.B. 1969. A matrix model for forest management. *Biometrics* 25: 309-15.

Van den Bosch, R. Hengeveld, and J.A.J. Metz, 1992. Analyzing the velocity of animal range expansion. *Journal of Biogeography* 19: 135-50.

Von Foerster, H. 1959. Some remarks on changing populations. In *The Kinetics of Cellular Proliferation* (F. Stohlman, Jr. ed.), 382-407. New York: Frame and Stratton.

Williams, D.W. and A.M. Liebhold, 1995. Detection of delayed density dependence: effects of autocorrelation in an exogenous factor. *Ecology* 76: 1005-8.

Wolda, H. 1989. The equilibrium concept and density dependence tests. What does it all mean? *Oecologia* 81: 430-2.

Wolda, H. 1991. The usefulness of the equilibrium concept in population dynamics. *Oecologia* 86: 144-5.

Wolda, H. 1995. The demise of the population regulation controversy? *Researches on Population Ecology* 37: 91-3.

Wolda, H. and B. Dennis, 1993. Density dependence tests, are they? *Oecologia* 95: 581-91.

Wolda, H., B. Dennis and M.L. Taper, 1994. Density dependence tests, and largely futile comments: answers to Holyoak and Lawton (1993) and Hanski, Woiwod and Perry (1993). *Oecologia* 98: 229-34.

Wood, S.N. 1993. How to estimate life history stage durations from stage structured population data. *Journal of Theoretical Biology* 163: 61-76.

Wood, S.N. 1994. Obtaining birth and mortality patterns from structured population trajectories. *Ecological Monographs* 64: 23-44.

Wood, S.N. and R.M. Nisbet, 1991. Estimation of Mortality Rates in Stage-Structured Populations. Berlin: Springer-Verlag.
Worner, S.P. 1992. Performance of phenological models under variable temperature regimes: consequences of the Kaufmann or rate summation effect. *Environmental Entomology* 21: 689-99.
Worton, B.J. 1995. Modelling radio-tracking data. *Environmental and Ecological Statistics* 2: 15-23.

Population and Community Ecology for Insect Management and Conservation, Baumgärtner et al. (eds)
© 1998 Balkema, Rotterdam, ISBN 90 5410 930 0

Measures of spatial pattern and spatial association for insect counts

Joe N. Perry
Department of Entomology and Nematology, Rothamsted Experimental Station, Harpenden, UK

ABSTRACT: This paper introduces new methods to measure the spatial association or dissociation between two populations measured simultaneously, or between the same population measured on two occasions. It is suitable for ecological data in the form of counts, where the two populations are sampled at identical specified locations. The methods form part of the SADIE system, that uses all the spatial information in the sample. Examples are given using data from the literature. Fortran software is freely available from the author.

1 INTRODUCTION

This paper is concerned with the problem of measuring and testing for spatial association between two populations for which the data are counts of individuals. Both populations are assumed to have been sampled at the same spatial locations, for which the two-dimensional coordinates are all known. As an example, Harkness & Isham (1983) reported the number of nests of each of two ant species, *Cataglyphis bicolor* and *Messor wasmanni* in a 4x8 grid of quadrats covering a 0.5ha field in Northern Greece (see their Tables 3 & 4). The two populations might be spatially dissociated, perhaps because of competition between nests of different species for the same limited resource, or positively associated, perhaps because of the need to occupy areas with similar microclimates, or they may occur at random with respect to one another. It is impossible to consider association in isolation from the spatial pattern of its two component populations. For example, the presence of a parasitoid population may clearly affect the spatial pattern of its host, which will attempt to relocate in refuges. This escape response by the host may then cause an alteration in the spatial pattern of the parasitoid, which strives to seek out the new locations of the host aggregations. This process is dynamic, and may result in a ceaseless shifting of positions of both species. Then again, there may be absolutely no direct effect of one species on another, yet the spatial pattern of both may be affected by some third species or by an environmental component that induces a degree of spatial association. Without further information, these two cases, of direct and indirect association, are indistinguishable in the data as defined. The null hypothesis is one of lack of any association or dissociation, i.e. that the location of the individuals of either one population are random with respect to those of the other. Whether or not this is the case, both populations may still exhibit strong spatial pattern individually, and these patterns may in any event be described and tested in each species in isolation from the other. Indeed, this should be done before the problem of spatial association is addressed so that the effect of spatial pattern for an individual population may be eliminated; this may be done by conditioning, as described more fully below.

Lack of space precludes a more detailed account of the principles underlying the methodology, which are given in more detail in Perry (1997a). Briefly, the technique forms part of a class termed SADIE (Spatial Analysis by Distance IndicEs) first introduced by Perry & Hewitt (1991). The basis of SADIE is to quantify the spatial pattern in a sampled population by measuring the total effort (in terms of the minimum distance moved), which the individuals in the observed sample must expend, to move to extreme arrangements, for example that in which the individuals in the sample are spaced as uniformly, or regularly, as possible. The degree of non-randomness within a set of data is quantified by comparing the observed spatial pattern with rearrangements in which the sampled counts are randomly redistributed amongst the units by permutation. SADIE has two advantages: a biologically more intuitive approach than traditional mathematically-based concepts involving sample variance and sample mean, and increased power due to the greater use of the spatial information in the sample.

It was developed specifically for counts, in contrast to geostatistics, that was developed for physical variables measured on continuous scales that display a stationary, stable covariance structure over a wide area. However, counts of individuals of a particular animal or plant species are not continuous, but discrete, are often, as here, distributed exceedingly patchily, and frequently comprise a majority of zero values. They are highly dynamic, often characterized by isolated clusters, which may be acting as metapopulations with varying degrees of inter-cluster dispersal; they therefore do not usually satisfy the assumptions required by geostatistical methods.

Perry (1995) gave notation and definitions for D, the distance to regularity, found by the transportation algorithm (Kennington & Helgason 1980) from the operational research literature. He showed how randomizations based on permutations of the counts could be used to derive from D an index, I_a, by which spatial non-randomness could be distinguished from departures from the Poisson distribution in the form of statistical variance heterogeneity. The latter is due merely to skewness in the frequency distribution of counts whereas the former arises from aggregation of those counts into clusters or from regularity of the counts. Values of I_a less than unity indicate regularity; values greater than unity indicate aggregation. The randomization distribution is used to define a specific test of the null hypothesis of a spatially random arrangement of the counts, based on P_a the proportion of the distribution that is as large or larger than the observed value D. Perry (1996) presented an algorithm that would allow a set of given counts to be redistributed by permutation amongst a set of given sample units to achieve any given degree of spatial pattern, where possible. He noted that the index based on distance to regularity could be sensitive to edge effects and showed how to constrain the simulated arrangements to allow for this. Perry (1997*b*) introduced an 'initial and final' (IAF) diagnostic plot to aid interpretation and studied the behaviour of two new indices and related tests. The first index, J_a, is based on the distance to crowding, C, the minimum distance that individuals in the sample must move to occupy a single sample unit, and has a related test of spatial non-randomness based on the proportion, Q_a, of randomized values as large or larger than C; the unit to which individuals move to compute C is termed the 'focus' of crowding. The second index, K_a, is based on distance to regularity, but the randomizations are constrained to study for edge effects; it has a related test based on the corresponding proportion, R_a. A short summary of these ideas, and an introduction to the problem of spatial association was given by Perry *et al.* (1996).

The techniques used in this paper extend somewhat those introduced in Perry (1997*a*), which are recalled very briefly in the next section. Perry (1997*a*) noted that techniques such as measuring the correlation coefficient between the two populations, operate merely on the lists of counts, discarding all the spatial information concerning their locations. He developed two different methods to improve this, both of which utilise spatial information from nearby but not necessarily exactly coincident units, in addition to that concerning the relative size of counts in coincident units. In his introduction to the papers in this session, Manly (1997) refers to two studies by which the methods for spatial association will be exemplified. The most detailed analysis will be of those data given in Table 1 of Manly (1997), for *Rubus fruticosus* collected by Hutchings (1978). Although both sets of data analyzed here involve a regular grid of sample units, the methods work perfectly well on irregularly spaced sample units.

2 MATERIALS AND METHODS

2.1 *Notation*

Let there be n sample units with coordinates (X_i, Y_i), $i = 1,...,n$, each containing a count, N_{i1u} from population one, and N_{i2u} from population two. Let the total counts for the two populations be $M_1 = \Sigma N_{i1u}$ and $M_2 = \Sigma N_{i2u}$. Both methods to be described require the counts of the two populations to be scaled so that each have the same total; this does not otherwise affect any of the calculations. Hence, the counts of population one are scaled to $N_{i1} = M_2 N_{i1u}$, $i = 1,...,n$, and of population two to

$N_{i2} = M_1 N_{i2u}$, $i = 1,...,n$, and both populations have a total count of M_1M_2 after scaling. The 'edge-displacement' for population one, the distance between the centroid of its counts $(\Sigma N_{i1}X_i/\Sigma X_i, \Sigma N_{i1}Y_i/\Sigma Y_i)$, and the centroid of the sample units $(\Sigma X_i/n, \Sigma Y_i/n)$, is denoted δ_1; a similar distance δ_2 is defined for population 2. A small value, ε, where usually $\varepsilon<<1$, is specified to control the spatial pattern of the randomizations, as described below; similarly, a small value ε_2 is specified to control their edge-displacement. The meaning of D, C, I_a, P_a, J_a, Q_a, K_a and R_a, was given in the Introduction; these values will be subscripted as required to refer to either of the individual populations, e.g. D_1 represents the distance to regularity of population 1.

2.2 *The index* I_t

The motivation behind this method is that the within-unit totals, $(N_{i1} + N_{i2})$, of both population's scaled counts, will tend to appear relatively aggregated over the sample units with a large distance to regularity, T, if the populations are associated, but relatively regular with a short distance to regularity, T, if dissociated, compared to randomised permutations of the original counts with the same degree of spatial pattern. Two forms of constrained randomization are adopted. First, constrained permutations of the observed counts from population one are made with counts of population two held at their original values; this yields a set of distances to regularity of the totals with that for the kth permutation denoted as $T_{k(2)1}$. There are two constraints on the permutations, imposed (Perry, 1997*a*) to condition on aspects of the observed spatial pattern, and thereby to reduce extraneous variation and provide fair comparisons; they are achieved using the algorithm described in Perry (1996). The first restricts the distance to regularity of the permutation, D^*_k, to be within a certain proportion of the observed value D_1, to satisfy: $1\text{-}\varepsilon < (D^*_k / D_1) < 1\text{+}\varepsilon$; the second restricts additionally the edge-displacement of the permutation, δ^*_k, to be within a certain absolute distance of the observed value δ_1, to satisfy: $|\delta^*_k - \delta_1| < \varepsilon_2$. Usually, at least 200 permutations are done, the resulting set of values of $T_{k(2)1}$ is then ordered, and the proportion, $P_{t(2)1}$ of its values that are as large as T is computed. This is used to perform the usual SADIE one-sided randomization test (Perry, 1995) for spatial randomness, in this case of the placing of the counts of population one with regard to those of population two, against the alternative that the observed arrangement of population one is either associated or dissociated with population two. An index of association, $I_{t(2)1}$, is formed from $T/E_{t(2)1}$, where $E_{t(2)1}$ represents the average value of $T_{k(2)1}$ over the randomizations. Values of $I_{t(2)1} > 1$ indicate association, values < 1 dissociation, while values close to unity indicate random placement of population one with regard to population two.

By analogy with the process above, constrained permutations of the observed counts from population two are made with counts of population one held at their original values; this yields the set $T_{k(1)2}$, another proportion $P_{t(1)2}$ and another index $I_{t(1)2}$. In principle, the inferences pertaining to each randomization process may be distinct; for example, it is possible for every individual of species 1 to have an individual of species 2 as its nearest neighbour, but for most nearest neighbours of species 2 to be individuals of their own species. However, in practice, the values of $I_{t(2)1}$ and $I_{t(1)2}$, and of $P_{t(2)1}$ and $P_{t(1)2}$ are usually similar, and are combined by arithmetic average to form a single index I_t and proportion P_t.

2.3 *The index* I_m

By contrast with the index I_t, above, where the information from coincident and nearby units is combined implicitly, an alternative method is now presented by which one index I_f is explicitly formed from non-spatial information in coincident units, a second index, I_z, is explicitly formed from information in nearby units, and an overall index, I_m, formed by combining these two. Their construction is best explained by reference to an IAF plot such as that of the newly emerged brambles (population one) in Figure 1*a*. An IAF plot graphically depicts the results from the transportation algorithm, that gives the optimal number of individuals required to move from each of the cells with initially more individuals than the sample mean, to cells with initially fewer, to achieve regularity. Notice that: (i) units with particularly large counts are identified by the emanation of several lines; (ii) for each sample unit there is either an outflow or an inflow; (iii) that this flow has an overall value that is either positive (outflow) or negative (inflow), denoted here for the ith unit of population one as f_{i1}; (iv) that the individual flows, either into or out of a unit, may be summed as vectors, again with positive

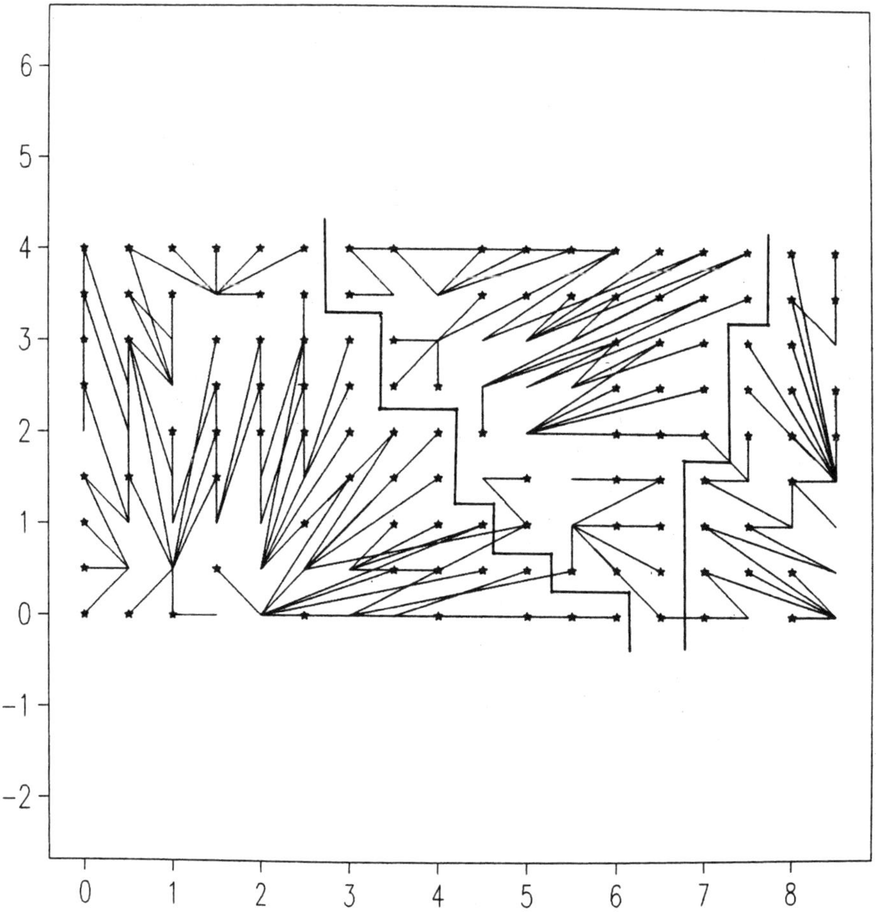

Figure 1. *a*. An IAF plot for the newly emerged brambles of Hutchings (1978). Units with inflows are indicated by *. The black lines partition units in the 18x9 grid into three regions within each of which the counts appear to form a single major cluster; for explanation of partitioning mechanism see text.

values for outflows and negative for inflows; (v) that the overall vector flow for unit i in population one has a strength, denoted here as F_{i1}, and a direction, θ_{i1}, where θ is measured clockwise from the vertical direction (y increasing); and (vi) the absolute value of F is always less than or equal to that of f, and only equal to it if all the individual flows are in exactly the same direction.

As an example, consider the count of three in the extreme lower-right corner (x=8.5, y=0, i=154) of the grid: $N_{154\,1u} = 3$ (Figure 1*a*), scaled by $N_{i1} = M_2 N_{i1u}$ to $N_{154\,1} = 195\text{x}3 = 585$. There were five outflows with values: 74.63 to (7, 0.5); 67.41 to (7, 1); 208.24 to (7.5, 0.5); 13.24 to (8, 0) and 13.24 to (8, 0.5). The total of these five flows, $f_{154\,1}$, is 376.76, and, of course, the remainder, after deduction of these four flows from 585, is, as required, 208.24, the sample mean of the scaled counts. Here, the vector strength of flow is easily calculated from its five components as $\boldsymbol{F}_{154\,1} = 373.43$, and the vector direction of the flow to be from (x=8.5, y=0) towards (x=6.45, y=1), i.e. in the direction of $\theta_{154\,1} = -64.04°$. Because the directions of flow nearly coincide, $\boldsymbol{F}_{154\,1}$ is not much less than $f_{154\,1}$.

Now consider two intermediate statistics that combine this information from the two populations. The first of these, denoted as ϕ, is defined as: $\Sigma f_{i1} f_{i2}$. If the two populations for a given unit both have counts that are greater than their respective population sample means they will both represent outflows and if counts that are both less they will both represent inflows, and in these cases their contribution to ϕ will be positive; but if one population represents an outflow and one an inflow then the contribution

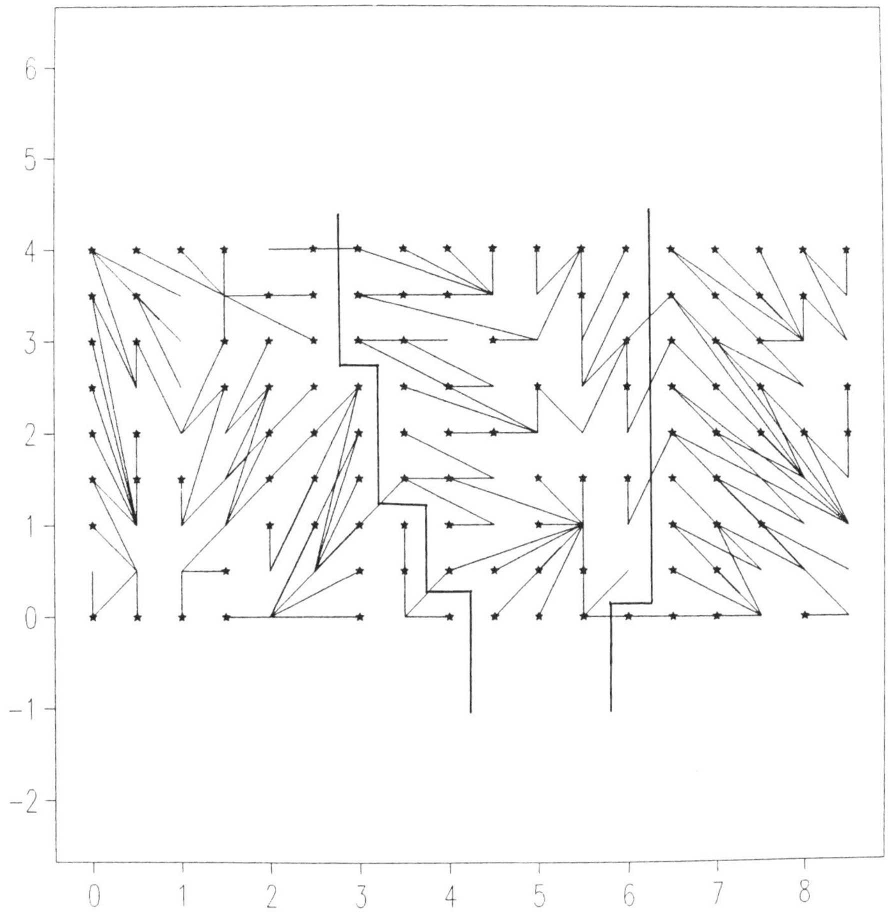

Figure 1. *b*. As for *a*, but for one year old brambles.

will be negative. Thus ϕ may be seen to comprise non-spatial information analogous to a correlation coefficient, and is likely to be positive if the two populations are associated, negative if dissociated; a value of zero is expected under the null hypothesis. Clearly, to be useable, ϕ will require some scaling; this scaling is used to produce the intermediate index I_f, described later.

The second intermediate statistic, denoted as ψ, is defined as: $\psi = \Sigma F_{i1} F_{i2} \cos(\omega_i)$, with $\omega_i = |\theta_{i1} - \theta_{i2}|$ if $|\theta_{i1} - \theta_{i2}| < 180°$, and $\omega_i = 360 - |\theta_{i1} - \theta_{i2}|$ if $|\theta_{i1} - \theta_{i2}| > 180°$; note that this notation supplants the slightly different version given in a preliminary form in Perry (1997*a*). If the two populations are strongly associated they will most likely have similar IAF plots, in which case for most units the values of θ for the respective populations will be similar, ω_i will be close to zero and ψ will be positive; if they are strongly dissociated then the IAF plots will be dissimilar, for many units ω_i will be close to 180°, and ψ will be negative; if they are random with respect to one another then a value of $\omega_i = 90°$ and $\psi = 0$ is expected. Clearly, plots of ω_i over the grid will be informative, particularly if there are large regions with similar ω_i values. However, the values of ω_i that together make up ψ require some initial scaling, and this is provided by the multiplication by the vector flows, but further scaling of ψ is also required to produce the intermediate index I_z, described later.

To extend the example above to exemplify the above statistics for two populations, consider the unit with a count of two in the extreme lower-right corner (x=8.5, y=0, i=154) of the grid of one year old brambles shown in Figure 1*b*: $N_{154\,2u} = 2$, scaled by $N_{i2} = M_1 N_{i2u}$ to $N_{154\,2}$ = 173x2 = 346. The total of the three outflows, $f_{154\,2}$, is 137.76, and the remainder, after deduction of these four flows from 346, is, as required, 208.24, the sample mean of the scaled counts (note that, as required, the sample means of the two scaled populations are identical). The contribution towards ϕ for this unit is therefore 51902.46. For population two, the vector strength of flow is $F_{154\,2}$ = 129.85, and the vector direction of the flow is from (x=8.5, y=0) towards (x=7.01, y=1), i.e. in the direction of $\theta_{154\,2}$ = -56.06°. Hence, ω_{154} = 7.97° and the contribution to ψ for this unit is 48021.51. As may be seen from the two IAF plots, the directions of the outflows are very similar for the two populations, so the value of $\cos(\omega_{154})$ = 0.990 is relatively large.

The simple process of scaling ϕ and ψ to obtain, respectively, I_f and I_z is achieved from randomizations; the randomizations used are precisely those described above. The constrained permutations of the observed counts from population one made with counts of population two held at their original values yield a set of values of ϕ and of ψ denoted, respectively, for the kth permutation, as $\phi_{k(2)1}$ and $\psi_{k(2)1}$. The means, $\mu(\phi_{k(2)1})$, $\mu(\psi_{k(2)1})$, and standard errors, $\sigma(\phi_{k(2)1})$, $\sigma(\psi_{k(2)1})$ of these permutations are noted. Each individual value is then standardised, so that each has an approximate standard normal distribution with mean zero and standard deviation unity, by: $\phi_{k(2)1}$ = $[\phi_{k(2)1} - \mu(\phi_{k(2)1})]/\sigma(\phi_{k(2)1})$ or by $\psi_{k(2)1} = [\psi_{k(2)1} - \mu(\psi_{k(2)1})]/\sigma(\psi_{k(2)1})$. The observed values of ϕ and of ψ are likewise standardized to yield the indices $I_{f(2)1}$ and $I_{z(2)1}$ respectively, by: $I_{f(2)1}$ = $[\phi - \mu(\phi_{k(2)1})]/\sigma(\phi_{k(2)1})$ or $I_{z(2)1}$ = $[\psi - \mu(\psi_{k(2)1})]/\sigma(\psi_{k(2)1})$. Under the null hypothesis both $I_{f(2)1}$ and $I_{z(2)1}$ have expected values of zero, positive values indicate association and negative values dissociation. It is possible to quantify the relative degree to which each unit contributes to the overall value of $I_{f2(1)}$ as $f_{i(2)1}$, where $f_{i(2)1}$ = $[f_{i1}f_{i2}$ - $\mu(\phi_{k(2)1})]/\sigma(\phi_{k(2)1})$; note that the average value of $f_{i(2)1}$ over the n sample units, $\Sigma f_{i(2)1}/n$, is equal to the index $I_{f2(1)}$. Contour plots of $f_{i(2)1}$ over the grid may be of use in interpreting regions of relatively high or low association or dissociation. Plots of the similarly defined $f_{i(1)2}$ will usually differ from plots of $f_{i(2)1}$ only marginally, by a very small constant value. Similarly, it is possible to quantify the relative degree to which each unit contributes to the overall value of $I_{z2(1)}$ as $z_{i(2)1}$, where $z_{i(2)1}$ = $[F_{i1}F_{i2}\cos(\omega_i)$ - $\mu(\psi_{k(2)1})]/\sigma(\psi_{k(2)1})$; and note that the average value of $z_{i(2)1}$ over the n sample units, $\Sigma z_{i(2)1}/n$, is equal to the index $I_{z2(1)}$.

The merging of the permutation statistics $\phi_{k(2)1}$ and $\psi_{k(2)1}$, and the indices $I_{f(2)1}$ and $I_{z(2)1}$, to form an overall index $I_{m(2)1}$ with a related test of association, is done as follows. Randomized values, $m_{k(2)1}$, are formed from $m_{k(2)1} = [\phi_{k(2)1} + \psi_{k(2)1}]/\sqrt{2}$, and these are compared against the index, $I_{m(2)1}$, defined as $I_{m(2)1} = [I_{f(2)1} + I_{z(2)1}]/\sqrt{2}$; once again, this method of combination ensures that the randomizations and the expected value of the index $I_{m(2)1}$ under the null hypothesis have zero mean and unit standard error, and that large values of the index indicate association and small values dissociation. The set of values

Table 1. Sample units for which the difference in count between age-classes is greater than unity. A '+' denotes that the count for one year old brambles was the higher; a 'x' that it was the lower. The same symbol enclosed by a circle indicates that the difference in count was larger than two.

y \ x	0.0	0.5	1.0	1.5	2.0	2.5	3.0	3.5	4.0	4.5	5.0	5.5	6.0	6.5	7.0	7.5	8.0	8.5
4					+													
3.5									x	⊕	+							+
3										x							⊕	x
2.5											x						+	
2			⊕	+									+					
1.5					x	x				+							⊕	x
1		+								⊕		⊕	+					+
0.5					⊗		⊗									⊕		
0							x									⊕		

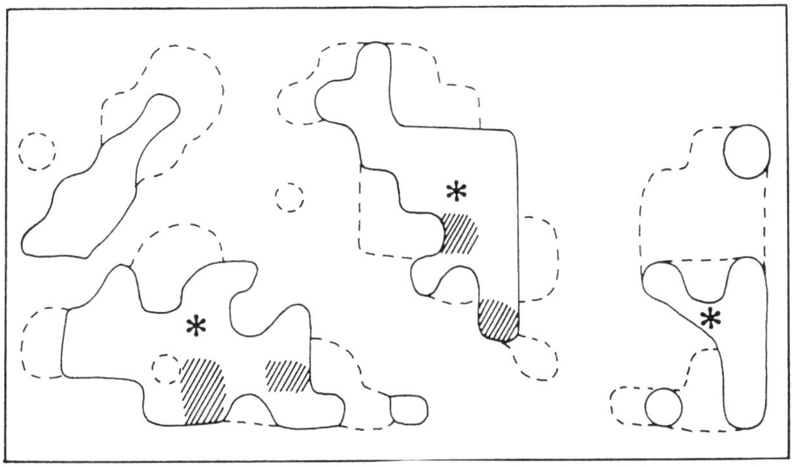

Figure 2.*a*. A simple contour plot of the newly emerged brambles. Units enclosed by dashed lines have a count of unity, units enclosed by solid lines have a count greater or equal to two, units with dark shading have counts greater or equal to five. The three units acting as the foci for the distance to crowding within the three partitions defined in Figure 1 are marked with a $*$; for further detain see text.

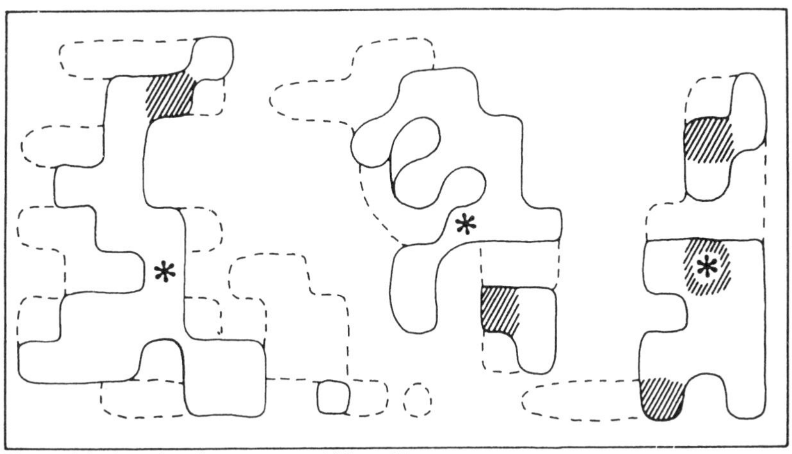

Figure 2.*b*. As *a*, but for one year old brambles.

of $m_{k(2)1}$ is ordered, and the proportion, $P_{m(2)1}$ of its values that are as large as $I_{m(2)1}$ is computed. This is used to perform the usual SADIE one-sided randomization test (Perry, 1995) for spatial randomness, in this case of the placing of the counts of population one with regard to those of population two, against the alternative that the observed arrangement of population one is either associated or dissociated with population two. As previously, the constrained permutations of the observed counts from population two made with counts of population one held at their original values yields the set $m_{k(1)2}$, another proportion $P_{m(1)2}$ and another index $I_{m(1)2}$ and, as long as the values of $I_{m(2)1}$ and $I_{m(2)1}$, and of $P_{m(2)1}$ and $P_{m(2)1}$ are similar, they are combined by arithmetic average to form a single index I_m and proportion P_m .

2.4 *Data*

The main data analyzed here are from Hutchings (1978), detailed in Table 1 of Manly (1997). Analysis begins with the spatial patterns of newly emerged and one year old brambles considered separately, before proceeding to consider association; for each of these stages novel diagnostic plots will be

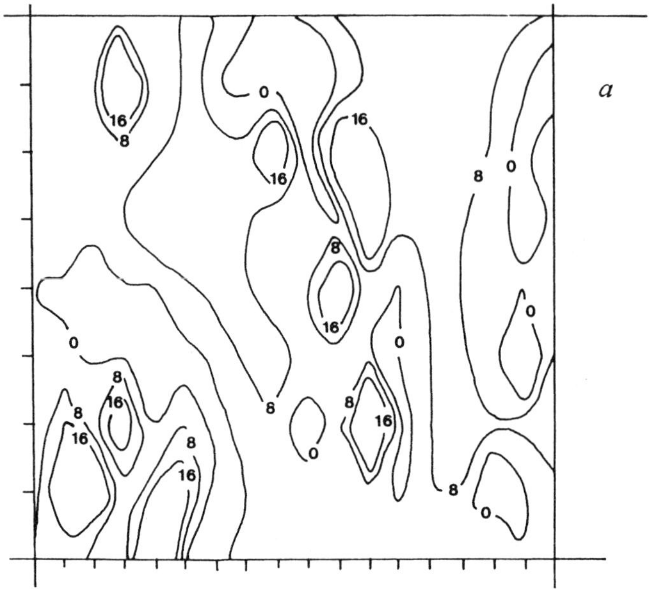
0
8
16
a

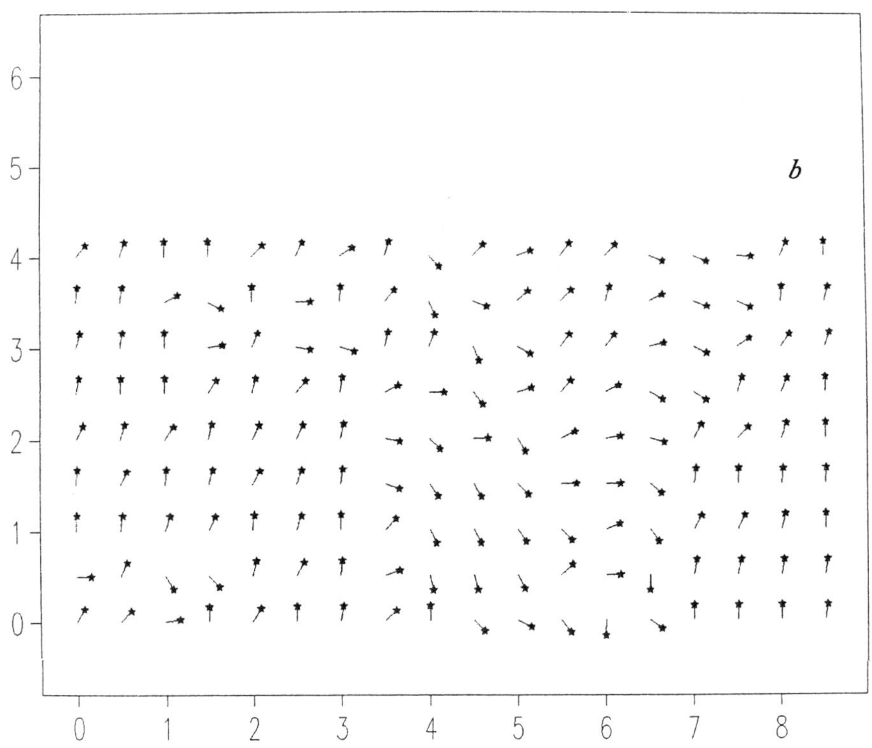
b
0
1
2
3
4
5
6
7
8
6
5
4
3
2
1
0

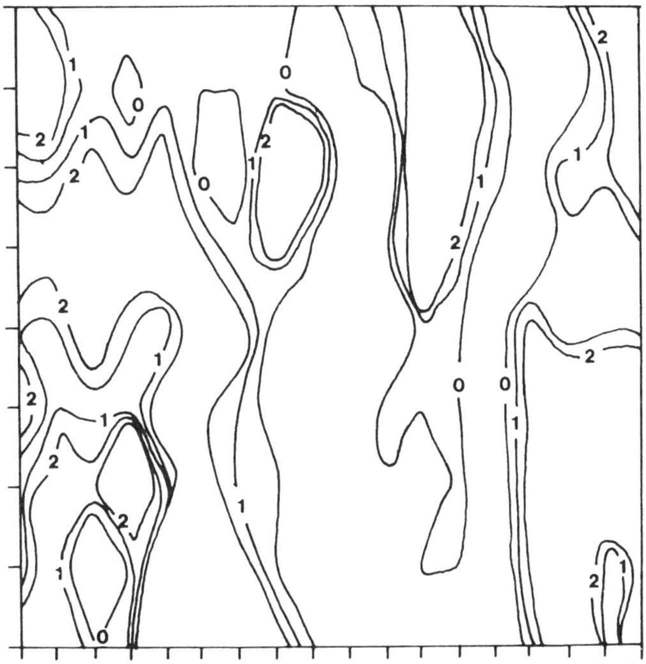

c

Figure 3. The contribution of individual units towards various statistics and indices of association. *a.* Contour plot of $f_{I(2)I}$, showing individual contributions towards $I_{f(2)I}$, a component index of association concerned with information in coincident units; axes scaled to same length. *b.* Plot of w_i, showing differences in those directional flows shown in Fig. 1 *a* and *b*; for each unit the value of w_i is depicted by the difference of the plotted line from the vertical with *y* increasing, i.e. w_i ranges from zero (vertical pointing up) through 90° (line points to the right) to 180° (vertical pointing down). *c.* Contour plot of $z_{I(2)I}$, showing individual contributions towards $I_{z(2)I}$, a component index of association concerned with information in neighbouring but non-coincident units; axes scaled to same length.

introduced for the first time. For each population, 250 randomizations were done. Brief details are also given of an analysis of the ants' nests of Harkness and Isham (1983) mentioned above, for which the rather small number of 100 randomizations were done for each population.

3 RESULTS AND DISCUSSIONS

3.1 *Newly emerged brambles - single population analysis*

The simplified coutour map of Figure 2a provides a visual summary of the data; 85 of the 162 units were occupied, with a sample mean of 1.07 brambles per unit. Clearly there are three major clusters in the data, although that on the left-hand side itself arguably might comprise two clusters. The three clusters were of different sizes, spaced at roughly equal intervals along the *x*-axis, although differing with regard to density in the y-direction. The value of I_a = 1.45 (P_a = 0.03) confirms that the data are significantly clustered, and that of R_a = 0.001 that this is not due to any edge effect. The small, non-significant value of J_a=1.04 indicates the presence of more than one cluster (Perry, 1997*b*). These clusters are visually clear in the IAF plot (Figure 1*a*), in which it is possible to categorise every unit into one of three sets that correspond in general location with the three clusters and their close vicinity, such that almost all the inflow units in a set receive flows from outflow units that are in the same set. In only four units is there any ambiguity in this partition of the grid; e.g. the unit at (6.5, 0) receives flows from units at (7.5, 0) and (5.5, 1) that are in different sets. Such inflow units are categorised as

Table 2. Indices and probability values for association in bramble data

Counts randomized: Counts fixed:	Newly emerged One year old		One year old Newly emerged		Average over randomizations	
	Index	Probability	Index	Probability	Index	Probability
Distance to regularity, (t)	$I_{t(2)1} = 1.30$	$P_{t(2)1} < 0.004$	$I_{t(1)2} = 1.27$	$P_{t(1)2} = 0.004$	$I_t = 1.28$	$P_t = 0.002$
Explicit flows in coincident units, (f)	$I_{f(2)1} = 8.04$	$P_{f(2)1} < 0.004$	$I_{f(1)2} = 7.71$	$P_{f(1)2} < 0.004$	$I_f = 7.88$	$P_f < 0.002$
Explicit directional flow in non-coincident units, (z)	$I_{z(2)1} = 1.00$	$P_{z(2)1} = 0.188$	$I_{z(1)2} = 0.831$	$P_{z(1)2} = 0.232$	$I_z = 0.917$	$P_z = 0.210$
Combined information, (m) from explicit methods (f) and (z)	$I_{m(2)1} = 6.40$	$P_{m(2)1} < 0.004$	$I_{m(1)2} = 6.04$	$P_{m(1)2} < 0.004$	$I_m = 6.22$	$P_m < 0.002$

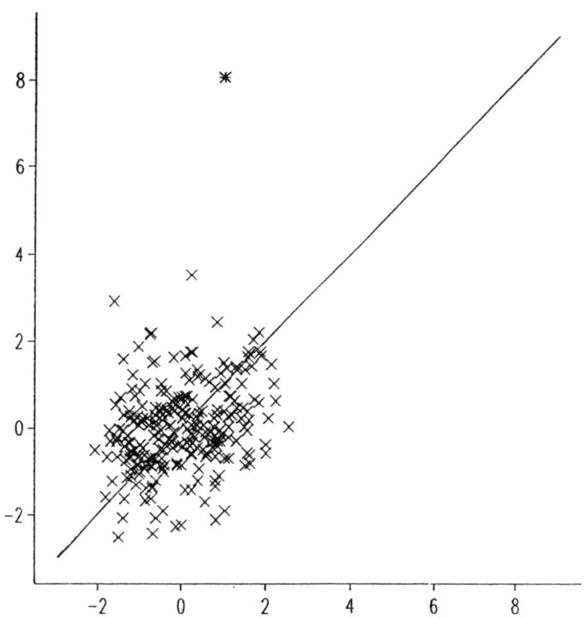

Figure 4 Graph of the 250 randomized standardized values $f_{k(2)1}$ versus $y_{k(2)1}$ (X), and, on the same axes, $I_{f(2)1}$ versus $I_{z(2)1}$ («); the equality line, $f_{k(2)1} = y_{k(2)1}$, is also shown for reference.

belonging to the set from which they receive the greatest outflow. In this case, since the scaled flow from the unit at (7.5, 0) was 34.91 and that from the unit at (5.5, 1) was 173.33, the unit at (6.5, 0) was assigned to the set to which the unit at (5.5, 1) belonged, i.e. that covering the middle of the grid. This process produced partitions (Figure 1*a*) of unequal size, 78, 55, and 29 units in the left, middle and right-hand partition, respectively, that resembled closely the shapes of the clusters shown in Figure 2*a*. Since the units formed almost completely separate, non-overlapping sets, in terms of their flows for distance to regularity, and since at this smaller scale each partition contained a single major cluster, it is reasonable to re-analyze them, separately, using the distance to crowding technique. As expected, the presence of single clusters at this smaller, within-grid scale, resulted in larger values of J_a that demonstrated strong spatial pattern; the three values of J_a were, respectively, 1.25, 1.45 and 1.11. The within-partition units that were the foci for crowding are shown in Figure 2*a*.

3.2 *One year old brambles - single population analysis*

Figure 2*b* provides a visual summary of the data; in comparison with those newly emerged, slightly more, 92, of the 162 units were occupied by one year old brambles, and the sample mean, 1.20, was slightly larger. There were still three major clusters in the data on the left, middle and right-hand sides of the grid, but their positions were noticeably changed; they appeared more equitable in size, and much more uniformly spread within the grid. The value of I_a=1.12 was much reduced, and not significant (P_a=0.21), although the spatial pattern appears clear. Possibly, this lack of power to detect non-randomness is an effect of scale, for when the units are partitioned into three sets using the IAF plot (Figure 1*b*), the values of J_a were, from left to right, 1.07, 1.23 and 1.18, the two larger being significant at the 5% level. The position of the foci for crowding (Figure 1*b*) emphasised, in contrast to the newly emerged brambles, how the clusters were more centred, especially in the *y*-direction. The three partitions were also of more equal size (64, 52 and 46 units, respectively) than for the newly emerged brambles.

3.3 *Spatial association between newly emerged and one year old brambles - preliminary analysis*

The above results demonstrate how, despite the clear and similar spatial pattern between the two age-classes, there were important differences too, so that while association was fairly evident, it was not uniformly large over the whole grid. Units of the grid where there was a relatively large difference between the two age-classes (Table 1) showed marked spatial pattern, counts for the one year old brambles being clearly larger towards the right-hand cluster, lower part of the middle cluster and middle of the left-hand cluster, but smaller at the bottom-right corner of the left-hand cluster (cf. also Figure 2*a*, *b*). The contour plot of $f_{i(2)1}$ (Figure 3*a*) reinforces these observations because the average value, around eight, indicates considerable overall association and for most units $f_{i(2)1}$ is positive, although the areas of the grid identified above have negative values, indicating the smaller regions of dissociation. By contrast, the plot of values of ω_i (Figure 3*b*) must be interpreted with reference to differences between the direction of flows in Figure 1*a*, *b*, which impart different information. This plot shows that for the majority of units on the left and extreme right-hand sides of the grid the directional flows were nearly identical, for those values of ω_i are close to 0° and even units that are the exceptions to this rule rarely show values of $\omega_i > 90°$. In the middle of the *x*-axis, however, especially for the smaller *y*-values, there was a large region of dissimilar directional flows, shown by many values of ω_i close to 180°. Perhaps it is inevitable, with the differences previously noted between the main clusters, that the main area where the directional flows would differ should be in the middle of the sample area. The contour plot of $z_{i(2)1}$ (Figure 3*c*) gives similar and little additional information; the mean value is close to unity, the majority of the units show positive values indicting overall association, and the main areas of negative contributions towards I_z clearly occur towards the centre of the *x*-axis.

3.4 *Spatial association between newly emerged and one year old brambles - formal analysis*

Both the I_t index and the I_m index are large and highly significant, showing positive overall spatial association, as expected. These values, and the individual components from the separate randomizations of newly emerged and one year old brambles are detailed in Table 2. For these data the association detected by the directional flow method (*z*) was, while still positive, less in degree than the flow in coincident units method (*f*), perhaps emphasising the importance of the differences between the two spatial patterns highlighted by the preliminary analysis above. The difference between the values of $I_{z(2)1}$ and $I_{z(1)2}$ is difficult to interpret without further biological information. If it represents a real difference then it would imply a slightly greater degree of affinity of newly emerged brambles for areas with one year old brambles than of one year old brambles for areas with newly emerged brambles; the difference is about 0.17 with a standard error of about 0.09, so it is just possible that the effect is real.

The interpretation of the combined coefficient, I_m, is aided by plotting the 250 randomized standardized values $\phi_{k(2)1}$ versus the randomized standardized values $\psi_{k(2)1}$, and on the same graph, $I_{f(2)1}$ versus $I_{z(2)1}$ (Figure 4). Note that, from the definitions of $m_{k(2)1}$ and $I_{m(2)1}$, these values may be seen visually as the projections of the respective points onto the equality line, plotted at 45° to the x- and y-axes. Hence, while the value of $I_{f(2)1}$ greatly exceeds its corresponding randomizations, $\phi_{k(2)1}$, that of $I_{z(2)1}$ does not ($P_{z(2)1} = 0.188$, see Table 2) exceed more than about 20% of its corresponding randomizations, $\psi_{k(2)1}$. However, the combined value, $I_{m(2)1}$, visualized as the projection of the asterisked point onto the equality line, clearly will greatly exceed ($P_{m(2)1} < 0.004$, see Table 2) all the values of $m_{k(2)1}$, which may be visualized as projections of the 250 individual points, labelled x, onto this line. Hence, in overall terms, the explicit combined index, I_m, agrees with the implicit index, I_t.

3.5 *Spatial association between* Cataglyphis bicolor *and* Messor wasmanni *ants' nests*

For these rather sparse data the following indices and probabilities were found: $I_{t(2)1} = 1.10$ ($P_{t(2)1} = 0.25$); $I_{f(2)1} = 0.99$ ($P_{f(2)1} = 0.52$); $I_{z(2)1} = 1.00$ ($P_{z(2)1} = 0.50$); $I_{m(2)1} = 1.00$ ($P_{m(2)1} = 0.51$); $I_{t(1)2} = 1.07$ ($P_{t(1)2} = 0.25$); $I_{f(1)2} = 0.99$ ($P_{f(1)2} = 0.58$); $I_{z(1)2} = 0.99$ ($P_{z(1)2} = 0.53$); $I_{m(1)2} = 0.99$ ($P_{m(1)2} = 0.55$). For these data the implicit method based on distance to regularity has perhaps slightly more power to detect any association arguably present. However, the value of I_t is nowhere near significant and the conclusion must be, without more detailed data, that there is no evidence that the two species do not place their nests randomly with respect to one another.

4 CONCLUSIONS

The concept developed in this paper for counts, of performing the randomization tests after conditioning on the observed arrangements, was suggested also by Harkness & Isham (1983) for mapped data, although they had no means to find such constrained arrangements Hence, their seeking of some ideal function that would measure the association between the nest locations of the two ant species: "Ideally this function for the observed nest processes would be compared with the same function calculated for two independent processes having the same distributions as the *Cataglyphis* and *Messor* nests ... we do not know what these marginal distributions are ... An alternative possibility would be to perform toroidal translations of the *Cataglyphis* data ... with the *Messor* data". The algorithm used here (Perry, 1996) goes a long way to providing such an ideal function for counts, without the need to use such a clumsy approach as toroidal translations, that are little more than a fudge for mapped data and completely unnecessary for counts.

This paper contains two recommended indices and related tests to use for formal analysis of association between two populations of counts sampled in the same locations. These use all the spatial information in the sample, in two different ways, both implicitly and explicitly. A wealth of diagnostic plots have been introduced to help interpret the data in a more informal way.

Versions of all the SADIE programs are now all available also as executable files and code, suitable for running on PCs with a 32-bit operating environment such as Windows95, under which they were developed, and WindowsNT. All SADIE software is available free by contacting the author via an email message to: joe.perry@bbsrc.ac.uk.

5 ACKNOWLEDGEMENTS

IACR Rothamsted receives grant-aided support from BBSRC. I am pleased to thank Niels Holst for his idea that IAF plots might form the basis for an index of association.

6 REFERENCES

Harkness, R.D. & Isham, V. 1983. A bivariate spatial point pattern of ants' nests. *Applied Statistics* 32: 293-303.

Hutchings, M. 1978. Standing crop and pattern in pure stands of *Mercurialis perennis* and *Rubus fruticosus* in mixed deciduous woodland. *Oikos* 31: 351-357.
Kennington, J.L. & Helgason, R.V. 1980. Algorithms for Network Programming. Wiley, New York.
Manly, B.F.J. 1997. Sampling and modelling of insect populations: Introduction. *This volume*.
Perry, J.N. 1995. Spatial aspects of animal and plant distribution in patchy farmland habitats. In: *Ecology and Integrated Arable Farming Systems*, pp. 221-242 (eds , D. Glen, M. Greaves & H.M. Anderson). Wiley, Chichester
Perry, J.N. 1996. Simulating spatial patterns of counts in agriculture and ecology. *Computers & Electronics in Agriculture* 15: 93-109.
Perry, J.N. 1997*a*. Spatial association for counts of two species. *Acta Jutlandica* (in press).
Perry, J.N. 1997*b*. Measures of spatial pattern for counts. *Ecology* (in press).
Perry, J.N., Bell, E.D., Smith, R.H. & Woiwod, I.P. 1996. SADIE: software to measure and model spatial pattern. *Aspects of Applied Biology* 46: 95-102.
Perry, J.N. & Hewitt, M. 1991. A new index of aggregation for animal counts. *Biometrics* 47: 1505-1518.

Population and Community Ecology for Insect Management and Conservation, Baumgärtner et al. (eds)

Simulation models of the dynamics of the sheep blowfly, *Lucilia sericata*

A. Fenton
School of Biological Sciences, University of Bristol, UK

ABSTRACT: Based on observed relationships between temperature and development rate, a series of deterministic and stochastic simulation models have been constructed to predict the seasonal pattern of abundance of the sheep blowfly, *Lucilia sericata*. These models have been used to explore a range of control strategies, showing that control methods targeting the adult flies, particularly early in the season should prove more beneficial in the suppression of the blowfly population than control aimed at the pre-adult stages. Furthermore, the models were used to show that given sufficient time, the population would reach a stable equilibrium determined by the number of suitable oviposition sites available to the gravid female flies. However, at the temperatures typically experienced in Britain, the population fails to achieve such a state before the initiation of diapause. Therefore, it is unlikely that the blowfly population is strongly regulated by such density dependent oviposition limitation.

INTRODUCTION

Computer simulation techniques are important tools which help in understanding the dynamics of biological systems and allow predictions of the way such systems will respond to external or internal pressures. In applied entomology, models which realistically simulate the behaviour of insect pest populations may allow predictions of the patterns of pest abundance to be made, and help identify novel control strategies.

The system under consideration here is that of the sheep blowfly *Lucilia sericata* Meigen (Diptera: Calliphoridae), which is a significant pest of sheep farms in Britain. A series of deterministic and stochastic simulation models have been developed to predict the local abundance of this species throughout the blowfly season on individual sheep farms in Britain (Wall *et al.* 1992a; Wall *et al.* 1993a; Fenton *et al.* 1997; Fenton & Wall 1997). Sensitivity analyses have been carried out on these models to identify the main factors affecting the predicted patterns of blowfly abundance, and determine the relative merits of a range of control strategies. Details of these models have been given previously (Wall *et al.* 1992a; Fenton *et al.* 1997), so the purpose of this paper is to summarise the model's construction and highlight some of the potential benefits of using such a model.

BLOWFLY LIFE CYCLE

The adult females lay egg batches of approximately 200 eggs in the wool of the sheep, selecting areas of high humidity, such as fleece soiled by faeces or urine (Cragg 1955). After hatching, the larvae feed on the epidermal tissues and skin secretions causing extensive tissue damage, which can lead to the death of the sheep (Guerrini 1988). When they have completed feeding, the third stage larvae drop to the ground where they undergo a period of dispersal as wandering larvae,

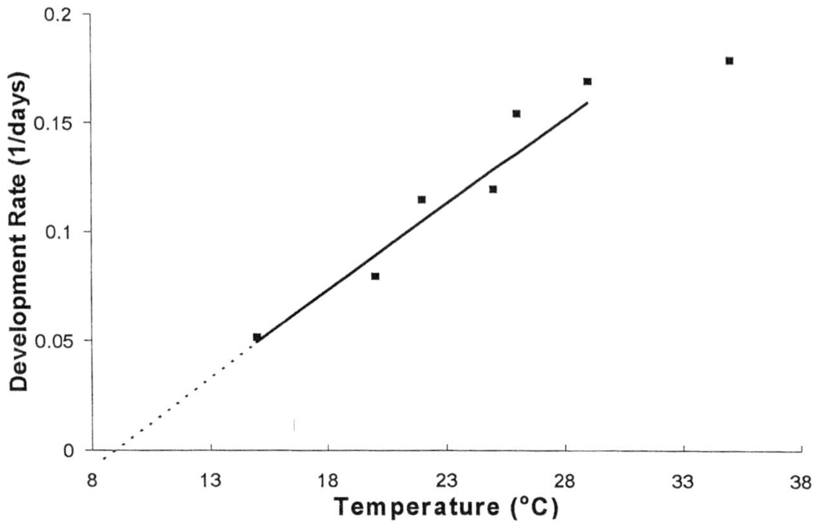

Figure 1. The rates of development (points) of third stage wandering larvae of *L. sericata* measured at a range of temperatures in the laboratory and the regression line through the linear portion of the relationship (solid line). The dotted line represents the extrapolation of the regression line to determine the base threshold temperature.

before burrowing into the soil to pupate. The emerging adult flies mate to begin the cycle once again. As the temperature and photoperiod decline in late August, females produce larvae that cease development and burrow into the ground where they overwinter as diapausing larvae (Saunders *et al.* 1985). During early spring the next year, the increasing temperature results in the diapausing larvae undergoing a period of post diapause development and then pupating, before emerging as the season's first cohort of adults.

QUANTIFYING INSECT DEVELOPMENT RATES

Temperature is of fundamental importance in determining the development rates of insects (Beck 1983; Wagner *et al.* 1984; Higley *et al.* 1986; Wall *et al.* 1992a). Hence, assuming ambient temperatures approximate insect body temperatures, it should be possible to describe the development of an insect population, given the temperatures experienced by the individuals in the population each day. A frequently used modelling technique involves the calculation of day degrees, which assumes that development rate is directly proportional to ambient temperature above a minimum threshold and that below this threshold no development occurs (Clements 1963; Moon 1983; Preuss 1983; Phelps *et al.* 1993).

To quantify the temperature-development rate relationship for *L. sericata,* individuals of each life cycle stage were reared under a range of constant temperatures (Wall *et al.* 1992b). A plot of mean development rates against temperature produces a sigmoidal pattern, with a linear relationship over the middle temperature ranges (Figure 1). By linear regression, this region can be extrapolated to where it crosses the x-axis, thereby providing an estimate of the base threshold temperature for that life cycle stage (Higley *et al.* 1986; Petitt *et al.* 1991; Bernal & Gonzalez 1993). The difference between this base temperature and the ambient temperature multiplied by the time taken to complete the life-cycle stage at that temperature, gives the number of day degrees required for complete development of that stage.

Given an estimate of the temperatures experienced by the population each day in the field, and the assumption that ambient temperature reflects the temperature experienced by the insects, the degree of development undergone by individuals of each life cycle stage in the population can be calculated. These day degree fractions are accumulated by members of the population daily until the full requirement has been attained, and they pass to the next stage.

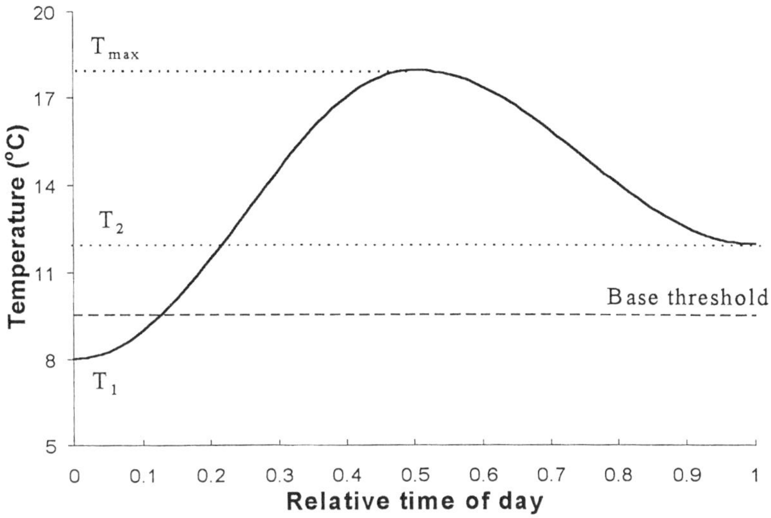

Figure 2. A sine wave representation of the daily temperature pattern (solid line), passing from a minimum (T1) at the start of the day, to a maximum (Tmax) at the middle, and a second minimum (T2) at the end of the day. The dashed line represents the base threshold temperature for third stage wandering larvae of *L. sericata*.

A reasonably accurate representation of the diurnal temperature pattern experienced in the field is a sine curve (Gettinby & Gardiner 1980), passing from a minimum, T_1, at the start of the day (time t=0), to a maximum T_{max} at the middle of the day (t=1/2), and returning to a second minimum, T_2, at the end of the day, where time t=1 (Figure 2). Details of the construction of the sine curves are given in Fenton *et al.* (1997).

For each life-cycle stage, the number of day degrees that is accumulated on any one day is given as the area below the temperature curve and above the base threshold for that stage. The calculated number of day degrees divided by the total number of day degrees required for completion of that stage gives the fraction of overall development undergone on that day. By this method, day degree fractions are accumulated even if the daily mean temperature lies below the base threshold, as long as the maximum temperature is above. Such conditions may occur in Spring and Autumn in Britain, so models which use only mean temperature values each day would underestimate the day degrees accumulated during these times of the year.

MODEL STRUCTURE

The models simulate the development of a single blowfly population on an individual sheep farm throughout the season; population movement such as immigration or emigration is disregarded. The population is represented as a collection of cohorts, where a cohort represents an identical group of individuals which start development at the same time. Hence, all the individuals arising from a single egg batch represent one cohort. The models iterate on a daily basis, starting on day 1, nominally 1 February, with an initial cohort of 100 larvae in a state of diapause. It is assumed that, before this time, temperatures are too cold for any post-diapause development to occur. After this point, the cohort undergoes development each day according to the temperatures it experiences, until sufficient day degrees have been accumulated to complete post-diapause development. The cohort then enters a pupal stage and, following completion of development of this stage, emerges as the season's first cohort of adults. After egg maturation, 50% of the adults oviposit the first egg batches of the season, laying 200 eggs each. To incorporate the effects of diapause in the model, larvae resulting from egg batches laid after day 210 (30th August) cease development to overwinter. Adult cohorts remain viable until the temperature drops below 10°C,

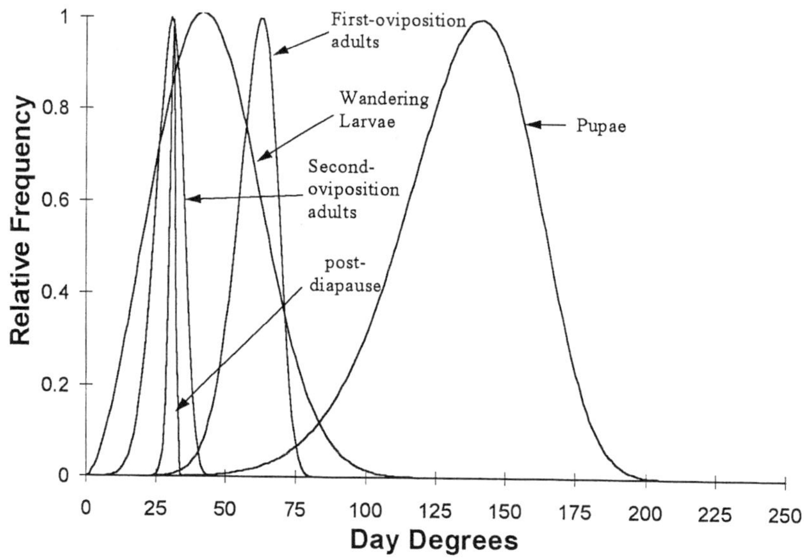

Figure 3. Weibull distributions (scaled to the same height) fitted to variations in development rate observed in constant temperature rearings for all life cycle stages of *L. sericata*.

when they are assumed to die. The model stops iterating once all cohorts are either dead or overwintering.

Mortality rates are assumed to be continuous and age-independent, so that the population age-distribution approximates a negative exponential (Wall *et al.* 1993b). The daily rate of *L. sericata* adult mortality in the field has previously been found to be temperature dependent (Wall 1993). Given the base temperatures for maturation of the ovaries and the day degree requirements presented in Table 1, at the temperatures experienced in Britain, this mortality rate typically allows approximately 15% of females to reach a first oviposition, 7% to reach a second, and 2.5% to reach a third. Pre-adult mortality rates in the field are unknown. However, as an arbitrary approximation, a constant proportion of individuals from all pre-adult cohorts are removed each day to produce an overall egg to adult mortality level of 50% per generation.

To simulate the effects of oviposition site deprivation on the population, it is assumed there is a maximum number of suitable oviposition sites available to the gravid female blowflies. This exact number of sites available does not qualitatively affect the predictions of the model, so an arbitrary maximum of 1500 suitable sites is assumed. A fixed proportion (10%) of this number of sites become available each day, up to the maximum number. Females unable to find a suitable site one day remain fully gravid until the next day, by which time more sites may have become available.

INTRODUCING STOCHASTICITY

The deterministic model assumes all individuals of a particular life cycle stage are identical in terms of their day degree requirements, with no accounting for variability between cohorts. However, in reality, insect populations of any one life-cycle stage typically show a positively skewed range of day degree requirements (Sharpe *et al.* 1977; Phelps *et al.* 1993). Variations such as these can affect both the patterns of abundance and the stability of the system to perturbations (Wagner *et al.* 1984; Dennis *et al.* 1986). The observed distributions in development rates for each life cycle stage can be described by a Weibull function (Fenton *et al.* 1997). The fitted Weibull distributions for each stage, scaled to the same height for comparison, are shown in Figure 3. Each cohort in the model is now randomly allocated day degree requirements for its life-cycle stage, generated from the appropriate Weibull distribution.

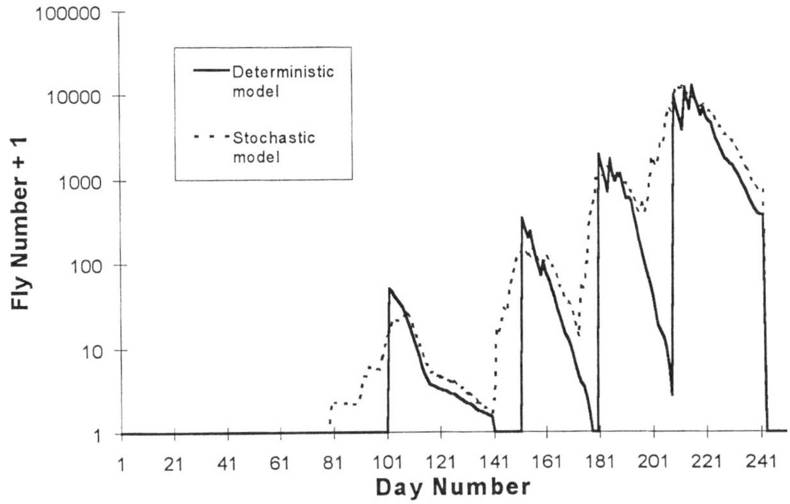

Figure 4. The seasonal pattern of *L. sericata* abundance as predicted by the deterministic and stochastic models, where day 1 is 1 February.

Since each run of this stochastic model produces a different pattern of abundance, the stochastic model is run a set number of times and the predicted population size on each day averaged across runs. By comparing the standard errors between sets of runs, it was previously decided that taking the average of 40 runs per treatment produced relatively consistent predictions (Fenton *et al.* 1997).

PREDICTED PATTERNS OF ABUNDANCE

The following simulations were carried out using temperature data recorded at ground level using an automatic data logger around the Bristol area in the south west of England in 1991. The output from the deterministic model predicts that, having completed post-diapause and pupal development, the adult flies first emerge on day 100 (Figure 4). The blowfly population then passes through a series of discrete generation waves, with fly numbers increasing exponentially each wave, peaking around days 150, 180 and 210. After day 210, the offspring from the fourth generation undergo diapause and no further population growth occurs. However, due to the variations in post-diapause and pupal development, the stochastic model predicts the first generation to emerge over a range of days, starting from day 80 (Figure 4). The subsequent generation waves are less distinct, with individuals of each generation present both earlier and later than the predictions of the deterministic model.

To test the accuracy of both model types, predicted population abundances have previously been compared to actual numbers of *L. sericata* caught on sticky targets baited with liver and sodium sulphide on sheep farms in Britain (Wall *et al.* 1993a; Fenton *et al.* 1997; Fenton & Wall 1997). Both models have been shown to provide an adequate prediction of fly abundance, with the deterministic model typically explaining around 60% of the variation in catch data, and the stochastic model 70%.

STRATEGIC CONTROL

To test the relative effects of different timings and targeting of control techniques, both models were initially run with 15% daily mortality applied to either the adults or the on-sheep stages over and above the background levels of mortality inherent in the models. Hence, out of those individuals surviving after the natural, daily mortality had been taken into account, a further 15% were removed each day as a result of control. Treatments were applied for periods of 30 days at one of 3 times of the season, timed to start just prior to the emergence of each generation wave, as predicted by the deterministic model. ie. days 100 to 130, 150 to 180 and 179 to 209. The level of control was then increased to 20% per day, and applied at the same times of the season.

Both models predicted similar results for both types, and all timings, of control application (Table 1). Adult control consistently produced a greater level of suppression than pre-adult control. For instance, the deterministic model predicts early season adult control to reduce the blowfly population by 90%, whereas early on-sheep treatment only reduces the population by around 70%. Furthermore, early season treatment of adults was far more successful in reducing the fly population than later treatments, which is in agreement with results found with previous models (Wall *et al.* 1993b). Mid season pre-adult control, however, suppressed the population to a greater extent than early season pre-adult control. Early on-sheep treatment still allowed many individuals to oviposit, thereby enabling the population to recover to levels observed in the absence of control, due to the number of oviposition sites available limiting maximum population size. However, targeting the offspring of the second generation of females does not allow sufficient time for the population to recover, and, hence, produces a greater level of suppression.

If the daily level of adult control is increased to 20%, the deterministic model predicts the blowfly population to become extinct after the first generation (Table 1). However, under the same level of control, the stochastic model predicts that the population is able to survive to pass through 4 full generations to the end of the season, albeit greatly suppressed. This is due to the variation in development rates between cohorts allowing some individuals to emerge either particularly early or particularly late and so, avoid the periods of control. Therefore, levels of suppression predicted by the deterministic model alone are unlikely to be realised in the field due to the inherent variability within the population.

LONG TERM BEHAVIOUR

The simulations presented so far were run until diapause interrupted further population expansion in late August. However, to explore the potential long-term behaviour of the blowfly system and help identify the main factors determining the pattern of *L. sericata* abundance, the stochastic model was allowed to run without the interference of diapause, for up to 300 days from 1 May.

The model was run with the maximum number of oviposition sites available to the blowfly population varied from 10 up to 1000, with 10% of this maximum number of sites becoming available each day. The temperature was kept at a constant mean value of 17°C each day. Under these conditions, the model was run 10 times for each treatment, with the resulting output being the average of these 10 runs. The predicted populations typically increased exponentially over

Table 1. Deterministic and stochastic model predictions of the effects of strategic control strategies targeting either adults or the on-sheep stages of the sheep blowfly *L. sericata*. The methods of control impose a stage-specific daily mortality of either 15% or 20% for periods of 30 days, where day 1 is 1 February. Results are expressed as the percentage change in the total number of flies present throughout the season (ie. the sum of fly numbers present on each day of the season) from a standard reference run of the appropriate model.

	Deterministic model		Stochastic model	
periods of imposed 15% daily mortality	on-sheep control	adult control	on-sheep control	adult control
days 100 - 130	73.9 %	95.6 %	55.1 %	89.9 %
days 150 - 180	76.4 %	80.9 %	81.0 %	72.4 %
days 179 - 209	68.8 %	72.1 %	67.0 %	68.2 %
periods of imposed 20% daily mortality				
days 100 - 130	91.8 %	*	85.7 %	95.8 %
days 150 - 180	96.2 %	89.2 %	96.7 %	79.0 %
days 179 - 209	86.9 %	80.9 %	88.4 %	81.3 %

* denotes population becoming extinct following treatment

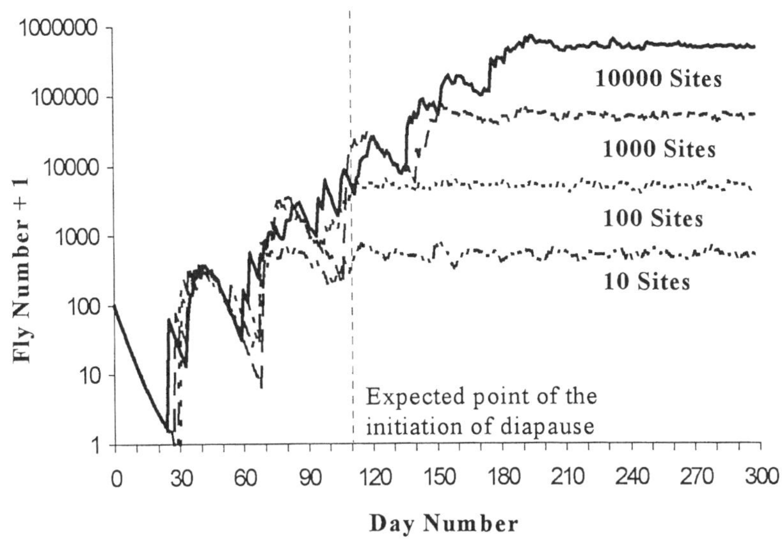

Figure 5. The effect of varying the maximum number of oviposition sites available to the *L. sericata* population as predicted by the stochastic model run beyond the point where diapause would normally inhibit population growth.

each generation until the number of suitable oviposition sites available became limiting. After this point the populations reached a stable equilibrium level (Figure 5).

There was a direct relationship between the number of oviposition sites available and the level at which the population reaches equilibrium. An increase in the number of sites allowed the blowfly population to pass through more generations before stabilising. However, at the temperatures used in these simulations (average seasonal temperature of 17°C), even with only 10 oviposition sites available, the population did not reach equilibrium before day 210, the point where diapause would normally be expected to inhibit further population growth.

DISCUSSION

Based on empirically derived relationships between temperature and development rate, a deterministic and a stochastic simulation model have been constructed to predict the abundance of the sheep blowfly, *Lucilia sericata*, throughout a season (Fenton *et al.* 1997). Realistic adult mortality rates are applied each day, and pre-adult mortality can be varied to further improve the accuracy of the models. These models typically predict the *L. sericata* adults to pass through four waves of emergence from early May to late September, with the final generation being limited by the onset of diapause. Similar deterministic models have been shown to significantly explain the pattern of blowfly abundance in the field (Wall *et al.* 1992a; Wall *et al.* 1993a).

The inherent variation in individual development rates within a population tends to blur the discrete generation waves predicted by the deterministic model (Wall *et al.* 1992a). The stochastic model, therefore, more closely resembles the pattern of blowfly abundance observed in the field (Fenton *et al.* 1997). Furthermore, since the stochastic model predicts the date of first emergence of the population over a range of days, the application of strategic control techniques just prior to the appearance of the first wave of adults can be timed more accurately in the field. Use of a deterministic model alone would result in some early developing flies being missed by such control efforts (Wall *et al.* 1993b).

The stochastic model also highlights the consequences large inter-cohort variation in day degree requirements may have on the future of the population, which will affect the accuracy of any prediction made by a model. At the start of the season, the random allocation of day degree requirements may result in the first cohort developing especially quickly. If this happens, the adults may emerge too early, when the temperatures are still too low to support them, and so the cohort dies (Fenton *et al.* 1997). This stresses the importance winter diapause has on synchronising the population at the same stage each year, and may explain the narrow range of variation in post-diapause development observed in the laboratory (Figure 3). Overwintering mortality is not explicitly included in the model, although, the fact that the blowfly population typically starts each year at low levels, despite the relatively high numbers present at the end of the season, suggests that there is a high degree of mortality occurring during diapause.

Both models presented show that the blowfly system is more sensitive to increases in adult mortality than increases in pre-adult mortality, especially early in the season. Furthermore, the stochastic model has previously been used to show that the population is relatively unstable, particularly early in the season when numbers are low (Fenton & Wall 1997), and that it may be possible to drive the population to extinction due to the effects of demographic stochasticity (Burkey 1995; Lindenmayer & Lacy 1995). Hence, as previous deterministic models have shown, strategic control targeting the adult flies early in the season should produce the greatest levels of suppression of the blowfly population. However, the stochastic model has the advantage of showing that the blowfly population can remain viable under levels of control that the deterministic model predicted sufficient to drive the population to extinction. Therefore, it is important to consider the effects of such variation when determining the levels and timing of control strategies.

In Britain, the development of *L. sericata* populations is inhibited by the initiation of diapause. In the present analysis, the simulations were run beyond the point where diapause would be expected to prevent further population growth. The results show that, given time, the *L. sericata* population will reach a state of stable equilibrium, due to the finite number of oviposition sites available to the population. This equilibrium state would continue indefinitely providing conditions remained the same.

The precise level attained at equilibrium appears to depend directly on both the number of oviposition sites available to the female blowflies and the rate at which new sites become available. Female *L. sericata* tend to oviposit in the breech area of sheep, preferentially selecting areas with high levels of faecal soiling (French *et al.* 1994; 1995). Clearly, increases in the rate of faecal soiling throughout a flock would increase the susceptibility of the sheep to strike by increasing the number of sites available to the female flies. Hence, worm control and techniques such as dagging, where faecally soiled wool is removed from the breech area, and tail docking, which reduce the incidence of faecal soiling could reduce both the number of suitable oviposition sites and the rate at which new sites occur. These management practices may prove to be effective contributory control methods, both directly reducing the incidence of blowfly strike and secondarily helping to suppress fly abundance. However, previous sensitivity analyses carried out over a typical blowfly season have shown that it is unlikely that blowfly populations are seriously affected by oviposition site limitation (Fenton & Wall 1997). Hence, although such management practices may reduce the immediate incidence of strike, they are unlikely to lead to a significant decline in the blowfly population throughout the season.

Any model based on the relationship between temperature and individual development rate depends on the accuracy with which the temperatures experienced by the target organisms is known (Moon 1983). The main problem with such a model is the accuracy with which temperature data obtained from weather stations reflects the actual temperature experienced by individuals in the field. Errors between actual and calculated rates of development would be greatest at extreme temperatures because this is where the temperature-development rate relationship deviates most from linearity (Clements 1963; Higley *et al.* 1986; Phelps *et al.* 1993). At the upper end of the temperature scale, this deviation starts to occur around 30°C for all stages of *L. sericata*. Since temperatures in Britain rarely exceed this, the error introduced at the higher temperatures is not too great. At the lower limit, the deviation tends to occur around 10°C, and so is usually only relevant during early Spring development, and would only affect those life stages exposed to the ambient temperature. Post-diapausing larvae, pupae, eggs and feeding larvae are all sheltered to some extent from the extremes of temperature that may occur in the field. Those stages which are exposed to the ambient temperature (adults, wandering larvae) are mobile, and may be able to find beneficial micro-climates where temperatures are nearer to optimal (Dallwitz 1984).

The above work has highlighted the use of deterministic and stochastic models in the prediction of the seasonal abundance of the sheep blowfly, *Lucilia sericata* on sheep farms in Britain. These models have been used to highlight the main factors affecting the stability ä explore the effects of various strategic control techniques. The stochastic model has been shown to have several advantages over the deterministic model, especially when considering the behaviour of the system under high mortality pressures. Such models can, therefore, prove to be invaluable tools in the fight against insect pest species.

ACKNOWLEDGMENTS

The invaluable support and advice of Drs. R. Wall and N.P. French and a LINK grant awarded by Bayer AG and the Ministry of Agriculture Fisheries and Food of the UK Government are gratefully acknowledged.

REFERENCES

Beck, S.D. 1983. Insect thermoperiodism. *Annual Review of Entomology* 28: 91-108.

Bernal, J and Gonzalez, D. 1993. Experimental assessment of a degree-day model for predicting the development of parasites in the field. *Journal of Applied Entomology* 116: 459-466.

Burkey, T.V. 1995. Extinction rates in Archipelagoes: Implications for populations in fragmented habitats. *Conservation Biology* 9: 527-554.

Clements, A.N. 1963. The physiology of mosquitoes. *International Monographs on Pure and Applied Biology* 17: 72-79.

Cragg, J.B. 1955. The natural history of sheep blowflies. *Annals of Applied Biology* 42: 197-207.

Dallwitz, R. 1984. The influence of constant and fluctuating temperatures on development rate and survival of pupae of the Australian sheep blowfly *Lucilia cuprina. Entomologia Experimentalis et Applicata* 36: 89-95.

Dennis, B., Kemp, W.P. and Beckwith, R.C. 1986. Stochastic model of insect phenology: estimation and testing. *Environmental Entomology* 15: 540-546.

Fenton, A. and Wall, R. 1997. Sensitivity analysis of a stochastic model of the sheep blowfly *Lucilia sericite. Journal of Applied Ecology* (in press).

Fenton, A., Wall, R. and French, N. 1997. Sensitivity analysis of deterministic and stochastic simulation models of populations of the sheep blowfly, *Lucilia sericata. Journal of Theoretical Biology* 184: 139-148.

French, N.P., Wall,·R. and Morgan, K.L. 1994. Lamb tail docking: a controlled field study of the effects of tail amputation on health and productivity. *The Veterinary Record* 134: 463-467.

French, N.P., Wall, R. and Morgan, K.L. 1995. The seasonal pattern of sheep blowfly strike in England and Wales. *Medical and Veterinary Entomology* 9: 1-8.

Gettinby, G. and Gardiner, W.P. 19870. Disease incidence forecasts by means of climatic data. *Biometeorology* 7: 87-103.

Guerrini, V.H. 1988. Ammonia toxicity and alkalosis in sheep infested by *Lucilia cuprina* larvae. *International Journal of Parasitology* 18: 79-81.

Higley, L.G., Pedigo, L.P. and Ostlie, H.R. 1986. DEGDAY: A program for calculating degree-days and assumptions behind the degree-day approach. *Environmental Entomology* 15: 999-1016.

Lindenmayer, D.B. and Lacy, R.C. 1995. Metapopulation viability of Leadbeater's possum, *Gymnobelideus leadbeateri*, in fragmented old-growth forests. *Ecological Applications* 5: 164-182.

Moon, R.D. 1983. Simulating developmental time of preadult face flies (Diptera: Muscidae) from air temperature records. *Environmental Entomology* 12: 943-948.

Petitt, F.L.., Allen, J.C. and Barfield, C.S. 1991. Degree day model for vegetable leafminer (Diptera: Agromyzidae) phenology. *Environmental Entomology* 20: 1134-1140.

Phelps, K., Collier, R.H., Reader, R.J. and Finch, S. 1993. Monte Carlo simulation method for forecasting the timing of pest insect attacks. *Crop protection* 12: 335-342.

Preuss, K.P. 1983. Day-degree methods for pest management. *Environmental Entomology* 12: 613-619.

Saunders, D.S., MacPherson, J.N. and Cairncross, K.D. 1985. Maternal and larval effects of photoperiod on the induction of larval diapause in two species of fly, *Calliphora vicina* and *Lucilia sericata*. *Experimental Biology* 46: 51-58.

Sharpe, P.J., Curry, G.L., DeMichele, D.W. and Cole, C.L. 1977. Distribution model of organism development times. *Journal of Theoretical Biology* 66: 21-38.

Wagner, T.L., Wu, H., Sharpe, P.J. and Coulson, R.N. 1984. Modelling distributions of insect development time: A literature review and application of the Weibull function. *Annals of the Entomology Society of America* 77: 475-487.

Wall, R. 1993. The reproduction output of the blowfly *Lucilia sericata*. *Journal of Insect Physiology* 39: 743-750.

Wall, R., French, N.P. and Morgan, K.L. 1992a. Effects of temperature on the development and abundance of the sheep blowfly *Lucilia sericata* (Diptera: Calliphoridae). *Bulletin of Entomological Research* 82: 125-131.

Wall, R., French, N.P. and Morgan, K.L. 1993a. Predicting the abundance of the blowfly *Lucilia sericata* (Diptera: Calliphoridae). *Bulletin of Entomological Research* 83: 431-436.

Wall, R., French, N.P. and Morgan, K.L. 1993b. Sheep blowfly population control: development of a simulation model and analysis of management strategies. *Journal of Applied Ecology* 30: 743-751.

Wall, R., Green, C.H., French, N.P. and Morgan, K.L. 1992b. Development of an attractive target for the sheep blowfly *Lucilia sericata*. *Medical and Veterinary Entomology* 6: 67-74.

Population and Community Ecology for Insect Management and Conservation, Baumgärtner et al. (eds)

Mathematical models for age-structured population dynamics: An overview

G. Di Cola & G. Gilioli
Dipartimento di Matematica, Università di Parma, Italy

J. Baumgärtner
International Centre of Insect Physiology and Ecology, Nairobi, Kenya

ABSTRACT: The overview on structured population models starts with the presentation of deterministic models (section 2). On this ground models with physiological variables (section 3) and, subsequently, stochastic models (section 4) are introduced. In the second section the continuous von Förster and the discrete Leslie model are shown to be the basic approaches to represent the dynamics of age-structured populations. The up-wind scheme of the von Förster model and the sojourn time models are related. For illustration purposes a sojourn time model and a Leslie model has been chosen. In the third section physiological variables are introduced into structured population models by considering a variable for mass or mass equivalents and/or by using physiological time and age rather than chronological time and age. Both possibilities have been considered in the illustrative example. In the fourth section macroscopic and microscopic stochastic models are introduced. The former are based either on diffusion processes or on time distributed delays. Microscopic models are often considered to be biologically more appealing than macroscopic models, they have recently received considerable attention in population ecology. To illustrate these approaches we primarily refer to structured population models developed by the authors.

1 INTRODUCTION

During the past two decades much progress has been made in the development of mathematical models for structured populations (e.g. Metz & Diekman, 1986, McDonald et al., 1989; Renshaw, 1992). During the same period, population ecologists have expressed an increasing interest in the analysis of population dynamics and in the design of population management systems based on the solid theoretical ground provided by mathematical models (e.g. Baumgärtner & Gutierrez, 1988; Di Cola & Gilioli, 1996; Gutierrez, 1996). However, our experience suggests that even more progress could have been made if theoretical and applied work would have been better coordinated than apparently has been done so far (Buffoni et al., 1995). The need to carry out interdisciplinary research has been recognized already at the beginning of this time period (e.g. Huffaker & Croft, 1976).

In this work we summarize the analysis of the mathematical models for the dynamics of age-structured populations in an attempt to further bridge the gap between theory and application. In general, we follow model development and model application through the following important steps: age-structured models, the incorporation of physiological variables, the addition of stochastic variability and the recent discussion of microscopic and macroscopic population models (Metz & Diekman, 1986; Munholland & Dennis, 1992). For the sake of simplicity, this presentation is limited to the models representing the local dynamics of single-species populations.

Furthermore, the analysis is restricted to the models representing the local dynamics of single-species populations including the temporal dynamics that depend on external abiotic factors but disregarding density dependent effects. However, the latter restriction is exclusively applied to the analytic part, whereas density-dependent effects do operate in some of the illustrative examples. In general, such models are particularly useful in theoretical demographic studies, but their applicability in population management is limited to a small temporal scale.

The above steps are expected to elucidate the common mathematical ground of widely used demographic models. This ground may facilitate the task of a population ecologist identifying and developing an adequate model. On each step representative examples are selected to illustrate the use of the models and to guide the reader looking for detailed information on model development and application.

2 DETERMINISTIC MODELS WITH AGE STRUCTURE

2.1 *Introduction*

Sinko and Streifer (1967) observed that nearly all mathematical models of single-species population assumed that the organisms were identical regardless of size and age. Among the exceptions there were the von Förster (1959) continuous and the Leslie (1945; 1948) discrete analogous model. In both models the changes in the density distribution $n(t,a)$ are related to chronological age a indicating that physiological qualities are generally disregarded.

This basis permitted the development of other less well known models that allow the introduction of physiological parameters (section 3) and stochastic development (section 4). Among them there are sojourn time models which have important analogies to the models derived from the von Förster equation.

2.2 *Model description*

2.2.1 The von Förster model

Von Förster proposed the following model for age-structured populations

$$\frac{\partial n}{\partial t}(t,a)+\frac{\partial n}{\partial a}(t,a)=-\mu(t,a)n(t,a) \tag{2.1}$$

with the boundary condition

$$n(t,0)=\int_0^{\infty}\lambda(t,a)n(t,a)da \tag{2.2}$$

and initial condition

$$n(0,a)=n_0(a) \tag{2.3}$$

where t is the chronological time, a is the chronological age, $n(t,a)\Delta t$ is the number of individuals at time t and chronological age a, and $\mu(t,a)$ is the specific (per capita) death rate of individuals, $\lambda(t,a)$ is the specific (per capita) fecundity rate of individuals.

2.2.2 The Leslie model

The Leslie model can be derived as a discrete analogue of the von Förster model for a population with periodical reproduction and overlapping generations. The solution of the equation (2.1) along the characteristic lines, in discrete terms, with $\frac{\Delta a}{\Delta t}=1$, is given by

$$n_i(t+\Delta t)=n_i(t)s_i(t) \tag{2.4}$$

with the boundary condition

$$n_0(t)=\sum_{i=0}^{m}\lambda_i(t)n_i(t) \tag{2.5}$$

where $\lambda_i(t)=\lambda(t,a_i)\Delta a$ is the fecundity rate of the i-th age class at time t, $s_i(t)=exp\left(-\int_{a_i}^{a_{i+1}}\mu(t,\xi)d\xi\right)\cong 1-\mu_i(t,a_i)\Delta a$ is the survival rate of the i-th age class at time t, $n_i(t)=n(t,a_i)$ and $a_{i+1}=a_i+\Delta a$, for $i=0, 1, \ldots, m$, $a_0=0$, m is the number of age classes obtained by subdividing the age. From equation (2.4) the widely used Leslie model can be obtained

$$\boldsymbol{n}(t+\Delta t)=\boldsymbol{F}(t)\,\boldsymbol{n}(t) \tag{2.6}$$

where $\boldsymbol{n}(t)=[n_0(t), n_1(t), \ldots, n_m(t)]^T$ and $\boldsymbol{F}(t)$ is the Leslie matrix (also known as the projection matrix)

$$\boldsymbol{F}(t)=\begin{bmatrix} \lambda_0(t) & \lambda_1(t) & \cdots & & \lambda_m(t) \\ s_0(t) & 0 & & & 0 \\ 0 & s_1(t) & \ddots & & \vdots \\ \vdots & & \ddots & \ddots & \\ 0 & \cdots & & s_{m-1}(t) & 0 \end{bmatrix}. \tag{2.7}$$

2.2.3 The up-wind scheme

The up-wind numerical scheme (Johnson, 1992) for equation (2.1) is described by the system of ordinary differential equations

$$\frac{dn_i(t)}{dt}+\frac{n_i(t)-n_{i-1}(t)}{\Delta a}=-\mu_i(t)n_i(t). \tag{2.8}$$

This scheme is equivalent to the following semidiscrete advection-diffusion model

$$\frac{dn_i(t)}{dt}+\frac{n_{i+1}(t)-n_{i-1}(t)}{2\Delta a}-\frac{\Delta a}{2}\left[\frac{n_{i+1}(t)-2n_i(t)+n_{i-1}(t)}{\Delta a^2}\right]=-\mu_i(t)n_i(t). \tag{2.9}$$

This can be interpreted as the discretization of a advective-diffusive equation with the diffusion coefficient $\left(\frac{\Delta a}{2}\right)$ explored further in section 4. The integration of equation (2.8), by means of forward Euler method, with $\Delta a = \Delta t$, yields

$$n_i(t+\Delta t) = n_{i-1}(t) + \left(s_i(t) - 1\right)n_i(t) \qquad \text{for } i = 0, 1, \ldots, m. \tag{2.10}$$

In matrix form we have

$$\boldsymbol{n}(t+\Delta t) = \boldsymbol{G}(t)\boldsymbol{n}(t) \tag{2.11}$$

where $\boldsymbol{G}(t)$ is like a Leslie matrix

$$\boldsymbol{G}(t) = \begin{bmatrix} \lambda_0(t) & \lambda_1(t) & \cdots & \cdots & \lambda_m(t) \\ 1 & s_1 - 1 & & & 0 \\ 0 & \ddots & \ddots & & \vdots \\ \vdots & & . & & 0 \\ 0 & \cdots & \cdots & 1 & s_m - 1 \end{bmatrix}. \tag{2.12}$$

The up-wind scheme of equation (2.1) appears to be less frequently used then Leslie's model. From a theoretical standpoint, however, it introduces a diffusion in population development (see section 4).

2.2.4 Sojourn time model

The development interval can be divided into a set of finite k sub-intervals (developmental stages or age classes). As a result, equation (2.1) can be approximated by a system of ordinary differential equations which transfers the model from a distributed parameter structure to a concentrated parameter structure (Cappio Borlino et al., 1991)

$$\frac{dn_i}{dt} = R_i - R_{i+1} - \mu_i(t)n_i(t) \tag{2.13}$$

where $i = 1, 2, \ldots, k$ is the subscript for the sub-interval, R_i represents the transfer rates among the different stages, R_0 is the reproductive rate, and $\mu_i(t)n_i(t)$ indicates the death rate in stage i. It can be assumed that

$$R_0(t) = \sum b_i(t)n_i(t). \tag{2.14}$$

The rates R_i can be approximated by

$$R_i \cong \frac{n_i}{T_i} \tag{2.15}$$

where T_i measure the average sojourn time, also called residence or maturation time, of each individual of the population in the stage i.

In such an approximation the different developmental stages are represented by a succession of homogeneous compartments, whereas the birth, development and death processes are described as input, output and connecting fluxes among the different compartments. By using equation (2.15) equation (2.13) becomes

$$\frac{dn_i}{dt} = \frac{n_{i-1}}{T_{i-1}} - \frac{n_i}{T_i} - \mu_i n_i \,. \tag{2.16}$$

If $T_{i-1} = T_i$ the model can be related to the up-wind scheme of equation (2.8). Moreover, the transformation of the equation (2.16) in discrete terms with respect to the time variable, as obtained by the Euler method, yields the following scheme

$$n_i(t+\Delta t) - n_i(t) = \Delta t\left[\frac{n_{i-1}(t)}{T_{i-1}} - \frac{n_i(t)}{T_i} + \mu_i(t) n_i(t)\right], \tag{2.17}$$

so that, with $\Delta a = \Delta t$,

$$n_i(t+\Delta t) = \left(\frac{\Delta t}{T_{i-1}}\right) n_{i-1}(t) + \left[s_i - \left(\frac{\Delta t}{T_i}\right)\right] n_i(t) \,. \tag{2.18}$$

In matrix form we have

$$\boldsymbol{n}(t+\Delta t) = \boldsymbol{G}(t)\boldsymbol{n}(t) \tag{2.19}$$

where $\boldsymbol{G}(t)$ is the matrix

$$\boldsymbol{G}(t) = \begin{bmatrix} \lambda_0(t) & \lambda_1(t) & \cdots & \lambda_m(t) \\ \dfrac{\Delta t}{T_0} & s_1(t) - \dfrac{\Delta t}{T} & & 0 \\ 0 & \ddots & \ddots & \vdots \\ \vdots & & & \\ 0 & \cdots & \dfrac{\Delta t}{T_{m-1}} & s_m(t) - \dfrac{\Delta t}{T_m} \end{bmatrix}. \tag{2.20}$$

Thus, the widely used sojourn time models and their time discretizations are similar to the up-wind scheme and the Leslie matrix model.

2.3 *Examples*

In this section we have presented models in which the per capita birth and death rates exclusively depend on age. These models have been extended to include density-dependencies and effects of variable environments (Gurney et al., 1983). Cavalloro et al. (1975) used a sojourn time model to estimate demographic parameters of *Bactrocera oleae* (Gmel.) populations.

The model on *Tribolium castaneum* Herbst (Desharnais & Liu, 1987) illustrates the use of the Leslie model and shows the effect of density-dependent effects on birth and death rates.

3 DETERMINISTIC MODELS WITH PHYSIOLOGICAL PARAMETERS

3.1 *Introduction*

The use of chronological age is justified as long the effects of external factors such as temperature or food are not explicitly addressed. If these factors are taken into account, the concept of physiological age, i.e. the level of development reached by an individual or by a cohort, has to be considered in model development. Of particular importance is the temperature which largely controls the development of poikilotherms. During development, physiological age augments by the summation of elementary developmental units which are obtained by expressing the developmental rates in function of external factors.

According to Schaalje and van der Vaart (1989) two classes of models have been used to represent physiological variables in population models. In *sojourn time models* all individual advance in "ages" at identical rates but change from one growth stage to the next at various "ages". If physiological age is considered, cohort development follows a characteristic line inclined by 45° on a Cartesian plain defined by physiological time and physiological age. Sojourn time models communicate little of individual development but more on how long individuals are sojourned in a stage under study. This model formally corresponds to the von Förster model of equation (2.1), in which chronological time and chronological age are substituted by physiological time and physiological age.

In the *developmental index models* all individuals change from one growth stage to another at the same "age", but advance in "age" at different rates. The model is called a developmental index model because the "age" of an individual carries information on the level of development. In contrast to the above sojourn time models, development index models require substantial changes in the von Förster model (see equation 2.1).

In this paper we focus on sojourn time models.

3.2 *Model description*

3.2.1 Generalized von Förster model

The development dynamics, in terms of physiological age (or developmental index) x, is described by the ordinary differential equation

$$\frac{dx}{dt} = v(t, x). \tag{3.1}$$

Accordingly, $v(t, x)$ is the developmental rate of individual organisms. We introduce a density function $n(t, x)$ stating that $n(t, x)\Delta x$ is the total number of individuals in a population between the physiological age x and $x + \Delta x$ at time t.

In many cases it can be assumed that the change in age with respect to time is an environmental variable-dependent rate function. For poikilotherms the temperature is such an environmental variable, and the physiological parameter x (developmental index) may change in time t with velocity v according to

$$\frac{dx}{dt} = v[\psi(t), x] \tag{3.2}$$

where $v(\cdot)$ is a temperature-dependent rate function, and $\psi(t)$ is the temperature at time t (Curry & Feldman, 1987).

Different models can be used to represent the development rate-temperature relationship. Among them there is the widely used linear model

$$v = c(\psi - \psi_0) \qquad \psi \geq \psi_0, \quad v = 0 \;\; \forall \;\; \psi < \psi_0$$

where c is a constant; the inverse of c is known as thermal constant and ψ_0 is the lower developmental threshold (Gilbert et al., 1976). Aging occurs in units of this thermal constant. For example, Hughes (1963) has expressed both time and age in physiological units of day-degrees and, thereby, created the ground for the models of Gilbert et al. (1976) and Gutierrez (1996). Nonlinear models to represent developmental rates-temperature relationships were proposed, for example, by Logan et al. (1976) and Sharpe & DeMichele (1977).

For a specified temperature regime $\psi(t)$ the extended von Förster model becomes for $t, x > 0$ (Curry & Feldman, 1987)

$$\frac{\partial n}{\partial t} + \frac{\partial}{\partial x}\left[v(\psi(t), x)n(t,x)\right] = -\mu(t,x)n(t,x) \tag{3.3}$$

with the initial condition

$$n(0,x) = n_0(x)$$

and boundary condition

$$v(t,0)n(t,0) = \int_0^{\infty} \lambda(t,x)n(t,x)dx$$

where $q(t) = v(t,0)n(t,0)$ is the inward flux of individuals. If instead of chronological time we use physiological time τ, then $dx/d\tau = 1$ and the equation can be reduced to the conventional original formulation of the von Förster model (2.1, 2.2, 2.3).

3.2.2 The generalized Leslie model

Also the Leslie method can be generalized to take into account physiological age x. To achieve the generalization we assume that equation (3.3) is applied to a population for which v, μ and λ are function of only physiological age. As a consequence the equation (3.3) is autonomous, and the solution trajectories exclusively depend on an initial physiological age distribution $n(0,x)$. Equation (3.3), upon application of the product rule for differentiation, becomes

$$\frac{\partial n(t,x)}{\partial t} + v(t,x)\frac{\partial n(t,x)}{\partial x} = -\left[\mu(x) + \frac{dv(t,x)}{dx}\right]n(t,x). \tag{3.4}$$

By the method of characteristics, equation (3.4) is equivalent to the set of differential equations

$$dt = \frac{dx}{v(t,x)} = \frac{dn(t,x)}{-\left[\dfrac{dv(t,x)}{dx} + \mu(x)\right]}. \tag{3.5}$$

Standard techniques provide, in discrete terms, the following solution to equation (3.5)

$$n(t+\Delta t, x_{i+1}) \cong n(t, x_i^*) s_i^*(t) \tag{3.6}$$

where $x_i^* = x_{i+1} - v_i(t)\Delta t$, $s_i^* \cong 1 - \left[\dfrac{v'_i(t) + \mu(x_i)}{v_i(t)}\right]\Delta x$, $v_i(t) = v(t, x_i)$ and $v'_i = \left.\dfrac{\partial v(t,x)}{\partial x}\right|_{x=x_i}$;

x_i^* can be interpolated between x_i and x_{i+1} as $x_i^* = \omega_i x_i + (1-\omega_i)x_{i+1}$ with $0 \le \omega_i \le 1$. Then, we can write

$$n_{i+1}(t+\Delta t) = \left[\omega_i n_i(t) + (1-\omega_i) n_{i+1}(t)\right] s_i^* \tag{3.7}$$

with boundary condition

$$v_0(t) n_0(t) = \sum_{i=1}^{k} \lambda_i(t) n_i(t),$$

where $q(t) = v_0(t) n_0(t)$ is the inward flux of individuals. In matrix form we have

$$\boldsymbol{n}(t+\Delta t) = \boldsymbol{F}^*(t)\boldsymbol{n}(t) \tag{3.8}$$

where $\boldsymbol{F}^*$ is a weighted Leslie matrix with s_i replaced by s_i^*.

This model can not be applied easily to the population processes, because the age-classes depend on both the developmental rate and time.

3.2.3 The up-wind scheme

The use of the up-wind scheme in modeling population dynamics with physiological variables was first proposed by McKendrick (1926) with reference to the cell population dynamics. As described in section 2 the up-wind scheme can be obtained by approximating the partial derivative $\left.\dfrac{\partial f}{\partial x}\right|_{x=x_i}$ by the backward difference $\dfrac{f_i - f_{i-1}}{\Delta x_i}$ in the equation (3.3) to give

$$\frac{dn_i(t)}{dt} + \frac{v_i(t)n_i(t) - v_{i-1}(t)n_{i-1}(t)}{\Delta x} = -\mu_i(t) n_i(t). \tag{3.9}$$

If we use the forward Euler method we have

$$n_i(t+\Delta t) = \frac{\Delta t}{\Delta x_i} v_{i-1}(t) n_{i-1}(t) + \left[s_i(t) - v_i(t)\frac{\Delta t}{\Delta x_i}\right], \tag{3.10}$$

where $s_i = 1 - \mu_i(t)\Delta t$ is the survival of the i-th age class at time t. If we take $\Delta x_i = v_i(t)\Delta t$ we obtain the matrix form

$$\boldsymbol{n}(t+\Delta t) = \boldsymbol{G}(t)\boldsymbol{n}(t). \tag{3.11}$$

The up-wind scheme is equivalent to

$$\frac{dn_i}{dt}+\frac{v_{i+1}n_{i+1}-v_{i-1}n_{i-1}}{2\Delta x_i}-\frac{\Delta x_i}{2}\left[\frac{v_{i+1}n_{i+1}-2v_in_i+v_{i-1}n_{i-1}}{\Delta x_i^{\,2}}\right]=-\mu_in_i \tag{3.12}$$

where $\frac{\Delta x_i v_i}{2}$ can be interpreted as a diffusion coefficient (see section 4).

3.2.4 Model for age and mass dependent growth (two-dimensional von Förster model)

Time and age variables alone are often not sufficient to adequately describe all population systems (Curry & Feldman, 1987). For example biomass-dependent growth must sometimes be considered, necessitating the inclusion of a biomass variable in the von Förster model in addition to time and age. Other physiological characters may also be represented in the models considered in this section.

Sinko and Streifer (1967) were the first to introduce, in addition to chronological age and chronological time, the mass w as a third variable into the von Förster model. Wang et al. (1977) substituted chronological age and chronological time by physiological age and physiological time and discretized the so extended Sinko-Streifer model (Gutierrez, 1996).

A more general description of the two-dimensional von Förster model has been proposed by Nisbet and Gurney (1982)

$$\frac{\partial n}{\partial t}(t,x,w)+\frac{\partial}{\partial x}(vn)+\frac{\partial}{\partial w}(gn)=-\mu(t,x,w)n(t,x,w) \tag{3.13}$$

with boundary condition

$$n(t,0,w_0)=\int_0^\infty\int_0^\infty\lambda(t,x,w)n(t,x,w)dxdw \tag{3.14}$$

where w_0 is the initial biomass of the newly born individuals (note that in this model the initial weight is the same for all individuals).

The analogous discrete of the model specified in equations (3.13) (3.14) in terms of the up-wind approximation is given by

$$\frac{dn_{ij}(t)}{dt}+\frac{v_{ij}(t)n_{ij}(t)-v_{i-1j}(t)n_{i-1j}(t)}{\Delta x}+\frac{g_{ij}(t)n_{ij}(t)-g_{ij-1}(t)n_{ij-1}(t)}{\Delta w}=-\mu_{ij}(t)n_{ij}(t) \tag{3.15}$$

where v is the development rate and g is the growth rate with reference to unit of biomass, with the boundary condition

$$n_{11}(t)=\sum_{i=1}^{I}\sum_{j=1}^{J}\lambda_{ij}(t)n_{ij}(t) \tag{3.16}$$

where $n_{ij}(t)=n(t,x_i,w_j)$, $v_{ij}(t)=v(t,x_i,w_j)$ and $g_{ij}(t)=g(t,x_i,w_j)$. In matrix form we have

$$\frac{d\boldsymbol{n}}{dt}=\boldsymbol{Fn} \tag{3.17}$$

where $\boldsymbol{F}$ is sum of a low-tridiagonal matrix and a dyadic matrix with the first row given by

$$[\lambda_{11}, \lambda_{12}, ..., \lambda_{1J}, \lambda_{21} ..., \lambda_{2J}, ... \lambda_{I1}, ... \lambda_{IJ}]. \tag{3.18}$$

As in the one-dimensional case this scheme displays diffusive properties in both age and biomass dimensions.

3.3 *Examples*

The population dynamics of cotton herbivores has been represented by a discretized extended von Förster model and coupled with the dynamics of cotton yield information (Wang et al., 1977). Thereby, physiological time and physiological age in day-degrees are used, and the dynamics of fruits is represented by a biomass-dependent three variable generalized von Förster model.

4 STOCHASTIC DEVELOPMENT MODELS

4.1 *Introduction*

The consideration of parameters with associated variability leads to various stochastic models. In this work we focus exclusively on the variability associated with developmental times.

Completion of development by a group of poikilotherms with identical age, i.e. a cohort, is distributed over a range of developmental times even if they experience the same environmental conditions. The dispersion of development is frequently ignored, with only a mean-value rate function being utilized and biological variability neglected (Curry & Feldman, 1987). If the range of the variation is significant in comparison to the mean, it becomes often important to take into account inter-individual variation in the rate of development.

In this section, two types of stochastic models, i.e. macroscopic and microscopic models (Munholland & Dennis, 1992), are presented. Life history data arising from single-generation populations, composed of single or multiple cohorts, are used to formulate macroscopic stochastic models representing gross or macroscopic changes in the population structure. The population attributes are usually represented by continuous or discrete distribution functions.

As an alternative, models describing the life history of an individual can be developed. These models are called microscopic because they cast the dynamics in more reductionistic terms by following individuals as discrete entities to construct the population dynamics. Often, microscopic models are considered to be biologically more appealing than those developed at the macro level (Munholland & Dennis, 1992).

Macroscopic models have also been addressed as *i*-state distribution models, while microscopic models have been named *i*-state configuration models (Metz & Diekman, 1986).

Among the macroscopic models we present reaction-diffusion models and time distributed delay models.

4.2 *Model description*

4.2.1 Macroscopic model

A) Diffusion models

Suppose that different individuals develop at different rates with mean $v(t, x)$ and variance $k(t, x)$. We assume that the stochastic process of change in development is Markovian, so that if the development level is known for any specific value of the time parameter t, that information is sufficient to predict the behavior of the process beyond that point (Bhat, 1984).

The dynamics of the population with death is described by the following forward Kolmogorov equation (Bellagamba et al., 1989)

$$\frac{\partial n(t,x)}{\partial t}+\frac{\partial}{\partial x}v(t,x)n(t,x)-\frac{1}{2}\frac{\partial^2}{\partial x^2}k(t,x)n(t,x)=-\mu(t,x)n(t,x), \tag{4.1}$$

where $\mu(t,x)$ is the death rate of an individual of age x at time t. The renewal process can be interpreted as the boundary condition

$$\left[v(t,x)n(t)-\frac{1}{2}\frac{\partial}{\partial x}(k(t,x)n(t,x))\right]_{x=0}=\int_0^\infty \lambda(t,x)n(t,x)dx, \tag{4.2}$$

and represents the inward flux $q(t)$ of individuals, where $n(t,x)$ is the density of the population and $\lambda(t,x)$ is the per capita fecundity rate. The initial condition is like (2.3).

Two numerical methods can be used to solve the equations (4.1), (4.2); the first approximation is obtained by combining the method of characteristics with finite element method (Buffoni & Di Cola, 1995; Buffoni et al., 1996) and using the Euler implicit time stepping. This procedure yields

$$\begin{aligned}&n_i(t+\Delta t)-n_{is}(t)-\frac{1}{2}\frac{\Delta t}{\Delta x_i^2}[k_{i+1}(t+\Delta t)n_{i+1}(t+\Delta t)-2k_i(t+\Delta t)n_i(t+\Delta t)+\\&+k_{i-1}(t+\Delta t)n_{i-1}(t+\Delta t)]=-\mu_i(t+\Delta t)n_i(t+\Delta t)\end{aligned} \tag{4.3}$$

with $q(t)=\sum_{i=1}^{k}\lambda_i(t)n_i(t)$ and where $n_i(t)=n(t,x_i)$, $\mu_i(t)=\mu(t,x_i)$, $\Delta x_i=x_{i+1}-x_i$, and $n_{is}(t)=n(t,x_i-v_i\Delta t)$.

The second method is directly based on finite element method and gives the following difference equations (Buffoni & Di Cola, 1995; Buffoni et al., 1996)

$$\begin{aligned}&n_i(t+\Delta t)-n_i(t)-\frac{\Delta t}{2\Delta x_i^2}[k_{i+1}(t+\Delta t)n_{i+1}(t+\Delta t)-2k_i(t+\Delta t)n_i(t+\Delta t)+\\&+k_{i-1}(t+\Delta t)n_{i-1}(t+\Delta t)]+\frac{\Delta t}{2\Delta x_i}[v_{i+1}(t+\Delta t)n_{i+1}(t+\Delta t)-v_{i-1}(t+\Delta t)n_{i-1}(t+\Delta t)]=\\&=-\mu_i(t+\Delta t)n_i(t+\Delta t).\end{aligned} \tag{4.4}$$

with $q(t)=\sum_{i=1}^{k}\lambda_i(t)n_i(t)$.

Both methods can be represented in a more simple form using the matrix notation.

B) Distributed delay models

The up-wind scheme (3.9) proposed by McKendrick (1926) provides the ground for the development of models which, for historical reasons, have been called box-cars (de Wit & Goudriaan, 1978) and time-distributed models (Manetsch, 1976). Here, we focus only on distributed delay models characterized by the assumption

$$v_i(t) = v(t) = \frac{k}{del(t)} \Delta x \qquad i = 0, \ldots, k, \tag{4.5}$$

where k is the number of age intervals and $del(t)$ is the expected value of emergence time. From (3.9) we obtain

$$\frac{dn_i}{dt} = \frac{k}{del(t)} [n_{i-1}(t) - n_i(t)] - \mu_i(t) n_i(t) \tag{4.6}$$

which, in terms of flux $r_i(t) = n_i(t) \frac{k}{del(t)}$, gives

$$\frac{d}{dt}\left[\frac{del(t)}{k} r_i(t)\right] = r_{i-1}(t) - r_i(t) - \frac{del(t)}{k} \mu_i(t) r_i(t). \tag{4.7}$$

This is similar to a model proposed by Manetsch (1976) and Vansickle (1977). In the case of time varying developmental times and mortalities the model does not lead to a definite distribution of emergence times.

However, when $del(t) = const$ (then $v(t) = const$) and $\mu_i(t) = const = \mu$ the density distribution $f(t)$ of emergence times of cohort individuals is a modified Gamma function (Vansickle, 1977; Plant & Wilson, 1986)

$$f(t) = \left(\frac{k}{del}\right)^k \frac{t^{k-1}}{(k-1)!} exp^{\left(-\frac{k}{del} t\right)} exp^{-\mu t}. \tag{4.8}$$

This shifted Gamma function is used for example in the Blythe et al. (1984) model which has found wide application. In absence of mortality $\mu = 0$ we obtain an Erlangian distribution of emergence times

$$f(t) = \left(\frac{k}{del}\right)^k \frac{t^{k-1}}{(k-1)!} exp^{\left(-\frac{k}{del} t\right)} \tag{4.9}$$

as observed by Manetsch (1976) and Severini et al. (1990).

Another stochastic interpretation of the up-wind scheme (3.9) proposed by McKendrick (1926) is given by the equation (4.10); in fact (3.9) can be regarded as the analogous discrete of the continuous equation

$$\frac{\partial n}{\partial t} + \frac{\partial}{\partial x}(nv) - \frac{\partial^2}{\partial x^2}\left(\frac{V}{2} n\right) = -\mu n \tag{4.10}$$

with $V = v\Delta x$. In fact, the (3.9) leads to a stochastic interpretation of the McKendrick scheme as the analogous discrete of the Kolmogorov equation for stochastic population development with mean development rate v and variance $v\Delta x$ (Buffoni & Di Cola, 1995; Buffoni et al., 1996).

4.2.2 Microscopic model

We assume that the development of an insect can be viewed as an accumulation of small developmental increments over time. For every small time interval Δt the developmental level changes by an amount Δx drawn from a continuous probability distribution. As Δt becomes smaller, we would expect both the mean and variance of Δx to get closer to zero. We assume in this case that the mean and variance of Δx can be approximated by linear function of Δt (disregarding terms of order greater than Δt), so that

$$E(\Delta x) \cong vDt \qquad \text{and} \qquad Var(\Delta x) \cong \sigma^2 \Delta t, \tag{4.11}$$

and we finally assume that $x(t)$ is composed of a sum of n such random increments, each drawn independently from a probability density function $d(\Delta t, \Delta x)$, so that

$$x(t) = \sum_{i=1}^{n} \Delta x_i \, . \tag{4.12}$$

A particularly convenient model for the development process $x(t)$ that satisfies the above conditions is the Brownian motion process with drift. The model is described by the following stochastic differential equation

$$dx(t) = v(x(t), t)dt + \sqrt{k(x(t), t)}d\omega \tag{4.13}$$

where $x(t)$ represents the developmental level, $v(x(t), t)$ is the expected value of developmental rate (drift), and $k(x(t), t)$ is the variance of this process (i.e., the biological variability in developmental rates). The renewal condition (boundary condition) of the process is given by

$$f(t,0) = \int_0^{\infty} b(x) f(t, x) dx \tag{4.14}$$

where $b(x)$ is the per capita birth rate in individuals of physiological age x, $f(t, x)$ is the age distribution of the population.

The stochastic differential equation (4.13) is discretized by the Euler method, so, in the time interval $[t, t + \Delta t]$ the developmental process of the j-th individual organism is described by

$$x_j(t + \Delta t) = x_j(t) + v(t, x_j(t))\Delta t + \eta_j \tag{4.15}$$

where the η_j are the random variables drawn from a normal distribution with mean 0 and variance $k(t, x_j)\Delta t$.

To take into account the effect of mortality on population dynamics a killing process is associated to the stochastic differential equation (4.13): given a death rate $d(t, x)$ of an individual of age x and at time t, in the time interval $[t, t + \Delta t]$ the individual has the probability $P(t, x)$ of dying

$$P(t, x) = d(t, x)\Delta t\,. \tag{4.16}$$

In summary, the life history of an individual can be simulated by a 'random walk' process to which death and reproduction are added. The developmental process is represented by means of equation (4.15), while the mortality process consists in an elimination of the individual on the basis of equation (4.16), and reproduction is simulated by creation of new individuals according to equation (4.14).

4.3 *Model for size-dependent stochastic growth*

In section 3 dealing with deterministic models, we argued that time and age variables alone are often not sufficient to adequately describe the population system, and that a mass or mass equivalent variable may be required (Curry & Feldman, 1987). The same statements is also applicable to stochastic models, but the presentation of these models goes beyond the scope of this overview, and hence is not pursued further.

4.4 *Examples*

In the past two decades many stochastic models for structured populations have been proposed and can be used for illustration purposes. Here, we present some examples which have been subjected to a detailed mathematical analysis by the authors.

First, we refer to macroscopic models. On the basis of the above specified first method to numerically solve the forward Kolmogorov equation (4.1) Barr et al. (1973) developed models for the dynamics of insect populations. Here, we refer to the example of Bellagamba et al. (1989) who modeled the dynamics of *Ceratitis capitata* (Wied.) populations using physiological time and age. Manetsch's (1976) time distributed delay model has successfully been applied to arthropod phenology forecasting in many pest management programs (Welch et al., 1978; Severini et al., 1990). Time invariant distributed delays models with losses enabled Gutierrez et al. (1984) to introduce time distributed development in the sojourn time model of Wang et al. (1977) (section 3). Subsequently, this model has found wide applications in ecological work (Gutierrez, 1996). On the basis of a time-varying distributed delay with losses, a macroscopic model was developed by Fouque and Baumgärtner (1996) and used to represent the fluctuations of the inland floodwater mosquito *Aedes vexans* (Meigen) in a temporary pool. Time-varying distributed delay with losses have been used in populations systems rèsearch (Baumgärtner et al., 1990).

Second, we refer to a microscopic model developed by Di Cola et al. (1996) for olive fruit fly *B. oleae* populations.

5 CONCLUDING REMARKS

We consider deterministic models as the first step in the development of structured population models. The continuous von Förster and the discrete Leslie models are considered as the first attempts to represent the dynamics of age-structured populations. The up-wind scheme of the von Förster's equation merits more attention in demographic work because it creates, by introducing artificial diffusion in population models, the theoretical basis for stochastic models. We also noticed that the up-wind scheme and the widely used sojourn time models are related. Consequently, the von Förster model can be considered as the basic equation for deterministic models on age-structured populations.

The von Förster model and the related equations appear to be valuable instruments for any study in population ecology. To make them applicable to particular population systems they can easily be extended, for example by considering density-dependent processes. An instructive example has been published by Desharnais and Liu (1987).

On the second step physiological elements are explicitly introduced into deterministic age-structured models. This can be achieved by introducing a variable for mass or mass equivalents as done by Sinko and Streifer (1967) and/or by substituting chronological time and age by physiological age and time, as done by Hughes (1963) and Gilbert et al. (1976). Schaalje and van der Vaart (1989) make a useful distinction between development index and sojourn time models. Sojourn time models, on the basis of the Sinko and Streifer (1967) model, with time and age expressed in physiological units have been discretized and successfully been applied to population and ecosystem dynamics (Wang et al., 1977).

On step three stochastic elements are added to physiologically based structured population models. This is an important step since many applied ecologists consider stochastic elements as essential qualities of population models. To discuss this step we found the separation between microscopic and macroscopic models very useful (Munholland & Dennis, 1992). Here, this terminology appeared more adequate than the corresponding characterization of individual based configuration and distribution models (De Angelis & Rose, 1992).

De Angelis and Rose (1992) compared the applicability of microscopic and macroscopic models to the dynamics of fish populations. They concluded that small populations, populations that are subjected to a high degree of temporal stochasticity in the environment, and populations in which environmental exposure and encounters with other individuals are likely to vary greatly within the population, are usually best described by microscopic models. Large populations in which all individuals are relatively similar are more likely to be satisfactorily modeled by macroscopic models. We suspect that these conclusions are valid not only for fish but also for insect population systems and recommend to carry out comparable work in entomology.

In macroscopic models stochastic elements can be introduced by means of the forward Kolmogorov (Barr et al., 1973) equation or by distributed delay processes (Manetsch, 1976; Vansickle, 1977; Severini et al., 1990). Welch et al. (1978) have compared these two analogous models and have found that the Kolmogorov equation is biologically more meaningful whereas the delay equation is easier to implement. Since the pioneering work of Welch et al. (1978) the two models have been intensively studied and a new comparative study should be undertaken.

The passage from age-structured deterministic models to physiologically based models with stochastic elements has enabled us to further clarify the theoretical ground for population model development. In our analysis of a poultry house mite system this ground facilitated model development and lead us to recognize the need to better coordinate theoretical and applied work with respect to spatial processes and population interactions (Buffoni et al., 1995).

ACKNOWLEDGMENTS

We thank Dr. A. P. Gutierrez, University of Berkeley, for stimulating discussions and useful suggestions.

REFERENCES

Barr, R.O., P.G. Cota, S.H. Gage, D.L. Haynes, A.N. Kharkar, H.E. Koenig, K.Y. Lee, W.G. Ruesink & R.L. Tummala 1973. Ecologically and economically compatible pest control. In Geier, P.W., L.R. Clark, D.J. Anderson, & H.A. Nix (eds.), Insects: studies in population management. *Ent. Soc. Australia, Memoirs* 1:241-264.

Baumgärtner, J. & A.P. Gutierrez 1988. Simulation techniques applied to crops and pest models. In Cavalloro, R. & V. Delucchi (eds.), Parasitis 88. Proceedings of a Scientific Congress, Barcelona, October 25-28. *Boletìn de Sanidad Vegetal*, fuera de serie, 17:175-214.

Baumgärtner, J., B. Wermelinger, U. Hugentobler, V. Delucchi, P. Baronio, E. De Berardinis, J.J. Oertli & C. Gessler 1990. Use of a dynamic model on dry matter production and allocation in apple orchard ecosystem research. *Acta Horticulturae* 276:123-139.

Bellagamba, V., G. Di Cola & R. Cavalloro 1989. Stochastic models in fruit-fly population dynamics. In Cavalloro, R. (ed.) *Proceeding of the CEE/IOBC International Symposium*, Rome 7-10 April 1987:91-98. Rotterdam: Balkema.

Bhat, U.N. 1984. *Elements of applied stochastic processes*. 2nd Edition. New York: Wiley.

Blythe, S.P., R.M. Nisbet & W.S.C. Gurney 1984. The dynamics of population models with distributed maturation periods. *Theoretical Population Biology* 25:289-311.

Buffoni, G. & G. Di Cola 1995. Discrete stochastic models in the dynamics of a size-structured population. In Kimmel, M. (ed.). *Proceedings of the 3rd International Conference on Mathematical Population Dynamics, Pau, June 2-5, 1992*: 212-21. Winnipeg: Wuerz Publishing Ltd.

Buffoni, G., G. Di Cola, J. Baumgärtner & V. Maurer 1995. A mathematical model for trophic interactions in an acarine predator-prey system. *J. Biol. Syst.* 3:303-312.

Buffoni, G., G. Di Cola & A. Ugolini 1996. Numerical methods for the solution of PDE describing the stochastic development of an age-structured population. In Di Cola, G. & G. Gilioli (eds.), *Computer science and mathematical methods in plant protection*: 12-21. Quaderni del Dipartimento di Matematica, Università di Parma, n. 135.

Cappio Borlino, A., G. Di Cola & G. Marras 1991. Mathematical modelling of natural population dynamics. *Memorie Istituto Italiano Idrobiologia* 49:127-162.

Cavalloro, R., G. Delrio & G. Di Cola 1975. Un modello matematico continuo per la stima dei parametri biologici di una popolazione di insetti polivoltini. *Redia* 56:375-385.

Curry, G.L. & R.M. Feldman 1987. *Mathematical foundations of population dynamics*. College Station: Texas A&M University Press.

DeAngelis D.L. & K.A. Rose 1992. Which individual-based approach is most appropriate for a given problem?. In DeAngelis, D.L. & L.J. Gross (eds.), *Individual-based models and approaches in ecology*: 67-87. New York: Chapman & Hall.

Desharnais, R.A. & L. Liu 1987. Stable limit cycles in laboratory populations of *Tribolium castaneum*. *J. Anim. Ecol.* 56:885-906.

Di Cola, G. & G. Gilioli 1996. *Computer science and mathematical methods in plant protection*. Quaderni del Dipartimento di Matematica, Università di Parma, n. 135.

Di Cola, G., G Gilioli & A. Cossu 1996. Models of stochastic development in insect populations. *Proc. XX International Congress of Entomology, Florence, August 25-31, 1996*.

Fouque, F. & J. Baumgärtner 1996. Simulating development and survival of *Aedes vexans* (Diptera, Culicidae) preimaginal stages under field conditions. *J. Med. Entomol.* 33:32-38.

Gilbert, N., A.P. Gutierrez, B.D. Frazer & R.E. Jones 1976. *Ecological relationships*. Reading: Freeman.

Gutierrez, A.P. 1996. *Applied population ecology: A supply-demand approach*. New York: Wiley.

Gutierrez, A.P., J. Baumgärtner & C.G. Summers 1984. Multitrophic models of predator-prey energetics: III. A case study an alfalfa ecosystem. *Can. Entomol.* 116:950-963.

Gurney, W.S.C., R.M. Nisbet & J.H. Lawton 1983. The systematic formulation of tractable single-species population models incorporating age structure. *J. Anim. Ecol.* 52:479-495.

Huffaker, C.B. & B.A. Croft 1976. Integrated pest management in the U.S.: progress and promise. *Environ. Health Perspect.* 14:167-183.

Hughes, R.D. 1963. Population dynamics of the cabbage aphid, *Brevicoryne brassicae* (L.). *J. Anim. Ecol.* 23:393-424.

Johnson, C. 1992. *Numerical solution of partial differential equations by the finite element method.* Cambridge: Cambridge University Press.

Leslie, P.H. 1945. On the use of matrices in certain population mathematics. *Biometrika* 35:183-212.

Leslie, P.H. 1948. Some further notes on the use of matrices in population mathematics. *Biometrika* 38:213-245.

Logan, J.A., D.J. Wollkind, S.C. Hoyt & L.K. Tanigoshi 1976. An analytical model for description of temperature dependent rate phenomena in arthropods. *Environ. Entomol.* 5:1133-1140.

Manetsch, T.J. 1976. Time-varying distributed delay models and their use in aggregative models of large systems. *IEEE Trans. Syst. Man Cybern.* 6:547-553.

McDonald, L., B. Manly, J. Lockwood & J.A. Logan 1989. *Estimation and analysis of insect populations.* Lecture Notes in Statistics, n. 55. Berlin: Springer.

McKendrick, A.G. 1926. Application of mathematics to medical problems. *Proc. of the Edinburgh Mathematical Society.* Vol. XLIV, Session 1925-26:98-130.

Metz, J.A.J. & O. Diekman 1986. *The dynamics of physiologically structured populations.* Lecture Notes in Biomathematics, n. 68. Berlin: Springer.

Munholland, P.L. & B. Dennis 1992. Biological aspects of a stochastic model for insect life history data. *Environ. Entomol.* 21:1229-1238.

Nisbet, R.M. & W.S.C. Gurney 1982. Modelling fluctuating populations. Chichester: Wiley.

Plant, R.E. & L.T. Wilson 1986. Models for age-structured populations with distributed maturation rates. *J. Math. Biol.* 23:247-262.

Renshaw, E 1992. *Modelling biological population in space and time.* Cambridge: Cambridge University Press.

Schaalje, G.B. & H.R. van der Vaart 1989. Relationships among recent models for insect population dynamics with variable rate of development. *J. Math. Biol.* 27:399-428.

Severini, M., J. Baumgärtner & M. Ricci 1990. Theory and practice of parameter estimation of distributed delay models for insect and plant phenologies. In Guzzi, R., A. Navarra & J. Shukla (eds.), *Physical climatology and meteorology for environmental application*: 674-719. Singapore: World Scientific Publishing.

Sharpe, P.J.H. & D.W. DeMichele 1977. Reaction kinetics of poikilotherm development. *J. theor. Biol.* 64:649-670.

Sinko, J.W. & W. Streifer 1967. A new model for age-size structure of a population. *Ecology* 48:910-918.

Vansickle, J. 1977. Attrition in time-distributed delay models. *IEEE Trans. Syst. Man Cybern.*, 9:535-638.

Von Förster, H. 1959. Some remarks on changing populations. In Stohlman, F. (ed.), *The kinetics of cellular proliferation*: 382-407. New York: Grune and Stratton.

Wang, Y., A.P. Gutierrez, G. Oster & R. Daxl 1977. A population model for plant growth and development: coupling cotton-herbivore interaction. *Can. Entomol.* 109:1359-1374.

Welch, S.M., B.A. Croft, J.F. Brunner & M.F. Michels 1978. PETE: an extension phenology modeling system for management of multi-species pest complex. *Environ. Entomol.* 7:482-494.

Wit, C.T. de & J. Goudriaan 1978. *Simulation of ecological processes. Simulation Monographs.* 2nd ed. Wageningen: Pudoc.

Population and Community Ecology for Insect Management and Conservation, Baumgärtner et al. (eds)

The impact of behavioural thermoregulation on reproductive rates in a grasshopper

J. Samietz & G. Köhler
Institute of Ecology, Friedrich-Schiller-University Jena, Germany

ABSTRACT: Oviposition rates of the European grasshopper *Stenobothrus lineatus* (Panzer) are clearly related to temperatures above a threshold of 23.5°C. Considering this temperature dependency, oviposition rates estimated for an experimental population on base of ambient field temperatures yielded much lower values in comparison to those actually determined in the field. These findings indicated that the thermoregulation of the females influences their reproductive rates in a decisive way. Field investigations showed that the body temperatures of *S. lineatus* females were much higher then the ambient air conditions raising intercepts of more than 20 K. Applying an empirical and a biophysical thermal model, the influence of passively acquired heat from environmental temperatures could be separated from active behavioural thermoregulation. The oviposition rates in the field can be estimated by simulating the body temperatures of the females on the base of weather data (temperature and sunshine periods). On the population level, the impact of behavioural thermoregulation, first by visiting sunflecks within the habitat and second by posturing towards the sun, could be quantified. In the Central European population studied between 1992 and 1995, the two modes of behavioural thermoregulation explained on average about 53% (basking in sunflecks) and 4% (posturing) of both the mean individual fertility and the population net reproductive rates.

1 INTRODUCTION

Between 1992 and 1995 the habitat utilization, population dynamics and life history of the temperate climate European grasshopper *Stenobothrus lineatus* (Panzer) was investigated. The resulting life-table data were examined by a key-factor analysis. Accordingly, the decrease from the potential fecundity of the females to the number of hatching first instar larvae, i.e. the oviposition and the egg storage during late summer and one-year hibernation, was found to be the most important factor in the population dynamics between 1992 and 1995 (Samietz et al. 1996). The contribution of the influences of both life-table compartments were unknown, but information about the fertility of the species in the field would allow us to evaluate their impact. As known for developmental processes in insects at all and in grasshoppers as well (e.g. Putnam 1963, Gage et al. 1976, Whitman 1986, Kemp & Dennis 1989, Carruthers et al. 1992), the oogenesis is temperature-dependent. We, therefore, assumed that this holds also for oviposition rates. Laboratory experiments were conducted to test this hypothesis and to quantify the influence of temperature on fertility of the grasshoppers (Samietz, in prep.). The results showed a lower temperature threshold of 23.5°C for successive ovipositions and, presumably, for the according development. Furthermore, the experiments demonstrated an almost linear temperature-dependency of oviposition rates above this threshold and below the maximum voluntarily tolerated temperature with a mean thermal constant of 450 Kelvin-hours (Kh) for each successive egg pod (Samietz, in prep.). These findings enabled us to estimate individual fertility and the total reproductive rates of the investigated field population given the weather- and habitat-dependent body temperature of the females.

Accordingly, the first purpose of the present study was to provide the data regarding field body temperatures of *S. lineatus*. Moreover, a second aim was to elucidate and quantify the causes determining temperature conditions of the species in the field. It is known for grasshoppers that the solar radiation energy plays a major role to increase body temperature above the ambient air temperature (e.g. Digby 1955, Stower & Griffiths 1966, Parker 1982, Chappell 1983, Kemp 1986, Wang & Chen 1989, Lactin & Johnson 1996, Carruthers et al. 1992). We supposed for *Stenobothrus lineatus* that these increases are not only caused by passive heat-gain – including the incident radiation energy flux – but also by active behavioural thermoregulation. This hypothesis may be tested in field experiments. However, to classify the influence of passive heat gain on one hand and active regulation on the other hand, we developed first a biophysical thermal model on the base of a heat-balance and second, an empirical model on the base of our measured data that describes the actually achieved temperature sums in the field. These models enabled us to quantify the influence of thermoregulation on reproductive rates and, furthermore, to partition the heat gain by active regulation into two classes of behavioural responses, i.e. visiting sunflecks in the vegetation and posturing towards the sun.

2 MATERIAL AND METHODS

2.1 *Body temperatures in the field*

During 52 surveys of 15 study days in July and August 1993, 1994, and 1995, we measured field temperatures of *Stenobothrus lineatus* on southernly exposed xerothermic grassland plots in the "Leutratal" nature reserve near Jena (Germany). The body temperatures of 461 individual females were measured by a "grab-and-stab" technique. First, every individual was caught by a sweep net or by hand, touching only the elytra but not the body. Second, a thermocouple microprobe (Testoterm, Lenzkirch/Germany, 97TE, NiCr-Ni, 0.5 mm) was injected quickly – within the next 5 s after capture – about 5 mm into the ventral pro-thorax towards the abdomen. Constant temperatures maintained for about 3 s were registered by a digital recorder (Testoterm 9010). The "grab-and-stab" procedure was criticized, however, mainly for reasons not valid for the investigated grasshoppers – e.g. quick heat-up of insects after flight or insufficient object-probe ratios (Stone & Willmer 1989, Heinrich 1993). Yet quickly used, it is the method of choice in grasshoppers of medium size.

The ambient temperature was gauged by a Testoterm probe as close as possible to the catching site of the animal. Simultaneously, horizontal and inclinative solar radiation flux densities Φ_R (in Wm^{-2}) were measured by a Thiess pyranometer (Thiess, Göttingen/Germany, CM11) and a Kipp & Zonen-radiation indicator (Kipp & Zonen, Delft/The Netherlands, CC20). In order to unify and calibrate habitat influences, the inclinative solar radiation flux was used to carry out further analyses of the data. Thermoregulation ability was obtained by linear regression analyses of body temperature T_b against the ambient temperature T_a performed for six ranges of solar radiation flux (0–199, 200–399, 400–499, 500–599, 600–799 and 800–1055 Wm^{-2}) separately. The slope of this regression is significantly smaller than one (isotherm) when the maintenance of T_b is ± independent of T_a and passive heating (see 2.2). It would be zero in the case of perfect regulation. Note that the analysis by linear regression assumes equal passive heating over the ambient temperatures tested and, therefore, it is valid exclusively for narrow ranges of solar radiation flux but will be biased with broader ranges.

2.2 *The thermal models*

Biophysical model: A simplified heat-balance model after Baumgärtner & Severini (1987) enabled us to approximate the operative environmental temperature T_e of the grasshopper females in the field. The heat-balance equation of a wet body is

$$a\,\Phi_R A - S\,[\Phi_H + \beta\,\Phi_V] = 0 \tag{1}$$

where Φ_R is the radiation flux density (in), Φ_H is the sensible heat flux density (out), Φ_V is the latent heat flux density (out), a is the absorption coefficient β is the resistance to evaporation, A is the body cross-section area, and S is the external body-surface area.

Baumgärtner & Severini (1987) applied this thermal balance in equation (1) with boundary layer theory and derive a simplified equation for the body-surface temperature T_s. In small insects, the body-surface temperature is usually equal to the internal body temperature (stated for grasshoppers: Chappell & Whitman 1990). Thus, we assumed T_e – the point of our interest – to be equal to T_s estimated by the model. Assuming that there are no transpiration and no forced convection it follows

$$T_e = T_s = T_a + a\,\Phi_R\,\frac{bA}{kS} \tag{2}$$

where T_a is the ambient air temperature, b is the body diameter and k is the thermal conductivity of the air (=0.026 $Wm^{-1}K^{-1}$). The absorption coefficient a was set to be 0.7 in the green grasshoppers under study. Digby (1955) approximated a by 0.7 in green *Carausius* and Chappell & Whitman (1990) gave the same value as a mean coefficient for grasshoppers in general. Anderson et al. (1979) also estimated the absorption of two grayish grasshoppers to be 0.7.

The body shape of female grasshoppers was considered to be a rotation ellipsoid with semi-major axis p=11.5 mm and semi-minor axis q=2.5 mm derived from the mean body length of 23 mm in Central Europe (Harz 1960) and a mean thoracic diameter of about 5 mm. As a result, the external body surface area was estimated to be 289 mm^2. The cross-section area along the major axis of the ellipsoid $A_{max} = \pi\, p\, q$ with respect to the angle α between the major axis and the heat source is

$$A(\alpha) = \pi p^*(\alpha)\, q\ . \tag{3}$$

The corresponding derivate projection of the semi-major axis is

$$p^*(\alpha) = \sqrt{q^2\cos^2(\alpha) + p^2\sin^2(\alpha)}\ . \tag{4}$$

Applied with equation (2), the extremes of equation (3) mark the possible temperature range of an individual in the sun due to behavioural regulation. The angle α=0° with $p=q$ refers to the minimum possible temperature in the sun corresponding to the minimum A. Accordingly, α=90° with $p=p^*$ refers to the maximum A and a maximum body temperature. Applying this elliptical model to behaviour, lateral basking – the adjustment of lateral orientation towards the sun – would have the same effect as dorsal basking. It should be noted that the latter ignores latent effects of the elytra either positively through heat gain as negatively through insulation.

Passive heat gain in the sun assumes a randomly chosen α between 0 and 90°. The according mean temperature intercept for passive heating ΔT_{pass} was estimated from equations (2) and (3) as follows:

$$\Delta T_{pass} = a\,\Phi_R\,\frac{b\,\pi q\,Ep^*(\alpha)}{k\,S} \tag{5}$$

where $Ep^*(\alpha)$ is the expected value of the derivate projection p^* with respect to α that being:

$$Ep^*(\alpha) = \int_0^{\frac{\pi}{2}} p^*(\alpha)\, P(\alpha)\, d\alpha \Big/ \int_0^{\frac{\pi}{2}} P(\alpha)\, d\alpha\ . \tag{6}$$

The probability $P(\alpha)$ is constant due to the required uniform random distribution to describe passive heating. The quotient in equation (6) reduced by the constant leads to

$$Ep^*(\alpha) = \frac{2}{\pi}\int_0^{\frac{\pi}{2}} p^*(\alpha)\, d\alpha\ . \tag{7}$$

Considering equation (4)

$$Ep^*(\alpha) = \frac{2}{\pi} \int_0^{\frac{\pi}{2}} \sqrt{q^2\cos^2(\alpha)+p^2\sin^2(\alpha)}\; d\alpha \tag{8}$$

The integral in equation (8) was numerically solved by applying a 32-point-Gauss formula and using the above-mentioned values for the semi-major and semi-minor axis of the ellipsoid model. The solution of (8) yields the expected value of p^*=7.744 mm with respect to passive heating. This value can be included in equation (5) to estimate ΔT_{pass} for any condition of Φ_R and a.

Empirical Model: Plotted over a broad range of potential solar radiation in the habitat, the relation between likely regulated body temperature and ambient temperature appears nonlinear. To describe this relationship, a sigmoid equation was used. We assumed that, whenever our weather station registered sunshine, the incident solar radiation flux in our optimum habitat was at least 600 Wm^{-2}. First, this approximation is useful because local weather stations make available sunshine periods rather then successive data of solar radiation. Second, during summer months within a southernly exposed slope, the time spans with potential Φ_R below 600 Wm^{-2} are relatively short and restricted to early morning and late evening – in our 15°-southernly sloped study plot before 06:00 in the morning and after 18:00 in the evening at the end of July (Samietz, unpubl.). Thus we expressed the body temperatures of *S. lineatus* females measured above 600 Wm^{-2} by

$$T_b = \frac{T_{max} - T_a}{1 + \left(\frac{T_a}{T_{infl}}\right)^s} + T_a \tag{9}$$

where T_{max} is the asymptotic maximum body temperature regulated by the animals under the specific radiation regime, T_{infl} is the ambient temperature at the inflexion point of the sigmoid curve, and s is the slope parameter. Considering the requirements of the field conditions, this function tends asymptotically towards the isotherm (T_b=T_a) with decreasing T_a. This implies a declining regulation performance near the lower activity threshold since, obviously, an animal cannot perform any kind of behavioural thermoregulation as long as T_b is below its thermal activity threshold. For this reason, earlier applications of symmetric logistic functions to describe $T_b(T_a)$ over broad ranges of radiation flux (e.g. Kemp 1986) do not comply neither with observations and nor with theory regarding the impossible asymptotic minima. The three unknown parameters of our equation (9) and their asymptotic variances were estimated iteratively by the Marquardt-Levenberg algorithm (Marquardt 1963, Press et al. 1986) using SigmaPlot program V. 2.01 (Jandel Scientific).

2.3 *Modelling field oviposition rates*

From the 1992 – 1995 field data ("Leutratal" nature reserve) we derived population curves of *S. lineatus* and estimated the number of mature females present on each day of the adult period (June to October) (Samietz 1994, Samietz et al. 1996, and unpubl.). The mean number of eggs laid by the population – necessary for estimation of population reproductive rates – is expected to be the product of the daily number of mature females, the daily oviposition rate of one female, and the number of eggs per pod (6.4 in the laboratory, Samietz, in prep.). The lower developmental temperature threshold for successive ovipositions is 23.5°C, and each consecutive pod is laid after a physiological time period of 450 Kelvin-hours (Kh) (Samietz, in prep.). The daily oviposition rate was then calculated as the quotient of daily temperature summation (TS_d) in Kh above the threshold and the thermal constant of 450 Kh. Summations were calculated for each hour and added to the required daily values. The ambient temperatures for each hour of the period under study were obtained from the Jena weather station as well as sunshine periods in tenths per hour (approx. 15,000 data sets). Three scenarios were calculated from these data in order to estimate the hourly temperature summations (TS_h) effective for oviposition:

(I.) First, we simulated the TS_h by the empirical thermal model that describes the actually acquired temperature sums in the field including likely variations of thermoregulation (equation 9). For Φ_R > 600 Wm^{-2} we estimated T_b for the sunny time span of each hour on the base of T_a and equation (9), and calculated the corresponding TS_h. For the remaining time span of one hour TS_h were calculated on the base of measured T_a. For verification purposes, the model was applied to the period of our field-oviposition experiment and compared with the corresponding empirically determined oviposition rates (cf. 2.4).

(II.) Second, the biophysical thermal model was applied to simulate passively achieved TS_h, without posturing to the sun, but considering the selection of sunflecks in the vegetation for basking whenever sun is shining. The fitted asymptotic maximum body temperature of 40.5°C, regulated by the animals under an optimum radiation regime (Φ_R > 600 Wm^{-2}), was used as an estimate of the voluntarily tolerated temperature. Thus, ΔT_{pass} estimated by the thermal model for the optimum habitat (according to the empirical model for Φ_R>600 and an average of 810 Wm^{-2}) was added to T_a. The resulting T_b* was truncated with the value of maximum voluntary tolerated temperature of 40.5°C to calculate TS_h for the sunny time span of each hour. As in the previous scenario TS_h for the remaining time span of each hour was calculated using T_a.

(III.) Third, the biophysical model was applied to simulate passive TS_h neither with posturing to the sun nor with selection of sunflecks in the vegetation for basking, i.e. without temperature increases through behavioural processes. Thereby, each sunny time span was reduced to its product with the probability of an animal being inactively in the sun. Presuming independence of animal locations from sunflecks, this probability equals the relative area of sunflecks within the occupied habitat. Within an optimal habitat, the ratio was 0.294 for 220 samples on 44 raster plots using a Decagon sunfleck ceptometer (Decagon Devices, Pullman, WA/USA) in 1995. For the resulting reduced "sunny" time span of each hour, TS_h was calculated from T_b* as mentioned in scenario II. Hence, for the remaining time span of the hour TS_h was calculated on the base of T_a.

2.4 *Actual field oviposition*

From 31 July to 9 August 1995, we conducted a field experiment to estimate the actual oviposition rates of *S. lineatus* in the "Leutratal" nature reserve. During the first day 60 females were marked individually by means of coloured numbered opalith plates glued on the pronota. In addition, we provided each individual with a small piece of self-adhesive reflecting tape (Scotchlite 7610 high gain, 3M Germany, Neuss/Germany) on both hind tibiae. Owing to the reflecting tape we could locate marked animals easily at night with a head lamp. The marked individuals were even visible at a distance of about 50 m from the observer. Thus the recapture probability of an individual during a night survey was above 90% in the present study and, in mobility studies, between 49 % and 97 % per survey (Samietz et al. 1996). Beginning in the evening of the first day we weighted the marked females daily during two hours after dusk using a Sartorius electronic balance (Sartorius, Göttingen/Germany, PT120-000V1). We considered that oviposition had occurred whenever we recorded a daily weight loss of at least 0.02 g (about 10% of initial weight). We recorded 85 interpod periods of 47 females and were able to estimate according oviposition rates. In order to avoid a possible pooling bias (Machlis et al. 1985), the individual means of the 47 females were used for further analysis of the data.

3 RESULTS

3.1 *Thermoregulation quantified*

In the sun, body temperatures of *S. lineatus* females in the field population are considerably higher than the ambient air temperatures T_a, reaching a maximum intercept of 21.9 K at a T_a of 19.7 °C and Φ_R=900 Wm^{-2}. The maximum body temperature was 42.6 °C. For narrow ranges of solar radiation flux, T_a plotted as a function of T_b exhibits a nearly linear relation. Figure 1 shows the corresponding plots for two opposite ranges of Φ_R. In both cases, the regression is significant (for statistics see Table 1). The left plot with an average Φ_R of 910 Wm^{-2} shows a nearly perfect maintenance of the body temperature independent of T_a. The values of T_b were between the isotherm and the maximum values predicted by the biophysical thermal model. They are, however, much higher than the expected passively achieved temperature without temperature increase by behavioural means (marked through the dashed lines). In both plots, the regression slope is significantly different from the

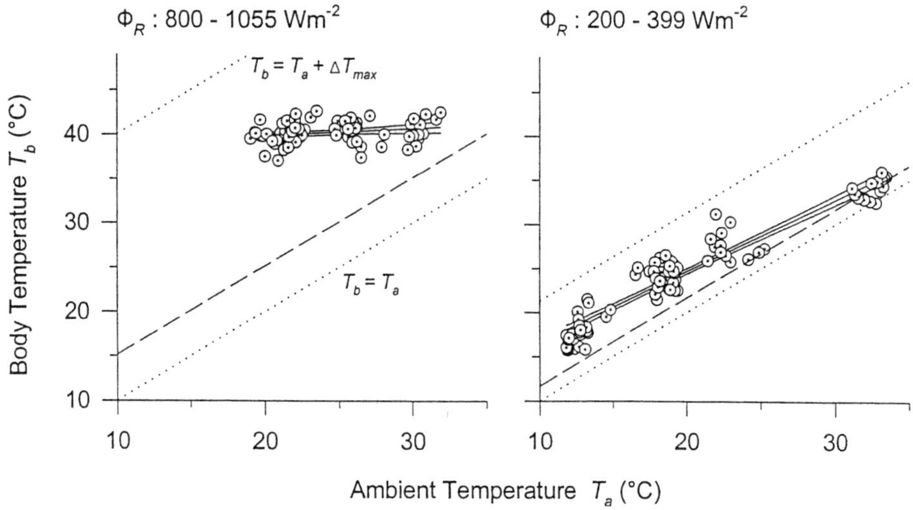

Figure 1. The body temperature T_b of *S. lineatus* females as function of the ambient temperature T_a in two exemplarily ranges of solar radiation flux Φ_R with linear regression lines and 95% confidence interval. Dashed lines refer to passively achieved T_b in the optimum habitat predicted by the biophysical thermal model neglecting behavioural thermoregulation. Upper dotted lines mark the higher limits of T_b attainable through perpendicular flanking towards the sun and lower dotted lines refer to the isotherm T_b=T_a. The area between both dotted lines marks the potential temperature range to regulate T_b through behavioural means at mean Φ_R.

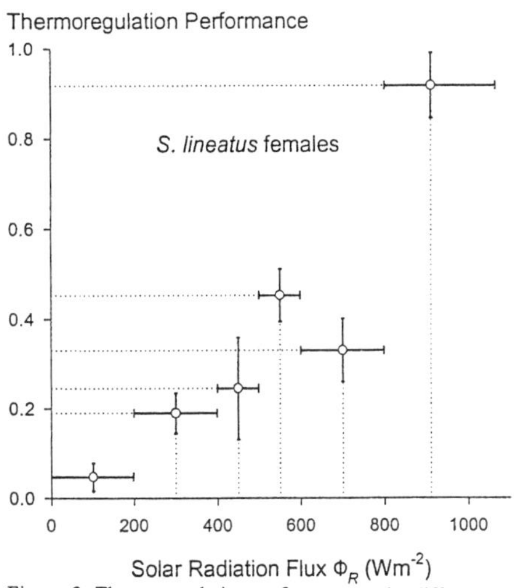

Figure 2. Thermoregulation performance – the difference between the isotherm's slope one (T_b=T_a) and the linear regression's slope of $T_b(T_a)$ – as a measure how females of *S. lineatus* maintain their body temperature T_b independent of the ambient temperature T_a. Data plotted with 95% CI (vertical bars) to average sample means and covered range of Φ_R (horizontal bars).

isotherm T_b=T_a (Table 1). Hence, thermoregulation can be demonstrated and quantified by the regression slope. The slopes of these linear regressions of $T_b(T_a)$ are provided in Table 1 including regression ANOVA and the results of the corresponding t-test with respect to H_0: slope=1. All regressions are statistically significant at P<0.01 and, hence, we have evidence for significant thermoregulation across all groups tested. In case of the lowest range of Φ_R between 0 and 199 Wm^{-2}, the regression slope is marginally but still significantly different from one. Furthermore, the slopes are significantly different among the results of the six regressions presented in Table 1 (comparison method after Sachs 1982: $F_{(5, 449)} = 965$, P<<0.001).

Table 1. Linear regression results of body temperature (T_b) as a function of ambient temperature (T_a) in narrow ranges of solar radiation flux density (Φ_R as average range mean, minimum, and maximum). N refers to sample size.

Φ_R			N	Regr.	SD	Regr. ANOVA		Test of H_0: slope=1	
mean	min	max		slope	(slope)	F	P	t	P
910	800	1055	78	0.08	0.04	4.8	0.032	-24.57	< 0.001
700	600	799	57	0.67	0.04	354.0	< 0.001	-9.26	< 0.001
550	500	599	41	0.55	0.03	335.1	< 0.001	-15.15	< 0.001
450	400	499	44	0.76	0.06	167.1	< 0.001	-4.18	< 0.001
300	200	399	103	0.81	0.02	1224.9	< 0.001	-8.23	< 0.001
100	0	199	138	0.95	0.02	3513.3	< 0.001	-3.01	0.003

To demonstrate how the animals maintain their T_b independent of T_a, we introduced the thermoregulation performances as the difference between one (the slope of the isotherm $T_a=T_b$) and the linear regressions slope of $T_b(T_a)$. In Figure 2 this measure is plotted vs. solar radiation flux Φ_R. Clearly, the thermoregulation performance of the females is strongly dependent on the incident solar radiation energy within their habitat. The latter correlation was remarkably high and statistically significant (r_S=0.94, P=0.017).

3.2 *Actual oviposition and prediction by the models*

The application of the empirical model (scenario I.) to the field data with $\Phi_R > 600$ Wm^{-2} resulted in T_{max}=40.5 ± 0.3 (SD) °C, T_{infl}=14.2 ± 0.3 °C, s=-6.4 ± 1.0. The coefficient of determination of this nonlinear fit is r^2=0.68. In Figure 3, the sequence of T_b estimated by this model is plotted with the curve of T_a for an exemplarily sunny summer day in Central Europe. The temperature summation effective for successive ovipositions is plotted on the second Y-axis for the ambient temperature, the empirical model (I) and the two scenarios on base of the biophysical thermal model (II. and III.). Calculated by the temperature summation, the corresponding theoretical oviposition rates (pods per day) during this representative day were estimated and are noted on the right margin.

The actual mean daily oviposition rate during the entire field experiment was 0.41±0.15 (SD) pods per day. When applying the ambient temperatures during the investigated time span to estimate physiological time, the temperature summation per day yields only 38 Kh. Accordingly, the expected

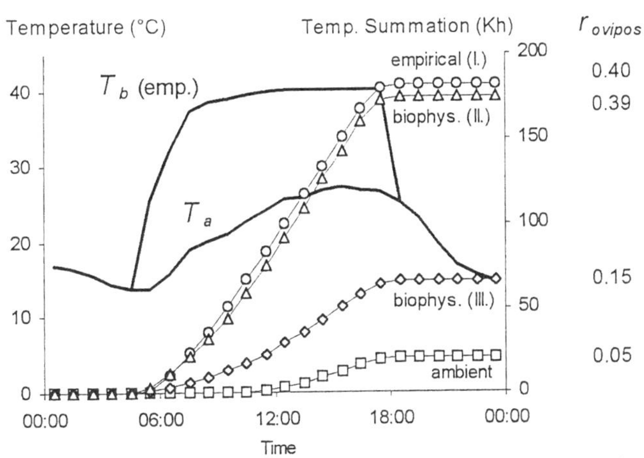

Figure 3. Ambient temperature (T_a) and estimated sequence of body temperature) on base of the empirical model (T_b(emp.) – bold lines) plotted for a exemplarily moderate sunny summer day in Central Europe (2 August 1995). Temperature summation effective for oviposition plotted for the ambient temperature (ambient), empirical model (I.), and the two scenarios on base of the biophysical thermal model (II. and III.); corresponding daily oviposition rates (r_{ovipos}, in egg pods per day) noted on right margin.

temperature summation per oviposition event in the field would be theoretically 93±24 (*SD*) Kh and, therefore, seriously lower than the mean thermal constant observed in the laboratory being 450 Kh per each egg pod (Samietz, in prep.). When applying the biophysical thermal model to simulate the temperatures without any kind of behavioural thermoregulation (scenario III.), the daily temperature sum is expected to be 76 Kh and the required summation per oviposition would be 188 ± 49 Kh. The results of scenario II. – allowing active basking in sunflecks – lead to a prediction closer to the laboratory data: a daily summation of 169 Kh and a sum of 417 ± 110 Kh per oviposition. The application of the empirical model to simulate field temperatures (scenario I.) shows a similar result and predicts 173 Kh per day during the investigated time span and a sum of 425 ± 112 Kh per egg pod laid.

The correspondence between the laboratory data and the actual field results under the applied scenarios was statistically evaluated by comparing overall linear regression slopes of oviposition rates as a function of daily temperature summations (nonintercept model) with regression slopes using the laboratory data. The variance ratio test (Zar 1984) was used to examine homogeneity of variances between the regressions. Due to heteroscedasticity, the test statistics had to be approximated by *t*-distribution applying a robust design to calculate degrees of freedom after Sachs (1992). The results confirmed the significant differences when using ambient temperatures ($t_{(15)}$=2.40, *P*=0.030; variance ratio=8.73 > $F_{(15,62,0.1)}$=1.60) and when employing scenario III. ($t_{(15)}$=3.54, *P*=0.003; variance ratio=4.27 > $F_{(15,62,0.1)}$=1.60). However, the application of the biophysical model with scenario II. ($t_{(15)}$=0.73, *P*=0.48; variance ratio=1.92 > $F_{(15,62,0.1)}$=1.60) and of the empirical model (scenario I.) showed no significant differences to the laboratory results ($t_{(15)}$=0.45, *P*=0.66; variance ratio=1.91 > $F_{(15,62,0.1)}$=1.60).

3.3 *Individual fertility and population reproductive rates*

To quantify the influence of thermoregulation on reproductive rates of *S. lineatus*, we applied the two scenarios with the biophysical thermal model (2.3, II. and III.) and the scenario with the empirical model (I.) to the population data and corresponding weather data recorded between 1992 and 1995. Results regarding individual fertility during these years are presented in Table 2. Applying the empirical model of actually achieved temperature sums, the mean number of eggs laid per female ranges from 17.6 to 30.5 with an average of 23.5. The individual number of eggs laid is much lower when applying scenario III.: the values range from 7.8 to 13.2 with an average of 10.1 during the study period. Therefore, the two investigated ways of heat-gain by behavioural thermoregulation contribute by 57.2% – 4.1% (posturing) plus 53.1% (visiting sunflecks) – to individual fertility of the females in the population of *S. lineatus*.

The results of the three scenarios on the population level are quantified and presented in Figure 4. The contributions of passive heat-gain to the population reproductive rates ranged from 34.9% in 1993 to 47.4% in 1994 with an average of 42.5 ± 5.3 (*SD*) %. The remaining temperature summation effective for reproduction, on average 57.5%, was contributed by the two investigated modes of thermoregulation combined. The regulation by posturing towards the sun played a minor role with an average of 4.0 ± 1.7% (1.9 – 6.0%). On the contrary, temperature summation by visiting sunflecks in the vegetation for basking is of major importance for explaining the population reproduction. This behaviour provides 53.5 ± 4.2 % (48.9 – 59.0%) of the effective summation during the four generations. The net reproductive rates R_0 (eggs/eggs) between the investigated generations reflect the same results (Figure 4): R_0 changes little if reducing the egg production by temperature summation through posturing. However, the differences are remarkable when removing the values contributed by both modes of thermoregulation, reaching the most serious decrease to nearly one third between 1992 and 1993.

Table 2. Scenarios to mean individual fertility of the *S. lineatus* females between 1992 and 1995 (lifetime number of eggs).

	1992	1993	1994	1995	Mean	*SD*
Including thermoregulation (scenario I. – empirical)	30.5	20.7	25.4	17.6	**23.5**	**5.6**
Without posturing towards the sun (scenario II.)	29.1	19.4	24.5	17.3	**22.6**	**5.3**
Without thermoregulation behaviour (scenario III.)	13.2	7.2	12.0	7.8	**10.1**	**3.0**

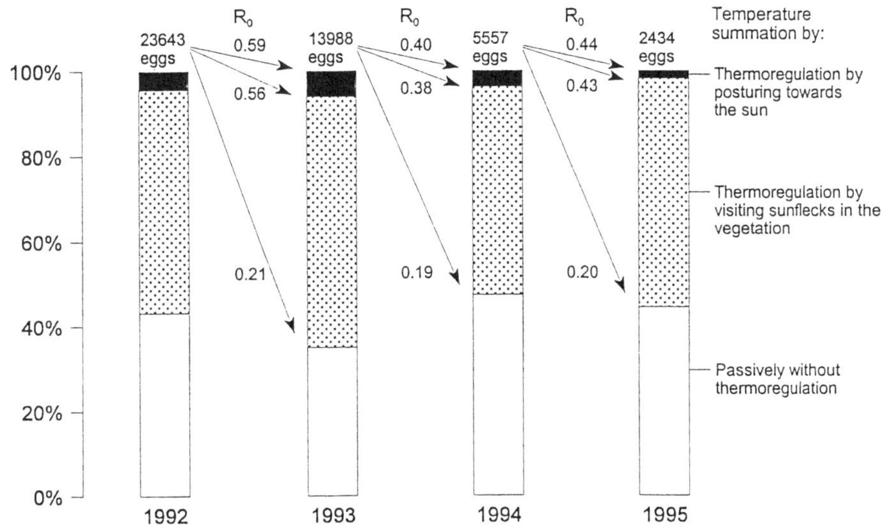

Figure 4. Temperature summation effective for reproduction of the *S. lineatus* population between 1992 and 1995 as cumulative percentages of values estimated by the empirical model that is the number of eggs noted on top of each bar. The population net reproductive rates (R_0) refer to the values of estimated egg production between the generations applying the empirical scenario I. (top), the biophysical thermal model with scenario II. (middle), and scenario III. (bottom) without thermoregulation.

4 DISCUSSION

It is generally known that the grasshopper metabolism and, thus, also development are temperature-dependent (e.g. Putnam 1963, Gage et al. 1976, Whitman 1986, Kemp & Dennis 1989). The fact that temperate climate grasshoppers increase their body temperature above ambient regimes by behavioural means to reach optimum thermal conditions is confirmed by many studies (Pepper & Hastings 1952, Anderson et al. 1979, Chappell 1983, Gillis & Possai 1983, Kemp 1986, Gillis & Smeigh 1987, Whitman 1987, Carruthers et al. 1992, Lactin & Johnson 1996). Also, the females of the grasshopper *S. lineatus* control their body temperatures through behavioural thermoregulation. The intercepts between ambient and body temperature reached values of more than 20K. These values are surprisingly high but lay inside the potential temperature range predicted by the biophysical model and, furthermore, they are comparable to extreme maxima recorded in other grasshopper species as well (e.g. Chappell 1983: 18K in *Melanoplus sanguinipes*; Kemp 1986: 18K in *Aulocara elliotti*).

Considering development and temperature increase by behavioural regulation, the great differences between fertilities observed in the field and in the laboratory are easy to explain. For example, Kriegbaum (1988) observed remarkably higher oviposition rates of *S. lineatus* in the field in comparison to the laboratory results of Richards & Waloff (1954) in the same species. Despite relatively high laboratory temperatures in relation to ambient condition in the field, laboratory studies are mostly conducted under low radiation conditions and, thus, without additional radiation heat-gain. As a further example, both Dingle et al. (1990) and Chappell (1983) stressed that *Melanoplus sanguinipes* would not complete its life cycle in higher altitudes up to 2,700 m without the radiation influence and likely thermoregulation.

However, only very few studies linked thermoregulation to the population ecology of the same species. Carruthers et al. (1992), for example, simulated growth rates and phenology of the grasshopper *Camnula pellucida* using ambient temperature on one hand, and ambient temperature plus solar heat gain on the other hand, to approximate body temperature of the individuals. They found that the simulated phenology on the base of ambient plus solar heat gain closely represented the field data. However, they did not separate the passively achieved operative environmental temperature from active thermoregulation. Whitman (1986 and 1988) emphasized the thermoregulation to be important for development of *Taeniopoda eques* but, in the same way, he did not separate active from passive heat gain. Furthermore, neither Carruthers et al. (1992) nor Whitman (1986 and 1988) quantified the benefit for succeeding generations.

Whereas the influences of passive heating and active regulation may be investigated by measuring operative environmental temperatures – e.g. by animal thermometers (Chappell 1983) – during short-term field experiments, population consequences can only be separated and investigated by a quantitative model. We approached this separation by the empirical and biophysical thermal model. Thus, we not only detected behavioural thermoregulation in *S. lineatus* but, for the first time, were able to estimate data regarding the influence on population level. Thereby, we get a quantitative understanding how a behavioural trait acts on individual fertility both as population reproductive success and, therefore, on the number of individuals in the following generation and possibly on the survival of a population. Moreover, these findings may allow to explain empirically discovered habitat requirements by functional connections.

ACKNOWLEDGEMENTS

We are grateful to J. Baumgärtner and K. Reinhardt for their helpful suggestions on the manuscript and to E. Finke for his comments on earlier drafts of this paper and, especially, for his aid with numerical solution of the integral. G. Jetschke kindly derivate the ellipse projection and J. Schumacher helped with some statistics. Furthermore, we would like to thank G. Kluge for making available weather data, D. Baumbach for partial entering weather data, and J. Klingelhöfer for his help with field works. The study was supported by German Research Foundation (DFG) grants Ko1494/1-1 and Ko1494/1-2 and partly by the German Ministry of Education and Research (BMBF) grant 0339521A.

REFERENCES

Anderson, R.V., Tracy, C.R. & Z. Abramsky 1979. Habitat selection in two species of short horned grasshoppers. The role of thermal and hydric stresses. *Oecologia* 38:359-374.

Baumgärtner, J., Severini, M. 1987. Microclimate and arthropod phenologies: the leaf miner Phyllonorycter blancardella F. (Lep.) as an example. In: Prodi, F., Rossi, F. & G. Cristoferi (eds): *Agrometeorology*. Fondazione Cesena Agricultura Publ. 498: 225-243.

Carruthers, R.I., Larkin, T.S., Firstencel, H. & Z. Feng 1992. Influence of thermal ecology on the mycosis of a rangeland grasshopper. *Ecology* 73:190-204.

Chappell, M.A. 1983. Metabolism and thermoregulation in desert and montane grasshoppers. *Oecologia* 56:126-131.

Chappell, M.A.& D.W. Whitmann 1990. Grasshopper Thermoregulation. In: Chapman, R.F. & A. Joern, *Biology of Grasshoppers*: 143-172. New York: John Wiley & Sons.

Digby, P.S.B. 1955. Factors affecting the temperature excess of insects in sushine. *Journal of Experimental Biology* 32:279-289.

Dingle, H., Mousseau, T.A. & S.M. Scott 1990. Altitudinal variation in life cycle syndromes of California populations of the grasshopper, Melanoplus sanguinipes (F.). *Oecologia* 84:199-206.

Gage, S.H., Mukerji, M.K. & R.L. Randell 1976. A predictive model for seasonal occurrence of three grasshopper species in Saskatchewan (Orthoptera: Acrididae). *Canadian Entomologist* 108:245-253.

Gillis, J.E. & K.W. Possai 1983. Thermal niche partitioning in the grasshoppers Arphia conspersa and Trimerotropis suffusa from a montane habitat in central Colorado. *Ecological Entomology* 8:155-161.

Gillis, J.E. & P.A. Smeigh 1987. Altitudinal variation in thermal behavior of the grasshopper Circotettix rabula (Rehn & Hebard) from central Colorado. *The Southwestern Naturalist* 32:203-211.

Harz, K. 1960. Geradflügler oder Orthopteren (Blattodea, Mantodea, Saltatoria, Dermaptera). In: Dahl, F., *Die Tierwelt Deutschlands und der angrenzenden Meeresteile nach ihren Merkmalen und nach ihrer Lebensweise*. 46. Teil. Jena: Fischer.

Heinrich, B. 1993. *The Hot-Blooded Insects - Strategies and Mechanisms of Thermoregulation*. Berlin: Springer.

Kemp, W.P. 1986. Thermoregulation in three rangeland grasshopper species. *Canadian Entomologist* 118:335-343.

Kemp, W.P. & B. Dennis 1989. Development of two rangeland grasshoppers at constant temperatures: development thresholds revisited. *Canadian Entomologist* 121:363-372.
Kriegbaum, H. 1988. *Untersuchungen zur "Lebensgeschichte" von Feldheuschrecken (Acrididae, Gomphocerinae): Fortpflanzungsstrategie und akustisches Verhalten im natürlichen Habitat.* Erlangen: Dissertation thesis Univ. Erlangen-Nürnberg.
Lactin, D.J. & D.L. Johnson 1996. Effects of Insolation and Body Orientation on Internal Thoracic Temperature of Nymphal Melanoplus packardii (Orthoptera: Acrididae). *Environmental Entomology* 25:423-429.
Machlis, L., Dodd, P.W.D. and J.C. Fentress 1985. The pooling fallacy: Problems arising when individuals contribute more than one observation to the data set. *Ethology* 68:201-214.
Marquardt, D.W. 1963. An Algorithm for Least Squares Estimation of Nonlinear Parameters. *Journal of the Society of Industrial and Applied Mathematics* 11:431-441.
Parker, M.A. 1982. Thermoregulation by Diurnal Movement in the Barberpole Grasshopper (Dactylotum bicolor). *American Midland Nataturalist* 107:228-237.
Pepper, J.H. & E. Hastings 1952. The effects of solar radiation on grasshopper temperatures and activities. *Ecology* 33:96-103.
Press, W.H., Flannery, B.P., Teukolsky, S.A. & W.T. Vetterling 1986. *Numerical Recipes.* Cambridge: Cambridge University Press.
Putnam, L.G. 1963. The progress of nymphal development in pest grasshoppers (Acrididae) of western Canada. *Canadian Entomologist* 95:1210-1216.
Richards, O.W. & P.D. Waloff 1954. Studies on the Biology and Population Dynamics of British Grasshoppers. *Anti-Locust Bulletin* 17:1-182.
Sachs, L. 1982. *Applied statistics.* Berlin: Springer.
Sachs, L. 1992. *Angewandte Statistik.* Berlin: Springer.
Samietz, J. 1994. *Untersuchungen zur Populationsgefährdungsgradanalyse zweier Heuschreckenarten (Saltatoria) in Halbtrockenrasen: Stenobothrus lineatus (PANZER) 1796 und Phaneroptera falcata (PODA) 1761.* Jena: Diploma thesis Univ. Jena.
Samietz, J. (in prep.). Temperature dependence and developmental thresholds of maturation and successive oviposition in a European grasshopper.
Samietz, J., Berger, U. & G. Köhler 1996. A Population Vulnerability Analysis of the Stripe-winged Grasshopper, *Stenobothrus lineatus* (PANZER) (Caelifera: Acrididea). In: Settele, J., Margules, C., Poschlod, P. & K. Henle (eds), *Species Survival in Fragmented Landscapes*: 299-311. Dordrecht: Kluver Academic Publishers.
Stone, G.N. & P.G. Willmer 1989. Endothermy and temperature regulation in bees: a critique of "grab and stab" measurement of body temperature. *Journal of Experimental Biology* 143:211-223.
Stower, W.J. & J.F. Griffiths 1966. The body temperature of the desert locust (Schistocerca gregaria). *Entomologia Experimentalis et Applicata* 9:127-178.
Wang, Z.-X. & Y.-L. Chen 1989. The influence of environmental temperature and humidity on the body temperature and water content of Chorthippus dubius (Zub.). *Acta Entomologica Sinica* 32:278-285.
Whitman, D.W. 1986. Developmental thermal requirements for the grasshopper Taeniopoda eques (Orthoptera: Acrididae). *Annals of the Entomological Society of America* 79:711-714.
Whitman, D.W. 1987. Thermoregulation and daily activity patterns in black desert grasshopper, Taeniopoda eques. *Animal Behaviour* 35: 1814-1826.
Whitman, D.W. 1988. Function and evolution of thermoregulation in desert grasshopper Taeniopoda eques. *Journal of Animal Ecology* 57:369-383.
Zar, J.H. 1984. *Biostatistical analysis.* Englewood Cliffs/New Jersey: Prentice-Hall.

Population and Community Ecology for Insect Management and Conservation, Baumgärtner et al. (eds)
© 1998 Balkema, Rotterdam, ISBN 90 5410 930 0

Extinction and the variability of populations

Brian H. McArdle
Biostatistics Unit, School of Biological Sciences, University of Auckland, New Zealand
Ian P. Woiwod
Entomology and Nematology Department, IACR, Rothamsted, Harpenden, UK

ABSTRACT: Generalized Linear Mixed Models are introduced as a way of estimating the percentage of a species' population that is in local decline. A technique is described so that, if this is high, an estimate can be calculated of the amount of time till a certain percentage of the population is below some critical local density. These techniques are illustrated using 2 moth species from the Rothamsted Insect Survey.

1 INTRODUCTION

It is a matter of increasing concern that an appallingly large number of species are in decline or are threatened with extinction. Although we are aware of the problem, there seems to have been remarkably little effort put into monitoring insect populations although they contain much of the earth's diversity and are vital components of many ecosystems. When we do have data on insects, it is nearly always in the form of densities measured at sampling stations rather than full population censuses - as is more often the case with declining vertebrates.

Dealing with densities can be considerably more complicated than with population size (McArdle and Gaston 1993). For example, when the population size is known then the rate of decline can be modelled, and the time to extinction (the point the model predicts a population of less than one) estimated by extrapolation. Such approaches often assume a constant rate of decline (e.g. Gaston and Nicholls 1995). However, if there is spatial variation in the local rate of decline, then the assumption of a constant average rate cannot be accepted. In figure 1 a population is modelled which consists of 100 subpopulations with constant (but different) rates of growth or decline (on a log scale). Growth rates are normally distributed with mean growth rate of -0.06 and standard deviation 0.06. The subpopulations start with a lognormal distribution of densities. Though the rate of decline is constant for each part of the population, it is clearly not constant for the population as a whole (the sum of the 100 parts of the population) - figure 1. If it were constant for the whole population then we would see a declining exponential curve. In fact, one might be tempted into suggesting density dependent dynamics were occurring, stabilising the population at lower densities. In fact there are no density dependent dynamics in this model. The population is simply the sum of 100 independent subpopulations with constant rates of growth or decline. In fact, because some of the populations have positive growth (the slopes come from a normal distribution), the population starts to increase again. When there is spatial variation in rates of decline, modelling the population's trajectory through time is complex unless we model the separate parts of the population, and predict the behaviour of the overall population from them.

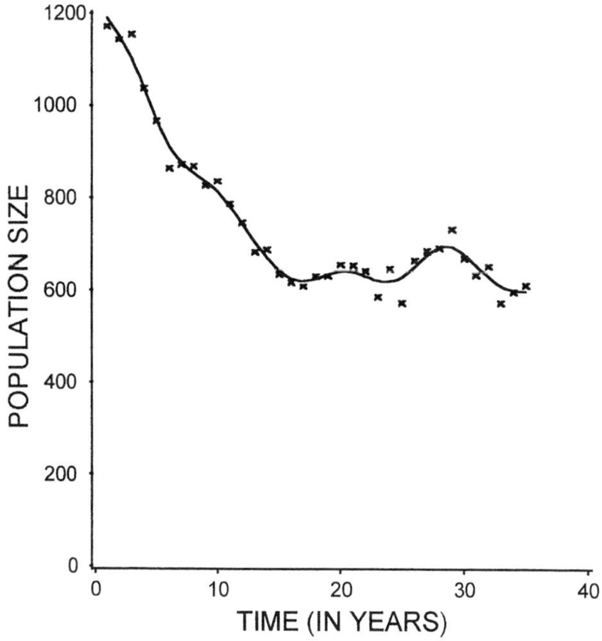

Figure 1. The trajectory through time of a population consisting of 100 strata with constant (but different) rates of growth or decline (on a log scale). The rates are normal with mean growth rate of -0.06 and standard deviation 0.06. The strata start with a lognormal distribution of densities. The figure shows the decline of the overall population (the sum of the 100 parts of the population). Though the rate of decline is constant for each part of the population it is clearly not constant for the population as a whole.

When we have measured only a sample of the possible parts of the population, perhaps as a series of density estimates, the situation becomes even more complex. We are now wanting to extrapolate in space as well as time. Even if all the sampled sites are in decline, some of the unsampled parts of the population may not be declining; they could be stable or even increasing, in which case the population will not go extinct, even though the sampled part of the population will. Estimating whether the population will go extinct, or making any statement about what proportion will, requires generalising from the sample of rates of decline to the statistical population of values. This in turn requires assumptions about the statistical or spatial distribution of rates of growth and decline. In this paper, rather than confront the problems of spatial modelling, assumptions are made about the statistical distribution.

One consequence of generalising to the whole population from only a part is that statements about the population as a whole must now be probabilistic. There can be no reliable estimate of time to extinction, since an unknown proportion of the population may in fact be stable. Also, since we are now having to model densities rather than population size, we can only model the expected density at a site, not the size of the population. This means it is difficult to specify a universal criterion for extinction. A very low density may not imply that extinction is close if the area of a sample is small relative to the size or mobility of the animal and the spatial extent of the population.

If we have the spatial extent of the population, then we can attempt to convert from the recorded densities to an estimate of population size. This would allow us to model the population and to get the information about extinction that we require. However, all too often the population of sites that we have sampled is from the set of sites that support the organism. The total area inhabited by the population may not be well known.

When we are unable to convert from density to population size we can still make statements like "greater than 90% of sites are in decline", "the time to 90% of the sites having an expected density of less than X is ...". These are more complex than "time to extinction" but are likely to be almost as useful if workers are not intimidated by their rigour. If we wished to make assumptions about the distribution of counts around each expected value we could rephrase the latter statement to "the time to 90% of the sites being X% certain to be empty is..." but there seems little point, it adds little except more assumptions.

2 STATISTICAL METHODS

Let us suppose that we have recorded the density, a count, of some insect at a number of sites regularly over time. In essence we are recording the trajectory of density through time for each site. We have therefore a Site by Time matrix.

In order to estimate the percentage of sites that are in decline we must estimate the mean and variance of the rates of decline found in different parts of our organism's range. In order to estimate the time for some percentage of the sites to reach a critical density we also need the mean and variance of the starting densities at the sites. We cannot model how long it takes unless we have information on where the sites are starting from. As a result the problem of modelling the decline of the total population can be broken into two parts:

1. estimating the rate of growth/decline and expected starting density for each site in the sample and,
2. estimating the mean and variance of the statistical population of rates of change and starting densities.

2.1 *Generalized Linear Models*

A constant rate of growth or decline gives a linear time trajectory if population density is plotted on a log scale (e.g. Gaston and Nicholls 1995). That is, on average the population density will grow or decline by a constant percentage in each time unit. It is therefore sensible to model the change of density through time on a log scale. There are two main ways of doing this: one is to log transform the data and use least squares regression assuming normal error, alternatively we can use a generalized linear model (McCullagh and Nelder 1989) with a log link and Poisson error (also known as log-linear models). If the numbers recorded on each sampling occasion were always relatively high, then both approaches would give much the same results. However if some of the counts are small then the differences would be larger. In particular if there are any zeros in the data, log transforming the data becomes impossible. Instead it would be conventional to use the $\log(N + c)$ transform, where c is an arbitrary constant, often 1. There are a number of reasons why this is less than ideal. Not least because it can predict negative densities (some others are described in McArdle, Gaston and Lawton 1990); additionally, at low densities the assumption of normal errors and constant variance becomes untenable.

An alternative approach increasingly advocated for ecologists for modelling monitoring data (e.g. Margules and Austin 1994) is the use of generalized linear models (sometimes known as GLIMs or GLzMs). I shall outline below the essential features but Crawley (1993) and Dobson (1983) are good introductions.

Generalized linear models are based on the premise that the purpose of any linear model (ANOVA and regression are linear models) is to model the expected, average, value (η_i) of the response variable (Y_i). The observed value Y_i is then a random value from some distribution (there is a choice) which has η_i as its expected value. For example, if we were dealing with say ANOVA or linear regression then the estimates of η_i would be better known as $\hat{Y}_i$ and Y_i would come from a normal distribution with variance σ^2 that would be estimated from the data. The importance of the generalized approach is that we are not restricted to normal errors or simple linear models. The relevant form for the handling of animal count data is the log-linear or Poisson model.

The count Y is modelled by the non-linear model

$$\eta_i = e^{\beta X_i + c}. \tag{1}$$

Which is equivalent to the log-linear model:

$$\log(\eta_i) = \beta X_i + c. \tag{2}$$

The variation of Y_i around the expected value η_i is treated as Poisson; i.e. Y_i is a Poisson count from a distribution with mean η_i. This is the model outlined by McArdle and Gaston (1995). From our perspective perhaps the most important feature of the Poisson model is that η_i is fitted to Y_i by maximum likelihood without ever requiring the Y_is to be log transformed (thus avoiding the necessity to fudge the zeros). If instead of using maximum likelihood we use a simple modification known as quasilikelihood (McCullagh and Nelder 1989) we can relax the Poisson assumption and model aggregated organisms.

We have now a modelling method that can allow us to estimate a constant rate of decline by fitting the model

$$\eta_i = e^{\beta t_i + c} \tag{3}$$

to the observed counts at times t_i. The rate of decline would be estimated by β, and the starting density (density at time 0) by e^c.

Unfortunately our troubles are not yet over. While our estimates $\hat{\beta}$ and c might be as good as we can get; unfortunately the standard errors and confidence intervals (and tests - should we be interested in those) will almost certainly be invalid since successive deviations (errors) from the fitted model will not be independent (Figure 2 shows this quite clearly). If on one occasion the density was higher than expected there is a high probability that it will be higher in the next time period. The errors are not independent - an essential assumption of the standard errors and tests for the classical approaches to linear models.

A further drawback is that we cannot estimate the proportion of all the sites that are in decline by simply fitting separate models to each site and then counting how many of the $\hat{\beta}$ values are negative. Because the $\hat{\beta}$ values are themselves subject to error (the unknown standard errors would tell us by how much) their distribution will be wider, with longer tails, than that of the βs. We would therefore consistently misestimate the true proportion. We shall see this in the examples.

We therefore need a technique that can handle autocorrelated errors and estimate the variances of the populations from which parameters come, and not merely the parameters themselves. Linear mixed model theory (Henderson 1984) can give us the technique we need.

2.2 *Linear Mixed Models.*

If we put aside the problem of autocorrelated errors for a moment, we can concentrate on the problem of dealing with a set of regression lines that are a random sample from a population of regression lines. Our problem restated is that we do not need the individual slopes and intercepts but the mean and variance of the population from which they come. This is essentially the difference between fixed and random effects and has been long known in ANOVA. In recent years, work that originated in animal breeding (Henderson 1984) to handle such models has found its way into more mainstream statistics.

While a detailed description is beyond the scope of this paper (an introduction can be found in McLean, Sanders and Stroup. 1991), a short outline is necessary. The traditional approach to random factors (and for the moment we shall be referring to ANOVA) does not distinguish between random and fixed effects during the parameter estimation phase of the model fitting (Searle, Casella and McCulloch 1992). They are only distinguished when the time comes to calculate standard errors or perform tests. For example take a mixed model where a set of treatments are all applied to the same set of randomly chosen animals. The effect of the animals and those of the interactions between treatment and animal are random effects, samples from populations of such effects. The model is usually written as:

$$Y_{ijk} = \mu + t_i + \alpha_j + \gamma_{ij} + \varepsilon_{ijk} \ , \tag{4}$$

where Y_{ijk} is the value measured on the kth sample taken from the jth animal receiving the ith treatment. The terms t_i, α_j, γ_{ij} and ε_{ijk} refer to the treatment, animal, interaction and sample error effects respectively; α_j, γ_{ij} and ε_{ijk} are random variates distributed normally with mean 0 and variances σ^2_α, σ^2_γ, and σ^2_ε, respectively. Though conceptually separate in this model, in fact the parameters t_i, α_j and γ_{ij} are estimated by fitting the matrix linear model:

$$\mathbf{y} = \mathbf{X}\boldsymbol{\beta} + \boldsymbol{\varepsilon} \ , \tag{5}$$

where $\mathbf{y}$ contains the Y values, $\boldsymbol{\beta}$ the parameter estimates for all the effects, random and fixed, and $\mathbf{X}$ codes for the ANOVA design.

In other words the fitting process does not distinguish between fixed and random effects. The approach now available is called Linear Mixed Modelling and explicitly separates the random from fixed effects during the fitting process. More importantly it estimates them in two separate steps, one for the fixed effects (e.g. t_i) and one for the random effects (α_j, γ_{ij} and ε_{ijk}).

Thus the model fitted is:

$$\mathbf{y} = \mathbf{X}\boldsymbol{\beta} + \mathbf{ZU} + \boldsymbol{\varepsilon}, \tag{6}$$

where $\boldsymbol{\beta}$ now refers to the parameters of the fixed effects and $\mathbf{X}$ the design for the fixed effects; $\mathbf{U}$ contains the estimates of the random effects and $\mathbf{Z}$ codes for their design. The important advance over the traditional linear model is that covariance matrices (containing σ^2_α, σ^2_γ and σ^2_ε) of $\mathbf{U}$ and ε can be estimated without ever actually estimating the parameters in $\mathbf{U}$. This is done using the estimation procedure REML (Restricted/Residual Maximum Likelihood - Robinson 1987). The estimates of the fixed effect parameters will generally be the same as for the more traditional Least Squares approach, but the elements of $\mathbf{G}$ (the covariance matrix of $\mathbf{U}$) and of $\mathbf{R}$ (the covariance matrix of ε) are estimated (provided these are of relatively simple structure) and their standard errors can be calculated. For example in the model above we have assumed independent random effects, so the covariance matrix $\mathbf{G}$ has no covariance entries and only two variances (those for α and γ); similarly $\mathbf{R}$ has only one variance (σ^2_ε). It is one of the more important advantages of this approach that it can handle situations where there are covariances (correlations) between either the random effects or the errors. This can allow, for example, the analysis of repeated measures data where the errors are serially correlated; but is flexible enough to handle other situations (such as those from animal breeding where the correlations are predefined by familial relatedness). This is not the place to go into the details of this process except to say that it has enormous potential for handling many otherwise difficult or intractable situations that restrict the use of standard Least Squares methods in ecology.

For most ecologists its advantages are:

Firstly, that it acknowledges that the individual parameters for a random effect are irrelevant and that all that needs estimating is the mean and variance. So that instead of losing a degree of freedom for each of the parameters only one is removed for each variance estimated (in our current example one for the distribution of α values and one for the γs).

Secondly, this method is actually modelling not just the variances but also the covariances between the random effects and can, if sufficient data is available, therefore accommodate and estimate correlations that may exist between sampling units.

2.3 *Using Linear Mixed Models to estimate the variances of slopes and intercepts.*

Just as ANOVA and regression are at root identical, both special cases of Linear Model theory, so the Linear Mixed Model theory applies to regression as well. A simple application would be the fitting of a set of regression lines that are a random sample from a population of regression lines. In this case the model could be initially envisaged as:

$$Y_{ij} = \tau + (\tau_i - \tau) + \beta X_{ij} + (\beta_i - \beta) X_{ij} + \varepsilon_{ij} \qquad (7)$$

The distribution of intercepts is modelled by an overall average mean intercept τ and α distribution of $(\tau_i - \tau)$ values with mean 0 and variance $\sigma^2_{(\tau_i - \tau)}$. The slopes are modelled with a mean of β and a variance of $\sigma^2_{(\beta_i - \beta)}$. Regardless of the number of lines fitted only four parameters are being estimated so only four degrees of freedom are being lost. This can be extremely important since the estimation of variances gives far less precise results than the estimation of means; so good variance estimates require many degrees of freedom. Optionally, and usefully, the correlation between the slopes and the intercepts can also be estimated (though this costs another degree of freedom).

2.4 *Generalized Linear Mixed Models (GLMMs).*

For our purposes neither Generalized Linear Models nor Linear Mixed Models on their own are enough. However, in recent years there have been a number of attempts to fuse the two approaches (Clayton 1992, Wolfinger and O'Connell 1993, Stroup and Kachman 1994, Lee & Nelder 1996). Of these, few have reached the computer packages available to ecologists. Recently however the macro GLMMIX, an implementation of Wolfinger's approach, has become available, distributed as an undocumented macro with the widely used package SAS. This allows the fitting of mixed models and the estimation of variance components within the framework of the Generalized Linear Model. The main restriction is that the random effects must be normally distributed within the linear part of the non-linear model. For example, take our model; we have a random (or at least representative) sample of sites, for each we wish to fit the model $\eta_{ij} = e^{\beta_j t_i + c_j}$, where j identifies the site. To use the Wolfinger approach to GLMMs we must assume that the β_j and the c_j are normally distributed with means β and c, and variances σ^2_β and σ^2_c and correlation $\rho_{\beta c}$. We are therefore assuming that the distribution of η at time zero (i.e. the distribution of e^c) is lognormal so that the distribution of observed counts Y over sites at time zero will be Poisson lognormal; a reasonable distribution for the description of spatial variation in count data (Cassie 1961). The distribution of rates of decline is assumed to be normal. These assumptions would have to be checked as a natural part of any analysis. A further feature of our model would be to assume that we have serially correlated errors within each site. If a site has higher than expected densities one year, it is more likely than not to have higher than expected the following year. If we ignored this assumption it could have major consequences on the validity of the standard errors. A simple correction for this would be to assume the errors follow an autoregressive order 1 process within a site (e.g. McArdle 1989). This assumes that the error depends on the previous time period only, not on any periods further back. This will usually be adequate for any species that do not cycle at the time scale being used. Thus seasonal cycles would be irrelevant for a study that took samples at the same time every year or pooled counts over a year. Given the relatively short duration of most monitoring studies (most have been going for <20 years) detecting such cycling from the data would be very difficult as time series typically have to be quite long in order to detect high order autocorrelation (McArdle 1989, Chatfield 1984). Knowledge of the animal's natural history would be the main defence against such a mistake.

2.5 *Estimating the proportion of a population in decline.*

Once the variance of slopes has been calculated then by assuming a normal distribution it becomes possible to estimate the proportion of the population that is in decline. A natural precaution at this point would be to check the assumption of normality. We do this by using a fixed effect model to estimate the individual slopes and intercepts and examine the normality of these values.

If we can accept the assumption of a normal distribution then it becomes possible to estimate the proportion of the population that is in decline. It is simply the probability of getting a value less than 0 from a normal distribution of mean β and a variance σ^2_β. The confidence interval on this estimate is slightly more difficult to calculate. One advantage of variance estimation by REML is that it provides standard errors for the estimates. Extensive simulations show that the standard deviation estimates are asymptotically normal, so the standard errors of the standard deviation estimates (obtained from the standard errors for variance by the delta method - Seber 1981) can be used to provide approximate confidence intervals. In this study we produce a confidence interval on the estimate of proportion in decline by simulating a population of means and standard deviations from normal distributions with the appropriate means and standard deviations (standard errors), and for each pair of values we calculate the proportion of the resulting distribution that is below zero. The confidence intervals are then given by the appropriate percentiles of these proportions. Simulations suggest that these work adequately even when the distribution of standard deviation estimates is not normal (i.e. when variances and sample sizes are small).

2.6 *Estimating the time for X% of the population to reach a critical density.*

Our ability to estimate the time to reach a critical density depends not only on the distribution of slopes and the intercepts but also on the correlation between the two. This correlation can be estimated as a covariance term in **G,** the random effects covariance matrix.

The time (t_{crit}) taken for the density of the population at a site to reach some critically low value (N_{crit}) is calculated as:

$$t_{crit} = (\ln(N_{crit}) - c)/\beta. \qquad (8)$$

The time for a proportion X of the population to reach N_{crit} is most easily estimated by simulation. Taking the bivariate normal distribution specified by the means and standard deviations of the slopes and intercepts and their correlation we can simulate a large number of regression lines and from them find the Xth percentile (t_{critX}) of the distribution of t_{crit}. Confidence intervals can be calculated in a similar way. By using the standard errors and the within sample correlation between estimates of slope and intercept, simulated sampling error can be added to the simulated population parameters thus allowing the estimation of the sampling distribution of t_{critX}, and so the 5th and 95th percentiles to give 90% confidence intervals. In simulations like these we use 90% confidence intervals as a matter of course since the behaviour of the extreme tails of the distributions may not conform perfectly to the normal distribution. As a result we can perhaps have more faith in the 90% than the 95% confidence intervals. Simulations suggest that these confidence intervals perform well even with relatively small numbers of sites (10) and small variances on the slopes.

3 EXAMPLES

There are very few insect databases with spatial replication and that are long enough to perform the type of analysis described here. One such is the moth data from the Rothamsted Insect Survey's light trap network, which has been used widely in spatio-temporal population dynamic studies (Woiwod and Harrington 1994). The data have been obtained over the last 30 years from a large number of standard Rothamsted type light traps (Williams 1948) which have been operated daily by volunteers throughout the British Isles.

We have used annual totals from these samples. These traps only catch small numbers and as species are differentially attracted to light these data provide only relative measures of population density, with zero catches indicating that densities have fallen below the recording threshold of the light traps. Also, as samples are small, only the commoner species tend to be recorded in sufficient numbers for analysis. Both these factors make the choice of critical density problematic, however the population trends in these common species are of major importance in any assessment of the effects of environmental change (Woiwod and Dancy 1987, Luff and Woiwod 1995).

For this study we have chosen two species that have relatively unbroken records from the same sites over a 14 year period (1977-1990). The species are both Geometridae; *Crocallis elinguaria* (L.), a univoltine species in Britain whose larvae are polyphagous on various trees and shrubs, and *Opisthograptis luteolata* (L.), a species with up to three overlapping generations in Britain, which is polyphagous on deciduous trees, especially Rosaceae. Altogether 31 different sites were used with a wide spatial spread. *O. luteolata* was recorded at all 31 sites, *C. elinguaria* only at 30. There were a few gaps in the time sequences (never more than one at a site) which were filled by imputation (Little 1987). We will examine the species separately.

3.1 *Crocallis elinguaria (L.).*

This species is clearly in decline (Figure 2). The GLMM fitting $\eta_{ij} = e^{\beta_j t_i + c_j}$ with AR(1) error gave the following estimates.

The autocorrelation was necessary in the model (as is clear in figure 2). The standard errors have been calculated by quasilikelihood, correcting for the presence of considerable extra-Poisson residual variance. The distribution of slopes and intercepts were not detectably different from normal (Figure 3). The percent of sites in decline was estimated as 98.7 (90% C.I. 90.4 - 100). Examining the slope estimates showed 86.6% (26 out of 30) were negative. As we explained earlier we would expect the parameter estimates to misestimate the true percentage because the variance of the distribution of slope estimates will be greater than that of the parameters. As a result we can expect in this case that when most of the sites are in decline then the percentage will be underestimated by the distribution.

Table 1. Parameter estimates (and standard errors) for fixed and random effects in a GLMM fitted to the *Crocallis elinguaria* data

Fixed effect parameter	Estimate	Standard error	Random Effect Parameter	S.D. estimate	Standard error
Mean site	2.69	0.17	Site	0.81	0.14
Mean slope	-0.065	0.011	Slope	0.03	0.01
Autocorrelation	0.36	0.062	Correlation (slope*site)	0.00084	0.0005
			Residual	4.48	0.43

The time to 90% of sites having less than a critical density of 1 animal per site per year (an arbitrary choice representing a very low density that might represent effective extinction) was estimated at 100.7 years (90% CI: 53.1 - 160.0). These figures of course can be used as indicative only.

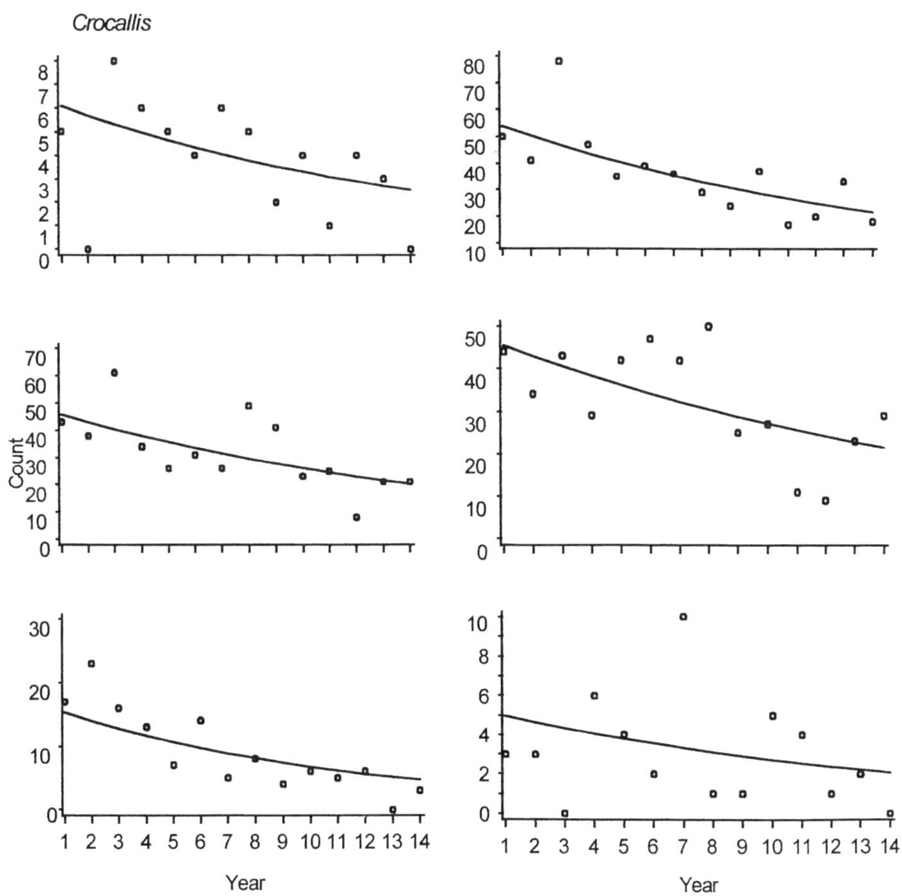

Figure 2. The trajectory through time for 6 randomly chosen sites for *Crocallis elinguaria* with fitted curves (a fixed effect log-linear model). Note the temporal autocorrelation that is clearly present.

Crocallis

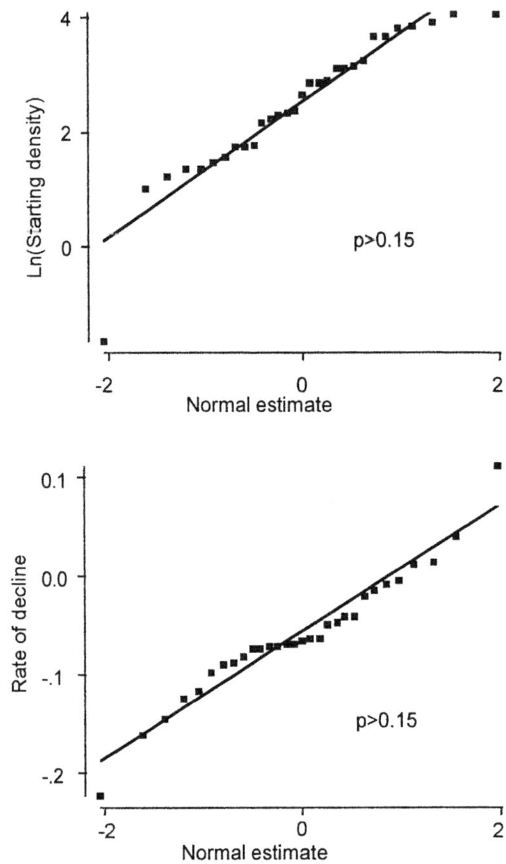

Figure 3. Normal probability plots for the parameter estimates from fixed effect log-linear models for *Crocallis elinguaria*.
a) intercepts (ln(starting density))
b) slopes(rate of growth/decline).

3.2 *Opisthograptis luteolata* (L.)

Figure 4 shows the trajectories through time for the same 6 sites as Figure 2. While some of the sites are decreasing, others seem to be showing no clear trend. The results of the GLMM are below. There does not appear to be any average decline apparent, indeed there may be an average

Table 2. Parameter estimates (and standard errors) or fixed and random effects in a GLMM fitted to the *Opisthograptis luteolata* data

Fixed Effect Parameter	Estimate	Standard error	Random Effect Parameter	S.D. estimate	Standard error
Mean Site	2.65	0.24	Site	1.26	0.19
Mean slope	0.034	0.01	Slope	0.04	0.01
Autocorrelation	0.20	0.061	Correlation (Slope*site)	0.0017	0.0008
			Residual	7.18	0.60

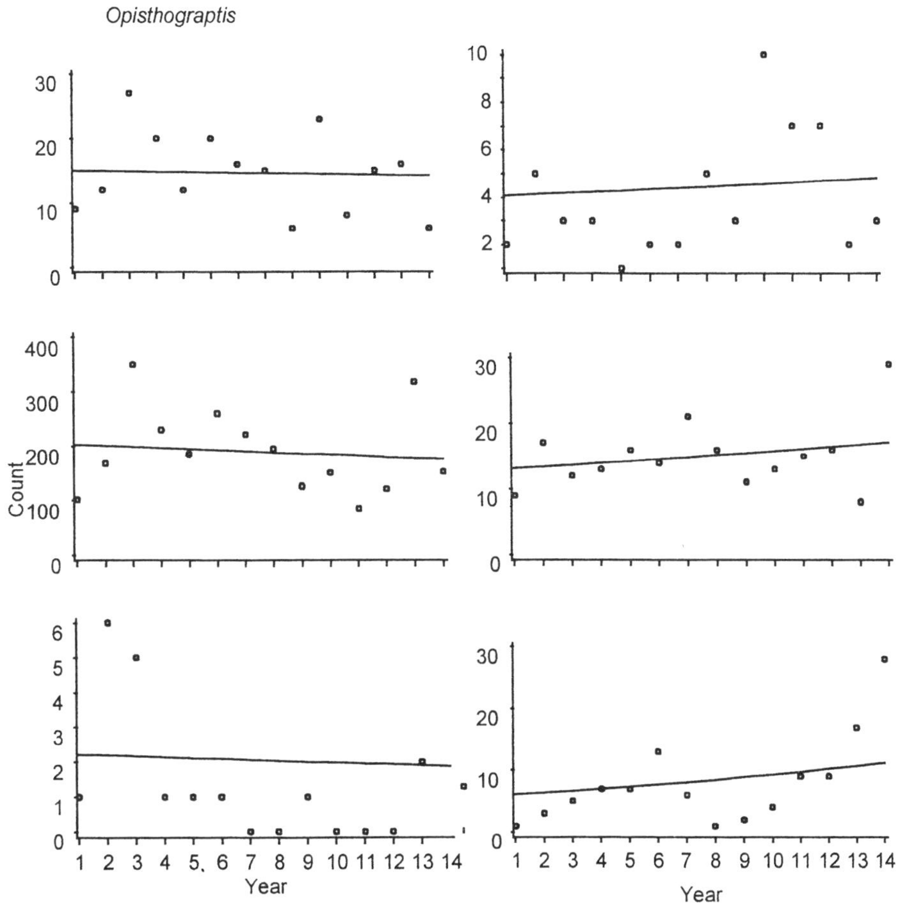

Figure 4. The trajectory through time for *Opisthograptis luteolata* at the same six sites as in figure 2. The fitted curves are from the fixed effect log-linear model.

increase. Once again there is strong autocorrelated variation around the fitted lines.

The estimate of the proportion of sites in decline is 20.2% (90% CIs 4.5 - 35.8). In this case there is clearly no point in estimating the time till 90% of the sites reach a critical density since most of the sites are not in decline. The data as analysed here shows no evidence of any danger to this species.

4 REFERENCES

Cassie, R. M. 1961. Frequency distribution models in the ecology of plankton and other organisms. *J. Anim. Ecol.* 31: 65-92.

Chatfield, C. 1984. *The Analysis of Time Series: an Introduction*, 3rd Edition, Chapman and Hall.

Clayton, D. 1992. Generalized linear mixed models in biostatistics. *The Statistician* 41: 327-328.

Crawley, M. 1993. *GLIM for ecologists*. Oxford, Blackwell.

Dobson, A. J. 1983. *An Introduction to Statistical Modelling*, Chapman and Hall, London.

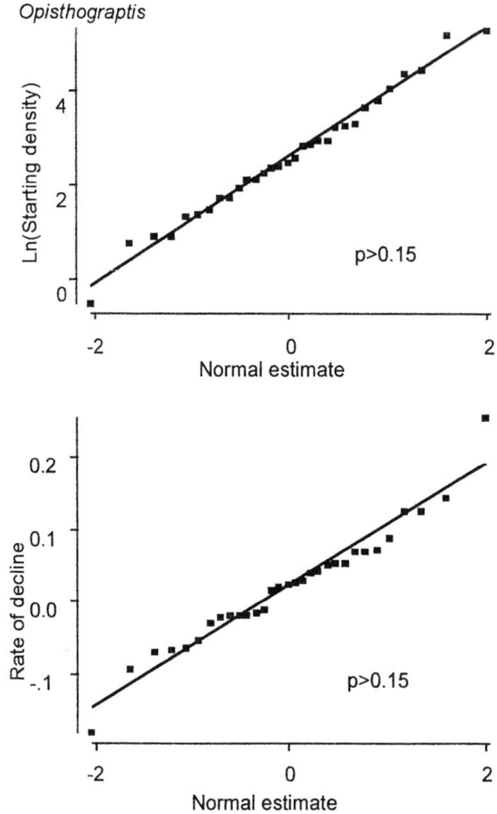

Figure 5. Normal probability plots for the parameter estimates from fixed effect log-linear models for *Opisthograptis luteolata*.
a) intercepts (ln(starting density))
b) slopes(rate of growth/decline).

Gaston, K. J. & A. O. Nicholls 1995. Probable times to extinction of some rarte breeding bird species in the United Kingdom. *Proc. Roy. Soc. Lond. (Ser. B)*. 259: 119-123.

Henderson, C. R. 1984. *Applications of Linear Models in Animal Breeding*. University of Guelph. Guelph, Canada.

Lee, Y. & J. A. Nelder 1996. Hierarchical Generalized Linear Models. *J. R. Stat. Soc. B* 58: 619-656.

Little, R. J. A., & D.B. Rubin, 1987. *Statistical Analysis with Missing Data*, New York: John Wiley and Sons.

Luff, M. L. & I. P. Woiwod 1995. Insects as indicators of land use change: a European perspective. In: R. Harrington & N. E. Stork (eds), *Insects in a changing environment*. London, Academic Press: 399-422.

Margules, C. R. & M. P. Austin 1994. Biological models for monitoring species decline: the construction and use of data bases. *Phil. Trans. R. Soc. Lond.* 344: 69-75.

McArdle, B. H. 1989. Bird population densities. *Nature* 338: 628.

McArdle, B. H. & K. J. Gaston 1993. The temporal variability of populations. *Oikos* 67: 187-191.

McArdle, B. H. & K. J. Gaston 1995. The temporal variability of densities: back to basics. *Oikos* 74: 165-171.

McArdle, B. H., K. J. Gaston & J. H. Lawton 1990. Variation in the size of animal populations: patterns, problems and artefacts. *J. Anim. Ecol.* 59: 439-454.

McCullagh, P. N. & J. A. Nelder 1989. *Generalized Linear Models*, Second Edition. London: Chapman and Hall..

McLean, R. A., W. L. Sanders & W. W. Stroup 1991. A unified approach to mixed linear models. *Am. Stat.* 45: 54-64.

Robinson, D. L. (1987). Estimation and use of variance components. *The Statistician* **36**: 3-14.

Searle, S. R., G. Casella & C. E. McCulloch 1992 *Variance Components*. New York: John Wiley and Sons

Seber, G. A. F.1982. *The Estimation of Animal Abundance and Related Parameters*. New York: Macmillan.

Stroup, W. W. & S. D. Kachman 1994. Generalized linear mixed models -- An overview. *Proceedings of the Kansas State University Conference on Applied Statistics in Agriculture*: 82- 98.

Williams, C. B. 1948. The Rothamsted light trap. *Proc. Roy. Ent. Soc. Lond. A*. 23: 80-85.

Woiwod, I. P. & K. J. Dancy 1987. Synoptic monitoring for migrant insect pests in Great Britain and Western Europe. VII. Annual population fluctuations of macrolepidoptera over Great Britain for 17 years. *Rothamsted Experimental Station. Report for 1986(Part 2)*: 237-264.

Woiwod, I. P. & R. Harrington 1994. Flying in the face off change: The Rothamsted Insect Survey. In: R. A. Leigh & A. E. Johnston *Long-term Experiments in Agricultural and Ecological Sciences*: 321-342. Wallingford: CAB International..

Wolfinger, R. & M. O'Connell 1993. Generalized linear mixed models: A pseudo-likelihood approach. *J. Stat. Comp. Simul.* 48: 233-243.

Population and Community Ecology for Insect Management and Conservation, Baumgärtner et al. (eds)
© 1998 Balkema, Rotterdam, ISBN 90 5410 930 0

Some problems and approaches to modelling insect phenology

S. P. Worner
Department of Entomology and Animal Ecology, Lincoln University, Canterbury, New Zealand

ABSTRACT: The development of models that predict the timing of important events in insect life cycles for control purposes has attracted considerable research attention in recent years. Approaches to phenological modelling range from simple physiological-time based models and static models, to more complex models that are process orientated. The use of hypothetical data to verify internal model assumptions in the absence of adequate observational data is illustrated. The implications of the rate summation or Kaufmann effect and the use of small and large samples of historical data for model validation are discussed. Predictive sample reuse methodology is used to validate a simple model based on long-term insect, climate and host-plant phenology data. Model validation indicated that possible interactions between host plant phenology, weather and insect population events merit close examination.

1 INTRODUCTION

A large body of theory and methods exists for modelling the dynamics and structure of populations of organisms. Classical models are based on the assumption that individuals in the population can be aggregated into one or several state variables that change with time. Empirical biologists are mindful, however, of the effect of spatial and temporal environmental discontinuities on population dynamics so that there are now many examples of models that combine analytical approaches with important environmental and biological components (e.g., DeAngelis & Gross 1992, Di Cola and Baumgärtner [this volume]).

Entomologists have long recognised the profound influence of temperature on the development rate, growth and fecundity of insects and the consequent effects on the timing of key events in insect life cycles. The temperature-time relationship and its effect on insect phenology have received most attention from applied entomologists who wish to use the relationship to construct prescriptive models that predict the timing and abundance of insect pests for purposes of control. Models that prescribe properly timed control strategies can reduce unnecessary insecticide application, reducing costs to the grower. Other important benefits include the delayed development of insecticide resistance, preservation of natural enemies, reduction of insecticide residues in crops, conservation of petroleum products and increased safety to farm workers, local residents and the environment. In addition, phenology models can improve the timing of sampling and monitoring programmes, the application of biological control procedures are often the basis of ecoclimatic assessment of potential establishment and distribution analyses of exotic pests (Sutherst & Maywald 1985, Worner 1988) and biological control organisms (Worner 1989, McClay & Hughes 1995). Furthermore, functions that model the relationship between insect development and temperature are important sub-components of larger numerical models that simulate insect population abundance.

Despite their wide use in applied entomology, there remain many problems associated with the derivation and use of such models. These problems are summarised and suggestions are made concerning how modelling the timing of key events in insect life cycles might be improved.

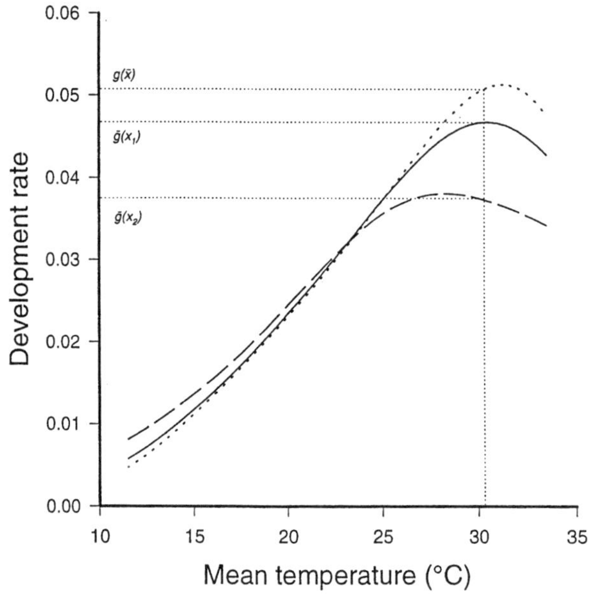

Figure 1. The effect of two sine-varying temperature regimes on a nonlinear function g. Under constant conditions (thick dotted line) development rate corresponds to that for the mean temperature ($g(\bar{x})$). For a sine-varying temperature regime of mean temperature ± 4 °C (solid line) the mean development rate corresponds to $\bar{g}(x_1)$. For a sine-varying temperature regime of ± 8 °C (dashed line) the mean development rate corresponds to $\bar{g}(x_2)$.

2 LINEAR VERSUS NONLINEAR MODELS

Three basic approaches to modelling insect phenology have been described. The oldest and most widely used model is the simple linear description of insect development in relation to temperature on which the thermal summation or degree day (DD) approach to timing prediction is based. Second, are the many different nonlinear functions that have been proposed to describe nonlinear development of insects. Nonlinear models require rate summation for timing prediction. Rate summation can be described as:

$$D = \sum r(T(t))dt$$

where development D is a function of temperature T, which in turn is a function of time t, and r is the development rate. The third approach consists of more complex "distribution models" (Logan 1988). Distribution models include both temperature-dependent development (described by either a linear or nonlinear function) and variation in development time (or rates) around the median individual. The biological validity (Gilbert & Raworth 1996) or usefulness of each approach continues to be debated (e.g., Hilbert & Logan 1983, Pruess 1983, Wagner et al. 1984, Higley et al. 1986).

3 FLUCTUATING TEMPERATURES AND THE RATE SUMMATION EFFECT

Much published research concerning insect development rates and phenology models shows that the effect of fluctuating temperatures and rate summation on nonlinear and linear model prediction has not been well understood. Some researchers have claimed that development accelerates under variable temperatures (Matteson & Decker 1965, Hagstrum & Hagstrum 1970); others report deceleration (e.g., Ellingsen 1969 [in Behrens et al. 1983]), or no difference (Morris & Fulton 1970, Humpesch 1982). In many instances, the explanation given for this effect is physiological. Figure 1 shows the effect of two sine-varying temperature regimes on a nonlinear function fitted to a well known data set for egg development of the melon fly, *Dacus cucurbitae* (Messenger & Flitters 1958). The function used is the popular biophysical model (Sharpe &

DeMichele 1977). Figure 1 illustrates that when the nonlinear rate of development of an insect species at constant temperatures is compared with predicted rates of development under an artificially varying temperature regime, acceleration of insect development at low temperatures and retardation at high temperatures occurs. Furthermore, the difference is accentuated by the diurnal range of the fluctuating temperature regime. This result applies to all nonlinear functions when the independent variable is varying in space and time (in general: $\bar{g}(x) \neq g(\bar{x})$ [Allen 1988]). In other words, for a nonlinear function (g), if the independent variable is varying in time or space, the mean of a function is not always equal to the function of the mean. The extent to which $g(\bar{x})$ and $\bar{g}(x)$ differ depends on the degree of nonlinearity of the function around the region of interest and the variability of the independent variable (in this case, temperature). In insect phenology this effect is known as the rate summation or Kaufmann effect (Ludwig & Cable 1933, Behrens et al. 1983, Ratte 1985, Worner 1992).

Clear explanations of the rate summation effect have been given by several authors, for example Pradhan (1945), Tanigoshi et al. (1976), Ratte (1985) and, more recently, Pedigo & Buntin (1994). Despite awareness of the existence of the rate summation effect among some scientists, many do not appreciate its consequences. Worner (1992) showed there is an interaction between the particular mathematical structure of a development rate model and rate summation. Lui et al. (1995) correctly pointed out that many studies to test model predictions under varying temperature regimes involve extrapolation of the rate function beyond the range of constant temperatures to which the function was originally fitted. But, extrapolation of a development rate function for prediction is necessary when it is well known that successful development can occur at temperatures that stray into regions that result in no development or cause high, even complete, mortality under constant temperature conditions. For example, in studies of the development of the melon fly under a constant temperature regime and varying temperatures normally experienced by the fly, Messenger and Flitters (1958, 1959) reported 0.3% hatch at 11.4°C constant temperature, yet at 11.3 ± 5.6°C and 11.2°C ± 8.3°C, 56.8 and 59.2% hatch (respectively) were recorded.

The selection of a development rate function for prescriptive use based on closeness of fit to developmental data from constant temperature experiments does not guarantee that it will give the best prediction under field conditions (that may involve extrapolation of the model). Three popular models gave similar fit to constant temperature development of the egg stage of the melon fly. However, under a variable regime of mean temperature ± 8.3 °C , the percent deviation of model predictions from observed (Messenger & Flitters 1959) development (Fig. 2) shows that for this data set, the Sharpe & DeMichele (1977) model gives better predictions of development under varying conditions.

Neither should selection of a nonlinear model be based on how well a model derived from constant temperature experiments predicts insect development measured under a narrow range of fluctuating field temperatures (as is typical when phenology models are validated over one or two years data for one or two sites). If the field temperatures vary within the more linear part of the development rate curve, then the nonlinear model has not been adequately tested. If proponents of the nonlinear approach believe that temperature extremes are important, then any model must be tested at those extremes.

Because insect development under varying temperature conditions is a function of the particular temperatures under which development occurs, a nonlinear model should not be fitted directly to development rates observed under variable temperature regimes. Worner (1992) showed that a function fitted to simulated insect development over sequences of naturally varying temperatures (that gave a range of mean temperatures for development) did not give acceptable predictions of development over the identical temperature regimes. Reasons for poor prediction are the rate summation effect, the fit of the model and possible bias in the diurnal temperature profile. Appropriate techniques that calculate an instantaneous development rate function from varying temperature development observations are available (Dallwitz & Higgins 1978, Allen 1988) but their adequacy is still not well established.

It may seem that because accelerated and decelerated development is observed in many experiments of insect development under variable temperature regimes that this is evidence that a nonlinear development model is the correct model to use for prediction. Such a conclusion may not be correct. Because the use of upper and lower thresholds introduce a form of nonlinearity to a linear function, in theory, a linear model will behave in the same way when temperatures fluctuate around those thresholds (Worner 1992).

Many entomologists believe that some unknown physiological mechanism acts in addition to the rate summation effect to advance insect development at low temperatures and retard

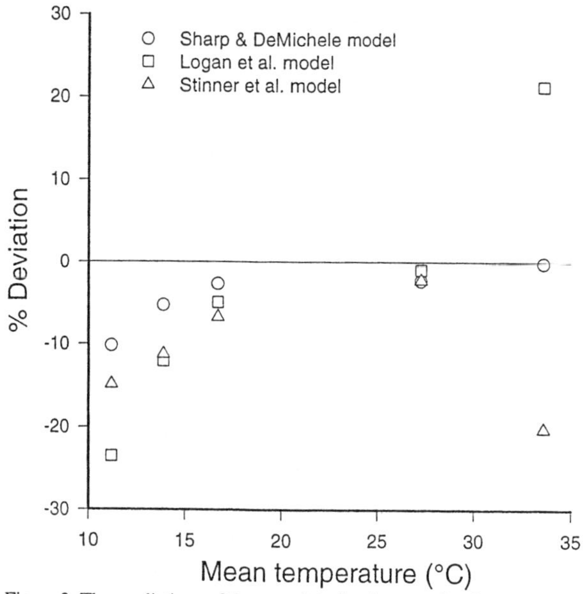

Figure 2. The predictions of three models fitted to egg development of melon fly under constant temperature conditions and their percent deviation from observed development of melon fly under a varying temperature regime of mean temperature ± 8.3 °C. Sharp & DeMichele model (1977), R^2=0.999; Logan *et al.*, (1976) model, R^2= 0.998 and the Stinner *et al.*, (1974) model, R^2 = 0.999.

development at high temperatures. As a consequence, there have been several studies where predictions of nonlinear models under varying conditions are compared to observed development under the same conditions to detect such an effect. More recently, Lui et al. (1995) carefully examined data from 26 insect species and concluded that rate summation could usually account for development at varying temperatures. Unfortunately, Lui et al. (1995) used the mean percentage deviation of the expected development from the observed development under varying temperatures to test for significant differences. But, because of large positive and negative deviations between observed and predicted development, some test based on absolute deviations would have been more appropriate.

4 DISTRIBUTION MODELS OF INSECT PHENOLOGY

The phenology models discussed so far are based on the development rate of the median or mean individual in a population. Distribution models (e.g., Logan et al. 1976, Sharpe et al. 1977, Welch et al. 1978, Régnière 1984, Dennis et al. 1986) include variability of development rates or time within the insect population at any particular temperature. The model proposed by Dennis et al. (1986) uses the logistic probability distribution function to describe this variability for r development stages of an insect population (see Dennis et al. 1986, Dennis & Kemp 1988 for details). If there are r development stages, the model has r unknown parameters. The r unknown parameters can be expressed as a_1, a_2, ...a_{r-1}, v. The quantity a_i, i=1, ..., r-1 can be interpreted as the time t (in terms of accumulated DD) at which half the individuals are in stage i or below, and v is a measure of the variability of development rates of the insects. Development is expressed in terms of accumulated DD. This model has been shown to fit observed data well (Kemp & Onsager 1986, Kemp & Dennis 1991). Yet, it is apparent from these studies (and the authors noted) that different parameter values (that may differ substantially) are obtained for different years. If this is true, the question is: Which parameters are used to make predictions for an independent data set, or is some average used? Parameters derived for two species of grasshopper over two seasons (1975 and 1976) at Roundup, Montana (Kemp & Onsager 1986) were used to predict key events for each species for a different season (1995) at a site (Billings, Montana) 80 km from where the original data were recorded (Table 1). Despite disparity in parameter values, the Julian dates for which the peak of the life stages were predicted during

Table 1. Accumulated degree days and Julian dates of the occurrence of key events for two grasshopper species for 1995, Billings Montana, using parameter estimates derived from two seasons' data recorded at Roundup, Montana in 1975 and 1976 (Kemp & Onsager 1986).

	1975 Parameters		1976 Parameters		Difference	
	DD	JD	DD	JD	DD	JD
Ageneotettix deorum						
50% instar 1	103	169	108	170	-5	-1
Peak instar 2	131	176	117	172	14	4
Peak instar 3	174	183	140	178	34	5
Peak instar 4	203	191	191	190	12	1
Peak instar 5	268	200	284	202	-16	-2
50% adult	312	206	356	211	-44	-5
Melanoplus sanguinipes						
50% instar 1	56	164	85	167	-29	-3
Peak instar 2	78	166	95	168	-17	-2
Peak instar 3	124	174	138	177	-14	-3
Peak instar 4	167	187	213	192	-46	-5
Peak instar 5	236	195	330	208	-94	-13
50% Adult	291	203	400	217	-109	-14

1995 were reasonably close with the exception of a 13 day and 15 day difference for peak instar 5 and 50% adult emergence, respectively, for *Melanoplus sanguinipes*. Differences between DDs accumulated on the predicted dates of key events are often large and some are large relative to differences between dates of occurrence, especially for peak instar 4 (46 DD) and 50% instar 1 (29 DD) for *M. sanguinipes*. The assumptions of the degree approach would require a greater agreement between the degree days accumulated for any percentile of the population between years. Kemp and Onsager (1986) suggested that differences in mortality from season to season may explain the differences in the fit of the original parameters (and that would account for the disparate predictions for the 1995 data). Of more concern, is the close agreement, in some instances, of the predicted days of occurrence despite sizable relative differences in accumulated DD. Apparently, high temperatures and a rapid rate of DD accumulation per day around the event of interest reduces the discrepancy between observed and predicted days. Does this mean that when phenology models are validated using small data sets that, occasionally, good agreement between predicted and observed dates may be attributed to rapid accumulation of rates or degree days rather than the adequacy of the model used?

5 USE OF HISTORICAL DATA

Regardless of the method used to build prescriptive models in entomology, large samples of historical insect development and climate data from the field that include extreme seasonal conditions are clearly needed for phenological (and population) model validation. Validation with independent data is the ultimate test of the assumptions of all models. Most scientists recognise the need for statistical rigour in sampling and experimentation. In particular, they acknowledge the need to obtain a representative sample of the system to be described or to test hypotheses over a range of conditions presented in the real world. Oddly, model validation procedures appear to have escaped such rigour. Any model is an abstraction of reality based on untested assumptions and, if constructed for prediction, its ability to do so is a hypothesis that must be tested using a representative sample of the real world.

Some investigators may argue that historical or long-term data (e.g., population events, population densities or weather variables observed in the field) present problems of precision and integrity. As the duration of the study increases, the quality of the data is likely to decrease if it has been collected for several different purposes (Pimm 1991). Regardless of this argument, all data possess flaws, and many short-term data sets are often collected by different people. An investigator always depends on the integrity, skill and intention of collaborators and technicians who collect the data. Despite problems of precision and integrity, large sets of historical climate and insect data at the very least provide a representative range of examples from the natural world we seek to explain.

Another criticism of the use of long-term insect and climate data is that historical weather data records (as normally collected) do not reflect insect microenvironments. Disparity in measurement of climate variables between an insect's microenvironment and macroenvironment, and spatial heterogeneity between sites are well recognised (e.g. Wellington 1954, Ferro et al. 1979, Baker 1980, Dobkin et al. 1987). These authors advocate more studies to determine the exact effect of macroclimatic variables on insect microenvironments. Clearly, microenvironmental studies will increase our understanding of the interactions between insect populations and the spatial and temporal heterogeneity of their environment. However, until such time as the inherent variability of measurement of heterogeneous microclimatic variables is resolved, scientists may have to depend on gross correlations between changes in the macroenvironment and insect microenvironment to develop practical predictive models.

6 POTENTIAL USE OF PLANTS AS INTEGRATORS OF WEATHER EFFECTS

Many phenology models in practical application suffer from an inherent weakness where they lack a reliable starting point (biofix) from which to start thermal or rate summation. Establishment of a reliable starting point can constrain either the application, or the predictive accuracy (or both) of phenological or population models. Given that the influence of host plant quality (and in some cases, host plant phenology) on insect population development has been well established in many studies, could the inclusion of relevant aspects of the phenology of the host plant(s) of some herbivorous insects improve the predictive efficiency of insect phenological models?

Despite abundant literature demonstrating or theorising close association between host plant nutritional quality and herbivore population growth and development (see Campbell 1989, Hunt et al. 1992, Thomas & Hodkinson 1991, De Barro 1992, Weibull 1993), plant phenology has been underutilised in population and phenology models for herbivorous insects. Figure 3 shows relationships between mean temperature over a phenophase interval defined by flowering phenology of two overwintering host *Prunus* species and the start of migration (defined as the second aphid caught in a suction trap) of *Phorodon humuli* versus the reciprocal of days over the interval, for two sites approximately 260 km apart in England. These relationships were highly significant despite the distances between the site where host plant phenology was recorded

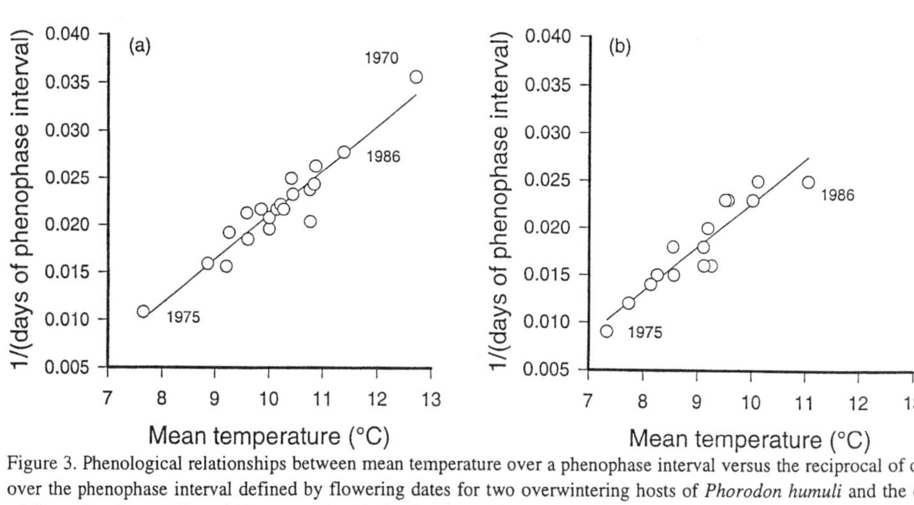

Figure 3. Phenological relationships between mean temperature over a phenophase interval versus the reciprocal of days over the phenophase interval defined by flowering dates for two overwintering hosts of *Phorodon humuli* and the date of first migration of the aphid at two sites in England. (a) Using 'Victoria' first flower; aphid migration data were recorded at Wye, Kent (1967-86), $r^2 = 0.91$, (b) using Myrobalan flowering over; aphid migration data were recorded at Rosemaund Experimental Husbandry Farm, Hereford (1972-1986), $r^2 = 0.90$ (From Worner et al. 1995).

Table 2. Predictive sample reuse (PSR), absolute mean error (ABSME) and root mean square error (RMSE) of residuals (in days) for prediction of the onset of *Phorodon humuli* migration at two sites in England.

Site		Prediction based on flowering phenology[1] (days)	Prediction based on mean calendar date (days)
Wye	ABSME	2.15	4.16
	RMSE	3.23	5.06
Rosemaund	ABSME	4.07	4.58
	RMSE	6.24	5.60

[1] For Wye, prediction was based on 'Victoria' first flower and for Rosemaund, Myrobalan flowering over.

(East Malling, Kent) and where the aphid data were recorded (Wye in Kent, 33 km from East Malling, and Rosemaund Experimental Husbandry farm in Herefordshire). Aphid migration data were obtained from the Rothamsted Insect Survey.

Clearly, for this aphid species, plant phenology can provide a meaningful starting point (biofix) for the integration of population development and the reciprocal of days over the phenophase interval provides a rate of approach to the event. Furthermore, the nature of this relationship indicates that the degree day approach to prediction can be used. Table 2 compares the absolute mean error (ABSME) and root mean square error (RMSE) of the predictive sample reuse (PSR) residuals using the host plant flowering - aphid migration model and thermal summation, with PSR errors obtained using average calendar date, to predict aphid migration at the two sites. Use of the average calendar date is the simplest approach to phenological prediction and in comparison gives relatively small prediction errors, especially the RMSE for the Rosemaund site. It is not surprising that the host plant flowering - insect migration model does not do well for the Rosemaund site considering the distance between Rosemaund and the site where host plant phenology data were recorded. Nevertheless, the host plant flowering - aphid migration model was also compared with two other methods that used historical data to predict insect phenology and again gave the lowest ABSME for the Wye site (Worner et al. 1995).

Much research clearly indicates that a complex interaction exists between temperature and photoperiod affecting both plant and insect phenology. These data show that use of plant phenology in insect phenology models may provide a means to integrate weather and possible photoperiodic effects on the induction of the developmental process of the insect population. Where data of host plant phenology are lacking, use of indicator non-host species for phenological prediction is possible (see Kemp et al. 1991).

I do not imply that such close associations as shown here exist for all herbivorous insects. Indeed, the uncoupling of phenological events between the host plant and the insect is an important possible explanation of shifts in density of some insect populations either between seasons on one host (Dobkin et al. 1987, Hunter 1990) or between hosts in a given season (Turgeon 1986). A close association between host plant phenology and insect phenology is most likely in those insects that spend most of their life cycle in or on a host plant. Nonetheless, subtle interactions between the host and those insect life stages that remain in close proximity may merit close examination.

7 STATIC MODELS OF INSECT PHENOLOGY

Multiple regression and multivariate methods are among the simplest approaches for predicting complex population processes. Use of these statistical methods has tended to be displaced by more complex modelling effort but, periodically, studies are reported (e.g., Price et al. 1985, Capinera & Horton 1989, Harrington et al. 1991) that continue to show the value of a purely statistical approach to generate interesting hypotheses. The potential use of host plant phenology in multiple regression models to predict 50% migration (as an indication of peak migration) of *P. humuli* at Wye and Rosemaund was investigated (Worner et al. 1995). Predictive sample reuse methodology showed that multiple regression models can give reasonable predictions for new data over a number of years. Statistical factors such as autocorrelation, collinearity and cross-correlations -- factors that are inherent in climate data used in such studies that can lead to erroneous interpretation of results -- were not significant statistical factors in the data used for these regression models.

The multiple regression approach indicates that insect modellers may need to consider weather conditions before the initiation of population events of interest (Capinera & Horton 1989, Harrington et al. 1991, Worner et al. 1995) as having a possible significant influence on the phenology of insect populations. Whether prior weather has a direct effect on the insect population (possibly causing differential survival of late or slow developing insects) and thus affects future population timing, or whether the effect is mediated by the host plant, should be determined by experimental studies.

8 CONCLUSIONS

The use of hypothetical data to verify internal model assumptions or to test alternative models in the absence of adequate observational data is widespread in current statistical research but seems underused in entomology. With careful thought such data can be devised. Alternatively, computer intensive methods such as randomization (e.g., the bootstrap and jackknife techniques) (Efron & Gong 1983, Manly 1990) and Monte Carlo methods may merit consideration in some instances.

Just as the power of electronic computation has resulted in the expansion of model development, it should also be used to expand methods of data analysis (Efron & Tibshirani 1991) and model validation. Statements such as "until appropriate validation data are collected [the model will not be truly validated]" have appeared in entomological literature for far too long. Few follow-up studies exist for published prescriptive models. One of the reasons may be the deficiency of long-term data for model validation. Clearly, funding for long-term studies is difficult to acquire. If recognised as essential to general insect population studies, one must ask why scientists have not insisted that long-term data be collected and why do we seem to have more models than there are data to test them?

Few comparative studies of the performance of different modelling approaches in answering specific questions in entomology exist in the literature. Comparative studies are sorely needed if progress is to be made towards increased availability of useful management models. Moreover, a combination of approaches may improve prediction. For example, Worner et al. (1995) found that for the aphid migration data discussed here, the simple average of the predictions of four phenological approaches reduced the ABMSE and RMSE for Wye to 1.7 d and 2.15 d, respectively, for the Wye site.

The type of data used to quantify the predictive relationship between host plant phenology and the phenology of *P. humuli* migration potentially can be greatly affected by sampling error and 'lurking variables', variables that impose pattern between unrelated variables (Barker & Sauer 1992). Despite this, the relationships should be intriguing enough to stimulate future studies of subtle phenological relationships between key host plants and phytophagous insects. Since it is important for insects in their evolution to optimise capture of resources, there is no reason why they should be constrained to use the well known cues of photoperiod and temperature to initiate life cycle changes. Clearly, more empirical studies (e.g., Gilbert & Raworth 1996) are required to attempt to derive the evolutionary and biological principles of insect growth and development.

REFERENCES

Allen, J. C. 1988. Averaging functions in a variable environment: a second-order approximation method. *Environ. Entomol.* 17: 621-625.

Baker, C. R. B. 1980. Some problems in using meteorological data to forecast the timing of insect life cycles. *EPPO Bull.* 10: 83-91.

Barker, R.J. & J.R. Sauer. 1992. Modelling population change from time series data. In *Wildlife 2001: Populations* (eds. D.A. McCullough and R.H. Barrett). Elsevier, New York.

Behrens, W., K-H. Hoffmann, S. Kempa, S. Gäßler & G. Merkel-Wallner. 1983. Effects of diurnal thermoperiods and quickly oscillating temperatures on the development and reproduction of crickets, *Gryllus bimaculatus*. *Oecologia* 59: 279-287.

Campbell, I. M. 1989. Does climate affect host-plant quality? Annual variation in the quality of balsam fir as food for spruce budworm. *Oecologia* 81: 341-344.

Capinera, J. L. & D. R. Horton. 1989. Geographic variation of the effects of weather on grasshopper infestation. *Environ. Entomol.* 18: 8-14.
Dallwitz, M. J. & J. P. Higgins. 1978. Users guide to DEVAR: a computer program for estimating development rate as a function of temperature. *CSIRO Div. Entomol. Rep. No. 2.*
DeAngelis, D.L. & L.J. Gross (eds.) 1992. *Individual-based models and approaches in ecology: Populations, communities and ecosystems.* Chapman & Hall, New York.
De Barro, P.J. 1992. The role of temperature, photoperiod, crowding and plant quality on production of alate viviparous females of bird cherry-oat aphid. *Entomol. Exp. Appl.* 65: 205-214.
Dennis, B., & W. P. Kemp. 1988. Further statistical inference methods for a stochastic model of insect phenology. *Environ. Entomol.* 17: 887-893.
Dennis, B., W. P. Kemp & R. C. Beckwith. 1986. A stochastic model of insect phenology: estimation and testing. *Environ. Entomol.* 15: 540-546.
Dobkin, D. S., I. Olivieri & P. R. Ehrlich. 1987. Rainfall and the interaction of microclimate with larval resources in the population dynamics of checkerspot butterflies (*Euphydryas editha*) inhabiting serpentine grassland. *Oecologia* 71: 161-166.
Efron, B. & G. Gong. 1983. A leisurely look at the bootstrap, the jackknife and cross validation. *Am. Stat.* 37: 36-48.
Efron, B. & R. Tibshirani. 1991. Statistical data analysis in the computer age. *Science* 253: 390-395.
Ellingsen, I. J. 1969. Effect of constant and varying temperatures on development, feeding and survival of *Adalia bipunctata* (Coccinellidae). *Norsk Entomol. Tidsskr.* 16: 121-125.
Ferro, D. N., R. B. Chapman & D. R. Penman. 1979. Observations on insect microclimate and insect pest management. *Environ. Entomol.* 8: 1000-1003.
Gilbert N. & D.A. Raworth. 1996. Insects and temperature - a general theory. *Can. Entomol.* 128: 1-13.
Hagstrum, D. W. & W. R. Hagstrum. 1970. A simple device for producing fluctuating temperatures, with an evaluation of the ecological significance of fluctuating temperatures. *Ann. Entomol. Soc. Am.* 63: 1385-1389.
Harrington, R., G. G. Howling, J. S. Bale & S. Clark. 1991. A new approach to the use of meteorological and suction trap data in predicting aphid problems. *EPPO Bull.* 21: 499-505.
Higley, L. G., L. P. Pedigo & K. R. Ostlie. 1986. DEGDAY: a program for calculating degree-days, and assumptions behind the degree-day approach. *Environ. Entomol.* 15: 999-1016.
Hilbert, D. W. & J. A. Logan 1983. Empirical model of nymphal development for the migratory grasshopper, *Melanoplus sanguinipes* (Orthoptera: Acrididae). *Environ. Entomol.* 12: 1-5.
Humpesch, U. H. 1982. Effect of fluctuating temperature on the duration of embryonic development in two *Ecdyonurus spp.* and *Rhithrogena cf. hybrida* (Ephemeroptera) from Austrian streams. *Oecologia* 55: 285-288.
Hunt, D. W. A., C. F. Drury & H. E. L. Maw. 1992. Influence of nitrogen on the performance of Colorado potato beetle (Coleoptera: Chrysomelidae) on tomato. *Environ. Entomol.* 21: 817-821.
Hunter, M. D. 1990. Differential susceptibility to variable plant phenology and its role in competition between two insect herbivores on oak. *Ecol. Entomol.* 15: 401-408.
Kemp, W. P. & B. Dennis. 1991. Toward a general model of rangeland grasshopper (Orthoptera: Acrididae) phenology in the steppe region of Montana. *Environ. Entomol.* 20: 1504-1515.
Kemp, W. P. & J. A. Onsager. 1986. Rangeland grasshoppers (Orthoptera: Acrididae): Modelling phenology of natural populations of six species. *Environ. Entomol.* 15: 924-930.
Kemp, W. P., J. S. Berry & J. M. Caprio. 1991. Use of ornamental lilac and honeysuckle phenophases as indicators of rangeland grasshopper development. *J. Range Manage.* 44: 583 586.
Logan, J. A. 1988. Toward an expert system for development of pest simulation models. *Environ. Entomol.* 17: 359-376.
Logan, J. A., D. J. Wollkind, S. C. Hoyt & L. K. Tanigoshi. 1976. An analytic model for description of temperature dependent rate phenomena in arthropods. *Environ. Entomol.* 5: 1133-1140.
Ludwig, D. & R. M. Cable. 1933. The effect of alternating temperatures on the pupal development of *Drosophila melanogaster* Meigen. *Physiol. Zool.* 6: 493-508.
Lui, S.-S., G.-M. Zhang & J. Zhu. 1995. Influence of temperature variations on rate of development in insects: analysis of case studies from entomological literature. *Ann. Entomol. Soc. Am.* 88: 107-119.

Manly, B. F. J. 1990. *Randomization and Monte Carlo methods in biology*. Chapman & Hall, London.

Matteson, J. W. & G. C. Decker. 1965. Development of the European corn borer at controlled constant and variable temperatures. *J. Econ. Entomol.* 58: 344-349.

McClay, A.S. & R.B. Hughes. 1995. Effects of temperatures on development rate, distribution and establishment of *Calophasia lunula* (Lepidoptera: Noctuidae), a biocontrol agent for toadflax (*Linaria spp.*). *Bio. Control* 5: 368-377.

Messenger, P. S. & N. E. Flitters. 1958. Effect of constant temperature environments on the egg stage of three species of Hawaiian fruit flies. *Ann. Entomol. Soc. Am.* 52: 109-120.

Messenger, P. S. & N. E. Flitters. 1959. Effect of variable temperature environments on egg development of three species of fruit flies. *Ann. Entomol. Soc. Am.* 52: 191-204.

Morris, R. F. & W. C. Fulton. 1970. Models for the development and survival of *Hypantria cunea* in relation to temperature and humidity. *Mem. Entomol. Soc. of Canada* 70: 1-60.

Pedigo, L. P. & G. D. Buntin. 1994. *Handbook of sampling methods for arthropods in agriculture*. CRC Press, Boca Raton, Florida.

Pimm, S. L. 1991. *The balance of nature: ecological issues in the conservation of species and communities*. University of Chicago Press, Chicago.

Pradhan, S. 1945. Insect population studies. II. Rate of insect development under variable temperature of the field. *Proc. Natl Inst. Sci. of India. Part B Biol. Sci.* 11: 74-80.

Price, J. R., J. E. Slosser & G. J. Puterka. 1985. Factors affecting survival of boll weevils in winter habitat in the Texas rolling plains. *Southwest. Entomol.* 10: 1-6.

Pruess, K. P. 1983. Day-degree methods for pest management. *Environ. Entomol.* 12: 613-619.

Ratte, H. T. 1985. Temperature and insect development. In *Environmental physiology and biochemistry of insects* (ed. K. H. Hoffman), pp. 33-66. Springer-Verlag, New York.

Régnière, J. 1984. A method of describing and using variability in development rates for the simulation of insect phenology. *Canadian Entomol.* 116: 1367-1376.

Sharp, P. J. H. & D. W. DeMichele. 1977. Reaction kinetics of poikilotherm development. *J. Theor. Biol.* 64: 649-670.

Sharp, P. J. H., G. L. Curry, D. W. DeMichele & C. L. Cole. 1977. Distribution model of organism development times. *J. Theor. Biol.* 66: 21-38.

Stinner, R. E., A. P. Gutierrez & G. D. Butler, Jr. 1974. An algorithm for temperature-dependent growth rate simulation. *Can. Entomol.* 106: 519-524.

Sutherst, R. W. & G. F. Maywald. 1985. A computerised system for matching climates in ecology. *Agric., Ecosyst. Environ.* 13: 281-299.

Tanigoshi, L. K., R. W. Browne, S. C. Hoyt & R. F. Lagier. 1976. Empirical analysis of variable temperature regimes on life stage development and population growth of *Tetranychus mcdanieli* (Acarina: Tetranychidae). *Ann. Entomol. Soc. Am.* 69: 712-716.

Thomas, A. T. & I. D. Hodkinson. 1991. Nitrogen, water stress and the feeding efficiency of lepidopteran herbivores. *J. Appl. Ecol.* 28: 703-720.

Turgeon, J. J. 1986. The phenological relationship between the larval development of the spruce budmoth, *Zeiraphera canadensis* (Lepidoptera: Olethreutidae), and white spruce in northern New Brunswick. *Can. Entomol.* 118: 345-350.

Wagner, T. L., H-I. Wu, P. J. H. Sharpe, R. M. Schoolfield & R. N. Coulson. 1984. Modeling insect development rates: a literature review and application of a biophysical model. *Ann. Ent. Soc. Am.* 77: 208-225.

Weibull, J. H. W. 1993. Bird cherry-oat aphid (Homoptera: Aphididae) performance on annual and perennial temperate-region grasses. *Environ. Entomol.* 22: 149-153.

Welch, S.M., B. A. Croft, J. F. Brunner & M. F. Michels. 1978. PETE: an extension phenology modelling system for management of multi-species pest complex. *Environ. Entomol.* 7: 487-494.

Wellington, W. G. 1954. Weather and climate in forest entomology. *Meteorol. Monogr.* 2: 11-18.

Worner, S. P. 1988. Ecoclimatic assessment of potential establishment of exotic pests. *J. Econ. Entomol.* 81: 973-983.

Worner, S. P. 1992. Performance of phenological models under variable temperature regimes: consequences of the Kaufmann or rate summation effect. *Environ. Entomol.* 21: 689-699.

Worner, S. P., S. L. Goldson & E. R. Frampton. 1989. Comparative ecoclimatic assessments of *Anaphes diana* (Hymenoptera: Mymaridae) and its intended host, *Sitona discoideus* (Coleoptera: Curculionidae), in New Zealand. *J. Econ. Entomol.* 82: 1085-1090.

Worner, S. P., G. M. Tatchell & I. P. Woiwod. 1995. Predicting spring migration of the damson-hop aphid *Phorodon humuli* (Homoptera: Aphididae) from historical records of host-plant flowering phenology and weather. *J. Appl. Ecol.* 32: 17-28.

Population and Community Ecology for Insect Management and Conservation, Baumgärtner et al. (eds)
© 1998 Balkema, Rotterdam, ISBN 90 5410 930 0

Quantitative analysis of gypsy moth spread in the Central Appalachians

Alexei A. Sharov
Department of Entomology, Virginia Polytechnic Institute and State University, Blacksburg, Va., USA

Andrew M. Liebhold
USDA Forest Service, Morgantown, W.Va., USA

ABSTRACT: Gypsy moth, *Lymantria dispar* (L.), is an introduced forest pest in North America that has been expanding its range to the west and south. In attempt to slow its spread, the USDA Forest Service has established several barrier zones in which isolated colonies are detected and eradicated. To evaluate the effect of barrier zones on the rate of gypsy moth spread, we measured the rate of spread as the average distance between population boundaries in consecutive years which were estimated using male moth counts in pheromone traps, egg mass counts, and defoliation maps in the central Appalachian Mts. in 1984-1995. In 1988-1992, the boundary of 1 moth per trap was on average 108.5 km from the boundary of defoliation, and male moth capture rate increased 10 times per 29 km perpendicular to the population front. Since 1990, the average rate of gypsy moth spread declined from 20.78 km/yr to 8.6 km/yr. This reduction of spread rate may have been due to eradication of isolated colonies in the study area that started in 1990. We developed a model of gypsy moth spread that considered establishment of isolated colonies beyond the moving population front. The model predicts that the barrier zone in the Appalachian Mts. should reduce the rate of gypsy moth spread by ca. 53% which is consistent with our data.

INTRODUCTION

The gypsy moth, *Lymantria dispar* (L.), was accidentally introduced to North America near Boston in 1869 and since that time it has been slowly expanding its range (Liebhold et al. 1989, 1992). The current distribution of the species includes most of the northeastern United States and parts of bordering Canadian provinces. Another, discrete population exists in Michigan that originated from a secondary introduction that was not successfully eradicated. The primary and secondary infested regions continue to expand.

Quantitative analysis of gypsy moth spread would be useful for: (1) proper timing of silvicultural measures which can reduce the adverse impact of gypsy moth defoliation (Gottschalk 1993), (2) planning sampling programs in areas at risk to defoliation, (3) planning areas for quarantine regulation, and (4) for planning and evaluating strategies to slow the spread of the gypsy moth (McFadden & McManus 1991).

In 1993 the USDA Forest Service initiated the Slow-the-Spread (STS) program, a pilot project designed to test the feasibility of slowing the spread of the gypsy moth over large regions (Leonard & Sharov 1995). The following 3 barrier zones were established along the advancing front of gypsy moth populations: (1) the Appalachian Mts. in Virginia and West Virginia, (2) northeastern North Carolina, and (3) the Upper Peninsula of Michigan. The strategy used in this project is to detect and eradicate (or suppress) isolated gypsy moth colonies that occur just beyond the expanding front of gypsy moth populations. Suppression of newly established colonies should reduce their growth and coalescence, and thereby reduce the rate of gypsy moth spread. A similar program, the Appalachian Integrated Pest Management (AIPM) project, was conducted from 1988-1992 in Virginia and West Virginia and was designed to suppress both high-density populations and isolated populations near the expanding front. Reduction of gypsy moth spread rate was one of the AIPM objectives which was adopted by STS. However, STS was designed to slow gypsy moth spread while using fewer pesticide applications than in the AIPM project.

In order to evaluate the effect of these projects on the rate of population spread it is important to have access to reliable methods for measuring population spread rates. Liebhold et al. (1992) suggested evaluating the rate of gypsy moth spread from the relationship between the year of establishment of quarantine regulations in a county and the distance of the county from the area infested in 1900. The estimated rate of spread varied from 2.82 to 20.78 km/yr. However the method of Liebhold et al. (1992) is not enough accurate for management purposes because it used a county-level spatial resolution and it may be biased due to subjective factors that affect the establishment of quarantine regulations.

More reliable estimates of the rate of spread can be obtained from geographically-distributed population samples. Gypsy moth populations are traditionally monitored using any of 3 methods: (1) aerial maps of forest defoliation, (2) counts of overwintering egg mass populations (Kolodny-Hirsch 1986), and (3) counts of male moths in pheromone-baited traps (Talerico 1981, Ravlin et al. 1987). Egg mass counts are the most reliable method for assessing densities of medium- and high-density populations and thus they are widely used for making decisions concerning suppression of outbreak populations (Ravlin et al. 1987). Counts of adult males are widely used to detect new isolated gypsy moth infestations because pheromone traps are highly sensitive at detecting low-density populations and this method is much less labor-intensive than egg mass sampling (Schwalbe 1981).

Sharov et al. (1995) proposed to estimate the rate of population spread as a distance between population boundaries in 2 consecutive years. A population boundary was defined as an isoline that separates areas with population density generally below and generally above a certain threshold. Threshold density is important when populations are evaluated on a continuous scale rather than as simply presence-absence. The first objective of this study was to implement this method for measuring the rate of gypsy moth spread and to evaluate the effect of population management on the rate of spread.

The second objective was to develop a model of gypsy moth population spread that could be used to test the effect of barrier zones on the rate of expansion of population range. Most models of population spread are based on diffusion equations. For example, the model of Skellam (1951) combines diffusion of organisms with exponential population growth. Reaction-diffusion models that are used for simulating population spread were reviewed by Hastings (1996). Marsula and Wissel (1994) used a diffusion-based model to simulate the management of a barrier zone designed to stop the spread of screwworm populations. However, simple diffusion models are not appropriate for simulation of the spread of gypsy moth populations (Liebhold et al. 1992). There is no continuous wave of population spread. Instead, discrete isolated infestations are established beyond the population front (McFadden & McManus 1991). These infestations apparently originate from egg masses and other life stages transported on human vehicles. Isolated infestations grow in size, coalesce and finally contribute to the expansion of the population front. Discontinuous long-distance dispersal is usually combined with local neighborhood dispersal that makes local infestations grow. Hengeveld (1989) suggested the term "stratified diffusion" or "stratified dispersal" to refer to a combination of different dispersal scales.

Shigesada et al. (1995) developed several models of stratified dispersal in animal populations. The scattered colony model assumes independent growth of a set of colonies which expand in size and simultaneously emit long-distance dispersal propagules. A more realistic coalescing colony model assumes that new colonies become established as a specific distance from the expanding population front. When their boundary reaches the population front, then they coalesce with the primary colony and their area is added to the infested area. Although models of Shigesada et al. (1995) generate reasonable rates of population spread, they are not realistic enough to be used for simulating the effect of barrier zones. For example, according to the coalescing colony model, eradication of offspring colonies can be restricted to a very small area because new colonies are produced at a fixed distance from the population front.

In this paper we developed a new model of stratified dispersal in which colonies can become established at varying distances from the population front. This model is biologically more realistic and provides more meaningful representation of barrier zone strategies.

MATERIALS AND METHODS

We used historical pheromone trap data (1980, 1984, 1988-1995) egg mass sample data (1988-1991) and defoliation maps in northwestern Virginia and southeastern West Virginia. Most of the data was collected during the Appalachian Integrated Pest Management (AIPM) project, 1988-1992, (Reardon 1991); and Slow-the-Spread (STS) project that started in 1993 (Leonard & Sharov 1995).

Most pheromone traps were the standard 1.9-liter milk-carton design (Schwalbe 1981). In the STS project, delta-traps were used in uninfested areas. All traps were baited with (+) disparlure impregnated dispensers. Traps were placed in regular grids; most grids traps were spaced by 2-3 km. The number of traps placed each year varied from 3,380 to 20,760.

Egg masses were sampled in 0.01 ha plots mostly in the area with male moth counts of 200-300 moths per trap and more. Each year from 11,650 to 45,860 samples were taken. The number of plots within a 1 km^2 area was determined using a sequential sampling plan (Fleischer et al. 1991) and generally ranged from 4 to 10. The number of new egg masses was counted on tree boles, branches and on the ground. Because we were interested in patterns of egg mass distribution larger than 1 km, egg mass counts were averaged in each 1 km^2 cell prior to analysis.

The location of traps and egg mass samples was recorded using Universal Transverse Mercator (UTM) (Snyder 1987) coordinates estimated from locations on standard 1:24,000 topographic maps published by the U.S. Geological Survey.

Preliminary analysis of male moth counts in 3 states (VA, WV, and NC) in 1995 indicated that the pattern of population spread in the Appalachian Mts. is different from that in the coastal plain and piedmont. Because most of data were collected in the mountains, we restricted the analysis to the mountain region (between the lines from Grant Town, Marlon Co., WV to Clendenin, Kanawha Co., WV and from Culpeper, Culpeper Co., VA to Sydnorsville, Franklin Co., VA).

Population boundaries were estimated as follows. The first step was to interpolate moth counts in pheromone traps using median indicator kriging with subsequent E-type estimation (Deutsch & Journel 1992). This method is not sensitive to data distribution and generates unbiased estimates (Deutsch & Journel 1992). Variograms were isotropic and were approximated by an exponential model with the relative nugget of 0.45 and effective range of 20.5 km for male moth counts. For egg mass counts parameters were: relative nugget, 0.58; and effective range, 17.8 km (Sharov et al. 1995). Median indicator kriging was performed on a lattice of 1 × 1 km cells. After E-type values were estimated, we delineated population boundaries using the following population thresholds: 1, 3, 10, 30, 100, and 300 moths per trap and 1, 3, 10, and 30 egg masses per 0.01 ha. Defoliation maps did not require kriging because the presence of defoliation was already coded for all cells in the map.

To estimate population boundaries we used the best cell classification method (Sharov et al. 1995). If a grid of cells is applied to the area, then a boundary line classifies cells into occupied and nonoccupied by a population. Some of these cells may be misclassified: they may be considered occupied whereas in reality they are not and vice versa. The method identifies a line that has no folds and that minimizes the number of misclassified cells. The weights assigned for cell misclassifications of the 1st and 2nd type (the 1st type of misclassification occurred when population was above the threshold but was classified as below the threshold; the 2nd type occurred when the population was below the threshold but it was classified as above the threshold) were: 1:1 for male moths, 3:1 for egg masses, and 30:1 for defoliation (Sharov et al. 1996a).

Boundary points were estimated in 5-km intervals. All population boundaries were functions in a rotated Cartesian coordinate system with x- and y-coordinates shown in Fig. 1. We used the same coordinate system for all boundaries so that they could be compared easily. The origin was selected arbitrarily so that x- and y-values were positive. The y-coordinate coincided with the

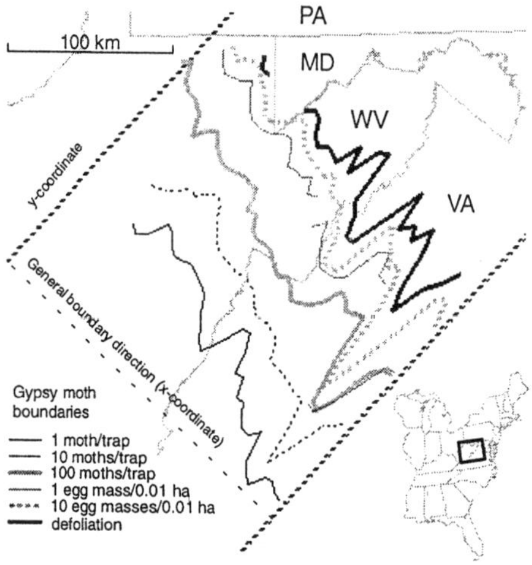

Fig. 1. Gypsy moth population boundaries estimated using best cell classification method from moth counts in pheromone traps, egg mass samples, and defoliation map in 1989.

border of the study area and had the azimuth of 42.7°. Average boundary location was estimated as the average y-value for all boundary points.

Population spread rate was measured using the following 2 methods: (1) as the difference between average boundary locations for the same populations threshold in 2 consecutive years and (2) as the average difference between local boundary points. These 2 measures would coincide if boundaries had no missing values. However, boundary lines were not always complete, and thus, these measures were different. Boundaries in 1984 had more than a half points missing, and this caused a considerable difference in estimates of spread rate obtained from 2 methods. Spread rates were estimated using male moth counts in pheromone traps for 6 population thresholds (1, 3, 10, 30, 100, and 300 moths per trap) and then averaged for all thresholds. Population monitoring was irregular before 1988; thus, there were no data between 1980 and 1984, and between 1984 and 1988. Thus, we estimated the average rate of population spread in each 4-yr interval as the distance between population boundaries at the beginning and end of the interval divided by 4.

To detect the reduction in the spread rate of gypsy moth populations we compared average spread rate before and after the start of pest management activity directed at slowing population spread. We expect that the reduction in population spread rate resulted mainly from eradication of isolated infestations just beyond the expanding population front, which started in 1990. Thus, we grouped spread rates into 2 categories: before 1990 and after 1990.

ANALYSIS OF GYPSY MOTH SPREAD

Average boundary values plotted against the distance along the y-coordinate (see Fig. 1) show the progression of the population front (Fig. 2). In 1988-1992, the distances between average boundaries estimated using different population thresholds generally remained the same from year to year (Fig. 2). Accumulated average distances between adjacent boundaries from 1 moth per trap to: 3, 10, 30, 100, 300 moths per trap, 1, 3, 10, 30 egg masses per 0.01 ha, and defoliation were 13.9, 28.4, 40.1, 54.7, 69.0, 92.4, 97.8, 104.6, 109.6, and 108.5 km, respectively. One order of magnitude increase in male moth counts occurred over a distance of ≈29 km across the population front. Egg mass boundaries were much closer to each other than male moth boundaries (Fig. 2). The average rate of population spread during that period was 10.4 km/yr. Thus, it took ≈11 yr for gypsy moth populations to grow from 1 moth per trap to the first defoliation.

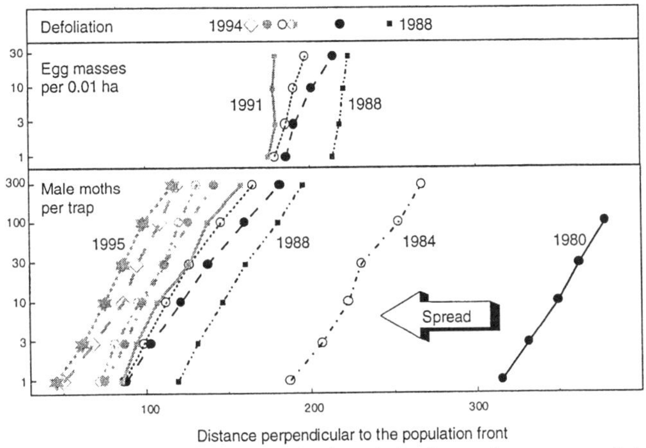

Fig. 2. Average gypsy moth population boundaries (y-coordinates averaged along each boundary) in different years and for different population thresholds (moth counts, egg mass counts, and defoliation). Distance perpendicular to the population front is measured from an arbitrarily selected point beyond the population front. Unlabeled lines represent data for the years (in chronological order) between those indicated on the graph.

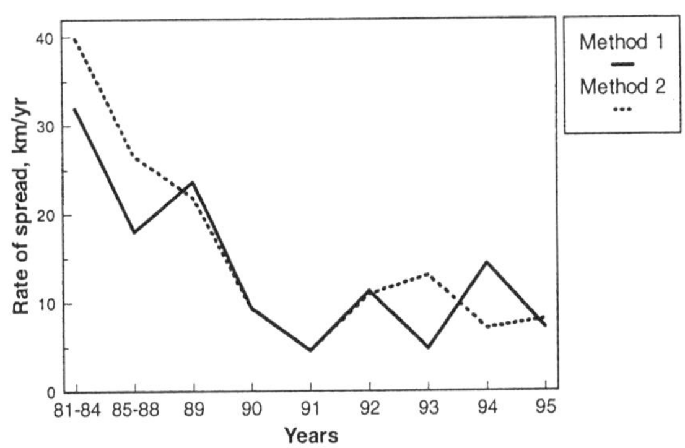

Fig. 3. The rate of gypsy moth spread estimated from 6 thresholds of moth counts as the difference between averaged boundaries (method 1) and as the average difference between local boundary points (method 2).

Population spread rates estimated using both methods declined during the census period (Fig. 3). In 1981-1990, the average rate of spread was 26.5 km/yr (average from 2 methods), whereas in 1991-1995, the average rate of spread was 8.6 km/yr. Liebhold et al. (1992) estimated the rate of uncontrolled population spread (without barrier zones) in areas with mean minimum temperature < 7°(C) (our study area satisfies this requirement) as 20.78 km/yr in 1966-1990 which is close to our estimates for 1981-1990. It is hard to tell which estimate of the uncontrolled spread rate is more accurate. Liebhold et al. (1992) used a less accurate method, but they analyzed a longer period. In this paper we will use the result of Liebhold et al. (1992) because it gives a more conservative estimate of the reduction in the rate of population spread (59% reduction).

MODEL OF GYPSY MOTH SPREAD

Our model of gypsy moth spread considers 3 zones at the expanding population front: the infested zone which is assumed to be a source of long-distance dispersal, the transition zone where new colonies become established, and the uninfested zone where no colonies exist (Fig. 4). We

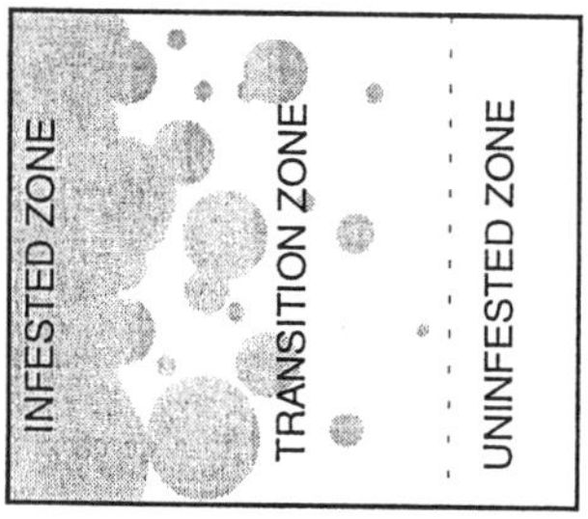

Fig. 4. Three zones at the expanding population front.

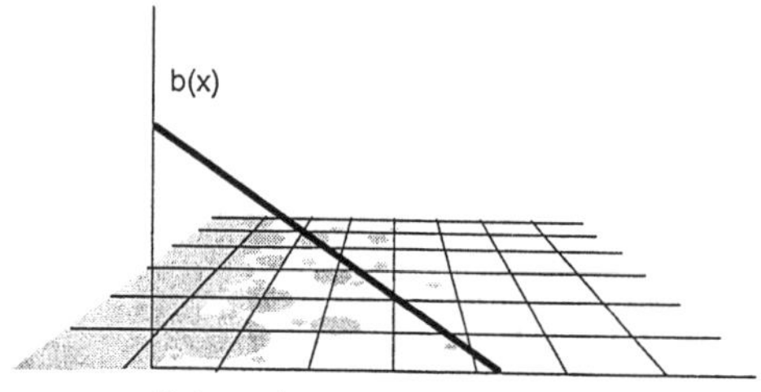

Fig. 5. The probability of colony establishment, $b(x)$, decreases with increasing distance, x, from the front of the infested zone.

assumed that the rate of colony establishment, $b(x)$, decreased with increasing distance, x, from the front of the infested zone (Fig. 5).

Average population density, N, is the mean number of individuals per unit area. The front of the infested zone is defined as a line where the average population density N reaches the carrying capacity, K. For simplicity, our model is built in one dimension which is considered perpendicular to the moving population front. The front of the infested zone is located at the point where $N = K$. Colonies grow with time and the value of N increases respectively. As a result, N becomes eventually $>K$, and the front of the infested zone moves forward to the point where the new value of $N = K$.

At distance x from the front of the infested zone, colonies of various age are present. The average population density can be estimated as the integral of the density of colonies of various ages multiplied by the number of individuals in these colonies. The number of colonies of age a at distance x is equal to the colonization rate at time $t-a$ where t is current time. We assume that the rate of spread has reached its asymptotic value, and therefore, it is constant. Then, at time $t-a$, the distance from the front was $x+av$; and colonization rate at that time (which equals to the number of colonies of age a at present time) was equal to $b(x+av)$, where $b(x)$ is the number of new colonies established annually at distance x from the population front, and v is the rate of spread. Then the average population density $N(x,v)$ at distance x from the front of the infested zone is:

$$N(x,v) = \int_0^\infty b(x+av)\cdot n(a)\cdot da \tag{1}$$

where $n(a)$ is the number of individuals in a colony of age a. The front of the infested zone is the point where the average population density reaches the carrying capacity, K:

$$N(0,v) = K. \tag{2}$$

Combining equations (1) and (2) we get a traveling wave equation:

$$\int_0^\infty b(av)\cdot n(a)\cdot da = K \tag{3}$$

This equation can be used for estimating the spread rate, v, if functions $b(x)$ and $n(a)$ are specified explicitly. The rate of colony establishment decreased linearly with the distance from the front of the infested zone (data supporting the linearity of this function is presented below). Thus,

$$b(x) = \max\left[0, c\cdot\left(1-\frac{x}{x_{\max}}\right)\right] \tag{4}$$

where x_{max} is the maximum distance at which colonies can become established (=width of the transition zone), and $c = b(0)$ is the maximum rate of colony establishment in the area adjacent to the infested zone. We assume that the number of individuals in a colony, $n(a)$, grow exponentially with colony age a:

$$n(a) = n_0 \cdot \exp(ra) \quad (5)$$

where n_0 is the initial number of individuals in a new colony, and r is the intrinsic rate of increase.

The traveling wave equation (3) combined with (4) and (5) yields the equation that can be used for estimating the maximum rate of population spread $v = v_{max}$ (without barrier zones):

$$\frac{cn_0 V}{r^2} \cdot \left[\exp\left(\frac{r}{V}\right) - \frac{r}{V} - 1 \right] = K \quad (6)$$

where variable $V = v_{max}/x_{max}$ represents the rate of population spread, v_{max}, relative to the maximum distance at which isolated colonies can be established, x_{max}. Equation (6) includes this relative population spread rate, V, only, and thus, spread rate v_{max} is always proportional to x_{max}. Equation (6) cannot be solved analytically and thus, we used numerical methods to estimate V.

The relative population spread rate, V, increased with increasing values of cn_0/K (Fig. 6). The value of cn_0/K is the density of organisms in newly established colonies just beyond the front of the infested zone (cn_0) relative to the total number of organisms at the front of the infested zone (K). It quantifies the intensity of colonization process. We expect this value to be very small (<0.1) in most dispersing populations because establishment of new colonies is a rare event and the initial number of individuals (n_0) is usually small.

The relationship between relative spread rate, V, and the intrinsic rate of increase, r, is almost linear (Fig. 6) in contrast to the diffusion model of Skellam (1951) which predicts that spread rate is proportional to the square root of the intrinsic rate of increase. This means that the intrinsic rate of increase has a much stronger effect on population spread rate in our model than in diffusion-based models.

Now we will assume that a barrier zone starts at distance x_0 from the front of the infested zone and extends to the end of the transition zone. The relative width of a barrier zone, W, is defined by its proportion of the width of the transition zone:

$$W = \frac{x_{max} - x_0}{x_{max}} \quad (7)$$

Eradication of all colonies in a barrier zone is equivalent to truncating the tail of the colonization function, $b(x)$, at $x \geq x_0$ (Fig. 7). This truncated function can be used in the traveling wave equation (3) to estimate the rate of population spread, v, in the presence of a barrier zone. We will consider reduction of population spread rate due to the barrier zone as a fraction of the spread rate without barrier zone, v_{max}:

$$R = \frac{v - v_{max}}{v_{max}} \quad (8)$$

The relationship between the reduction in the rate of population spread, R, and the relative width of the barrier zone, W, was studied using the equation:

$$(1 - R + QW) \cdot \exp\left[\frac{Q(1-W)}{1-R}\right] + R - \exp(Q) = 0 \quad (9)$$

where $Q = rx_{max}/v_{max}$. Equation (9) was derived from 2 traveling wave equations (3) with non-truncated and truncated $b(x)$ functions. Equation (9) cannot be solved analytically, and thus, we used numerical methods to estimate R as a function of W for different values of parameter Q (Fig. 8).

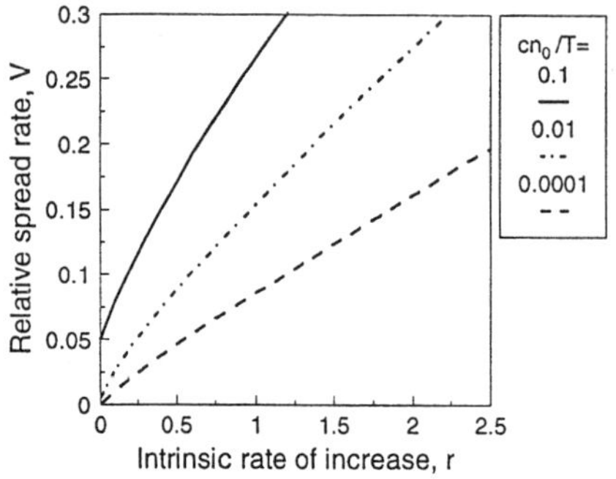

Fig. 6. Relative population spread rate ($V = v_{max}/x_{max}$) predicted by the model as a function of intrinsic rate of increase, r, and parameter cn_0/T.

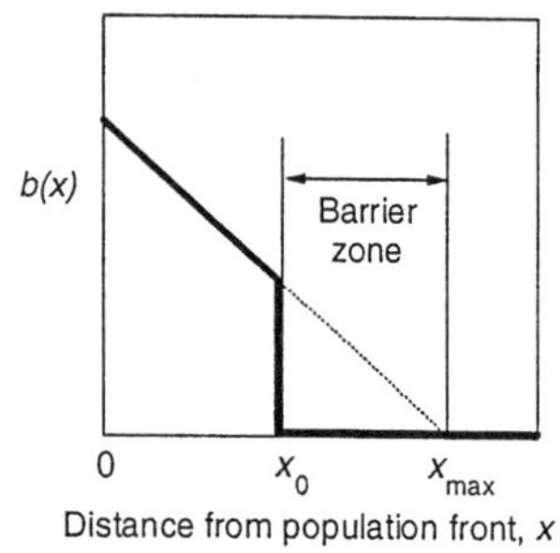

Fig. 7. The function of the probability of colony establishment, $b(x)$, becomes truncated beyond distance x_0 from the front of the infested zone if a barrier zone is set in the area from x_0 to x_{max}.

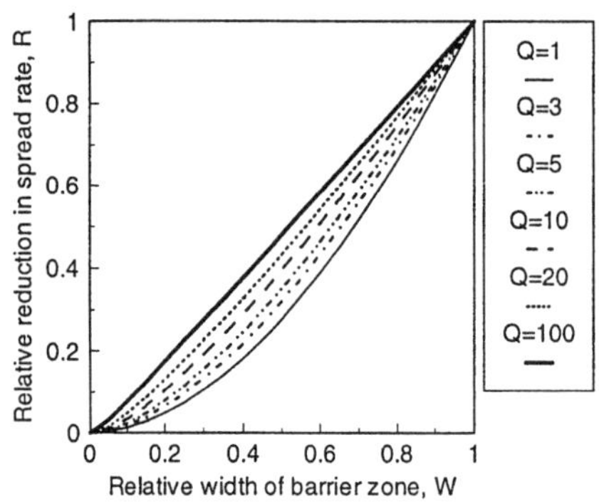

Fig. 8. Relative reduction in spread rate, $R = (v_{max} - v)/v_{max}$, predicted by the model.

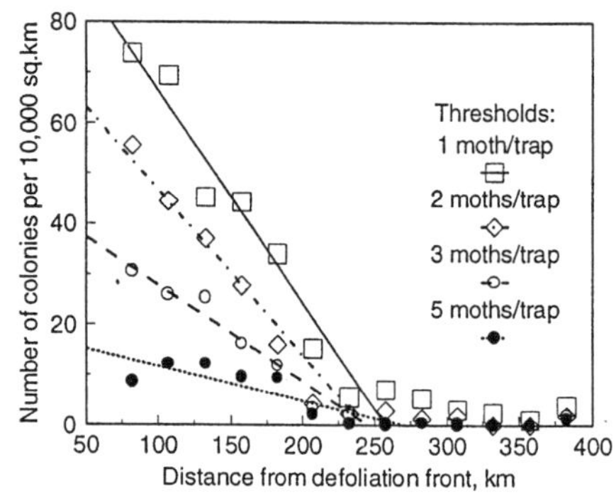

Fig. 9. Number of gypsy moth colonies detected at different distances from the front of the infested zone using 1, 2, 3, and 5 moths per trap as detection thresholds.

Parameter Q can be interpreted as follows. The ratio x_{max}/v_{max} equals the amount of time a location exists in the transition zone (without any population management). Then, Q is the net rate of increase, r, multiplied by time in the transition zone. Thus, $\exp(Q)$ is the ratio of population numbers in a colony established at the front of the transition zone when it reaches the infested zone to the initial population numbers when it was established. Unless population numbers increase at least 3 times in isolated colonies in the transition zone, population spread will be continuous and dynamics would closely follow that of a simple diffusion. Thus, the value of Q should be ≥ 1. The predicted efficiency of a barrier zone in slowing population spread increases with increasing values of parameter Q (Fig. 8). Thus, the higher is the intrinsic rate of increase, r, the greater is the percent reduction in the rate of population spread caused by a barrier zone.

Model parameters: c, x_{max}, n_0, r, and K were evaluated as follows. We considered defoliating populations as the major source of long-distance dispersal which leads to establishment of isolated colonies. Thus, the front of the defoliated area was considered as the front of the infested zone in our model. The maximum distance from the population front, x_{max}, at which colonies can become established was estimated from male moth counts in grids of pheromone-baited traps in 1984 and in 1988-1995. In order to count isolated colonies, we selected traps that caught N or more moths and then grouped these traps if they were separated by less than 4 km (4 km was selected because it was larger than the maximum intertrap distance in most grids). These groups were counted in each year. Colonies may coalesce closer to the population front. Thus, we restricted the counting of colonies to the area beyond the 1 moth per trap boundary. The number of detected colonies was divided by the area covered by pheromone traps that was estimated as the area of the 4-km buffer zone set around all traps used in the analysis. Data from all years were pooled together.

Previously we found that the boundary of 10 moths per trap is the most stable in space and time and therefore, it is most reliable for monitoring of population spread (Sharov et al. 1996b). The average distance from the 10 moth per trap boundary to the area where defoliation first occurs was ≈80 km. Thus, the distance from a colony to the defoliation front was estimated as the distance from the center of mass of a cluster of traps to the 10 moths per trap boundary plus 80 km.

Gypsy moth colonies detected using pheromone traps cannot be directly interpreted in the model because the majority of them disappear naturally over time, whereas colonies in the model grow continuously. However, we can assume that the number of colonies that will eventually become established is proportional to the number of detected colonies.

The number of colonies detected decreased almost linearly with increasing distance from the population front (Fig. 9). Lower thresholds, N, resulted in a larger number of colonies detected. However, the maximum distance from the defoliation front did not change considerably with male count threshold and was ≈250 km. Thus, we assume that $x_{max} = 250$ km.

We could not determine parameter c from Fig. 9 because we did not know what proportion of detected colonies would become established. Thus, we decided to rely on the number of eradicated colonies in the STS project area in VA and WV. The number of treatment blocks was 5, 12, and 7, and the project area was 12,443; 12,649; and 10,147 km^2 in 1993-95, respectively. This makes the rate of colony establishment in the STS area equal to 0.00068 per km^2.

Parameter c equals the rate of colony establishment just beyond the front of the defoliation area. The average distance from the center of the STS project area in VA and WV to the boundary of 10 moths per trap was ≈70 km in 1993-95. Thus, the distance from the STS area to the defoliation front was ≈150 km. Assuming the linear function $b(x)$ (eqn. 4), parameter c was estimated as $0.00068 \cdot x_{max} / (x_{max} - 150) = 0.0017$.

The initial population numbers n_0 in isolated colonies was not known. We doubt that a colony may start from 1 egg mass because of the low mating probability (Sharov et al. 1995a). Our expectation of the average initial number of egg masses that may start a new colony is $n_0 = 5$.

The carrying capacity, K, was taken equal to 200,000 egg masses per 1 km^2 because we previously showed that the defoliation boundary (detected at ≈30% level) was mostly located between the boundaries of 10 and 30 egg masses per 0.01-ha plot (Sharov et al. 1995).

The intrinsic rate of population increase, r, can be estimated from egg mass counts. However, egg masses can be counted with acceptable accuracy only in moderate- or high-density populations. Thus, they could not be counted in isolated low-density colonies beyond the population front. Furthermore, we think that extrapolation of r-values from high- to low-density populations is dangerous because population dynamics at low densities may differ substantially from population dynamics at higher densities (Campbell & Sloan 1978, Elkinton & Liebhold 1990). Thus, we adjusted parameter r until the model gave a rate of population spread that matched rates observed in nature.

We assume that the initial rate of spread $v_{max} = 20.78$ km/yr at $t = 0$ represents uncontrolled population spread. Parameter $r = 1.706$ was estimated numerically using equation (6). This means that isolated populations increase their numbers $\exp(r) = 5.51$ times per year.

Now we can substitute parameter values into eqn. (9) and estimate the effect of the STS barrier zone in VA and WV on the rate of gypsy moth spread. Currently, the barrier zone starts at a distance of ≈100 km from the defoliation front, and its width is also ≈100 km. The width of the transition zone is $x_{max} = 250$ km and thus, the distant portion (50-km band) is not covered by the barrier zone. However, isolated colonies are detected and eradicated even beyond the barrier zone by a different agency (USDA Animal and Plant Health Inspection Service). Thus, we can consider that the width of the barrier zone is 150 km. Then, $W = 150/250 = 0.6$. Parameter $Q = rx_{max} / v_{max} = 1.706*250/20.78 = 20.52$. Substituting these values into equation (9), we predict relative reduction of population spread rate $R = 0.543$ (54.3%).

The actual rate of gypsy moth spread in the central Appalachian Mountains has declined by 59%, which is close to that predicted by the model.

DISCUSSION

Studies of biological invasions were always based on qualitative (presence-absence) data (Elton 1958, Hengeveld 1989). Extensive historical data on gypsy moth abundance gave us a unique opportunity to study the spread of this species at a quantitative level over a large area. We have developed new methods for analysis of the progression of population boundaries estimated at various thresholds: 6 thresholds of male moth counts in pheromone traps, 4 thresholds of egg mass counts, and defoliation. Our methods are based on estimation of population boundaries (Sharov et al. 1995), which are mathematical functions in a rotated coordinate system. This is a powerful quantitative method that can be used for documenting and predicting the rate of spread of any introduced exotic organisms.

The average locations of population boundaries estimated from various thresholds form a profile of the traveling wave of the spreading population (Fig. 2). The distance from the boundary of 1 moth per trap to defoliation was 108.5 km, and it took ≈11 yr for a population to build up until defoliation first happened. Log-transformed population numbers increased linearly in space (Fig. 2) which is the result of the exponential population growth. Male counts in pheromone traps are indirect measures of population numbers and they are not necessary proportional to the population density. Several factors may distort proportionality: moth dispersal, saturation of pheromone traps (Elkinton 1987), and competition between traps and natural females for male attraction. However, trap saturation becomes important only at high capture rates, and competition between traps and natural females was shown to be non-significant (Elkinton & Carde 1984). Thus, we think that moth counts in traps are more or less proportional to the density of males in the area, but the density of males does not necessary correspond to the density of reproducing populations because of possible long-distance dispersal of male moths. Mark-recapture experiments were used to analyze short-range dispersal of male moths (Elkinton & Carde 1980), however long-distance dispersal requires additional studies.

The rate of spread of gypsy moth populations was reduced by 59% after 1990, the year when eradication of small isolated colonies has started beyond the expanding population front. The model of stratified dispersal suggests that eradication of isolated colonies may be the cause of the observed reduction in the rate of population spread.

However, we should be cautious in interpreting these data. First, the barrier zone strategy was implemented consistently only in the STS project that started in 1992. In the previous AIPM project, eradication of small isolated colonies was conducted simultaneously with intensive suppression of outbreak populations. Suppression activity may have also contributed to the reduction of gypsy moth spread rate in the area. Second, population spread rate exhibited considerable fluctuations in time (Fig. 3). Thus, the accuracy of estimated reduction in the spread rate of gypsy moth populations is not high.

It is interesting to compare our results with historical data on gypsy moth spread. A barrier zone was established in 1923 from Canada to Long Island along the Hudson River Valley to prevent the spread of gypsy moth to the west (McManus & McIntyre 1981). This zone was managed until 1941 when it finally became generally infested. Later in 1953-1958, another barrier zone was operated along the Adirondack Mountains and Allegheny plateau. These barrier zones did not stop the advance of population front, but spread rates were very low (2.82 km/yr) during that time as compared with the periods from 1900 to 1915 (9.45 km/yr) and from 1966 to 1990 (20.78 km/yr) (Liebhold et al. 1992). Despite of limitations of the method used by these authors to estimate the rates of population spread, results suggest that the barrier zone along the Hudson River may have significantly contributed to slowing gypsy moth spread.

ACKNOWLEDGMENTS

We thank E. Anderson Roberts for providing access to data-base of gypsy moth counts, Shelby J. Fleischer and Frank C. Curriero for discussion of the manuscript. This research was funded by the Grant No. 95-37313-1892 from the USDA, Cooperative State Research, Education, and Extension Service.

REFERENCES

Campbell, R.W. & R.J. Sloan, 1978. Numerical bimodality among North American gypsy moth populations. *Environ. Entomol.* 7: 641-646.

Deutsch, C.V. & A.G. Journel, 1992. *GSLIB. Geostatistical software library and user's guide.* Oxford University, Oxford.

Elkinton, J.S., 1987. Changes in efficiency of the pheromone-baited milk-cartoon traps as it fills with male gypsy moths (Lepidoptera: Lymantriidae). *J. Econ. Entomol.* 80: 754-757.

Elkinton, J.S. & R.T. Carde, 1980. Distribution, dispersal and apparent survival of male gypsy moths as determined by capture in pheromone-baited traps. *Environ. Entomol.* 9: 729-737.

Elkinton, J.S. & R.T. Carde, 1984. Effect of laboratory-reared female gypsy moths, *Lymantria dispar* L. (Lepidoptera: Lymantriidae), on the capture of males in pheromone-baited traps. *Environ. Entomol.* 13: 1377-1385.

Elkinton, J.S. & A.M. Liebhold, 1990. Population dynamics of gypsy moth in North America. *Annu. Rev. Entomol.* 35: 571-596.

Elton, C.S., 1958. *The ecology of invasions by animals and plants.* London, Methuen.

Fleischer, S.J., F. W. Ravlin & R.C. Reardon, 1991. Implementation of sequential sampling plans for gypsy moth (Lepidoptera: Lymantriidae) egg masses in eastern hardwood forests. *J. Econ. Entomol.* 84: 1100-1107.

Gottschalk, K.W., 1993. *Silvicultural Guidelines for forest stands threatened by the gypsy moth.* USDA For. Serv. Gen. Tech. Rep. NE-171.

Hastings, A., 1996. Models of spatial spread: is the theory complete? *Ecology* 77: 1675-1679.

Hengeveld, R., 1989. *Dynamics of Biological Invasions.* Chapman & Hall, London, New York.

Kolodny-Hirsch, D.M., 1986. Evaluation of methods for sampling gypsy moth (Lepidoptera: Lymantriidae) egg mass populations and development of sequential sampling plans. *Environ. Entomol.* 15: 122-127.

Leonard, D.S. & A.A. Sharov, 1995. Slow The Spread project update: Developing the process for evaluation. Pp. 82-85. In: S.L.C. Fosbroke and K. W. Gottschalk (eds.) *Proceedings. USDA Interagency Gypsy Moth Research Forum 1995.* USDA For. Serv. Gen. Tech. Rep. NE-213.

Liebhold, A.M., J.A. Halverson & G.A. Elmes, 1992. Gypsy moth invasion in North America: a quantitative analysis. *J. of Biogeogr.* 19: 513-520.

Liebhold, A., V. Mastro & P.W. Schaefer, 1989. Learning from the legacy of Leopold Trouvelot. *Bull. Entomol. Soc. Amer.* 35: 20- 21.

Marsula, R. & C. Wissel, 1994. Insect pest control by a spatial barrier. *Ecol. Modell.* 75/76: 203-211.

McFadden, M.W. & M.E. McManus, 1991. An insect out of control? The potential for spread and establishment of the gypsy moth in new forest areas in the United States. Pp. 172- 186. In: Y.N. Baranchikov, W.J. Mattson, F.P. Hain & T.L. Payne, (eds.) Forest insect guilds: Patterns of interaction with host trees USDA For. Serv. Gen. Tech. Rep. NE-153.

McKendrick, A.G., 1926. Applications of mathematics to medical problems. *Edinburgh Math. Soc. Proc.* 44: 98-130.

McManus, M.L. & T. McIntyre, 1981. Introduction. Pp. 1-7. In: C.C. Doane & M.L. McManus (eds.) *The gypsy moth: Research toward integrated pest management.* USDA Tech. Bull. 1584.

Ravlin, F.W., R.G. Bellinger & A.E. Roberts, 1987. Gypsy moth management programs in the United States: status, evaluation and recommendations. *Bull. Entomol. Sic. Am.* 33: 90-98.

Reardon, R.C., 1991. Appalachian gypsy moth integrated pest-management project. *For. Ecol. and Manage.* 39: 107-112.

Schwalbe, C.P., 1981. Disparlure baited traps for survey and detection. Pp. 542- 548. In: Doane, C.C. and M.L. McManus (eds.) *The gypsy moth: Research toward integrated pest management.* USDA Tech. Bull. 1584.

Sharov A.A., E.A. Roberts, A.M. Liebhold & F.W. Ravlin, 1995. Gypsy moth (Lepidoptera: Lymantriidae) spread in the Central Appalachians: Three methods for species boundary estimation. *Environ. Entomol.* 24: 1529-1538.

Sharov, A.A., A.M. Liebhold & E.A. Roberts, 1996a. Gypsy moth (Lepidoptera: Lymantriidae) spread in the Central Appalachians: Comparison of population boundaries obtained from male moth capture, egg mass counts & defoliation records. *Environ. Entomol.* 25: 783-792.

Sharov, A.A., A.M. Liebhold & E.A. Roberts, 1996b. Methods for monitoring the spread of gypsy moth (Lepidoptera: Lymantriidae) populations in North America. *J. Econ. Entomol.* (in press).
Shigesada, N., K. Kawasaki & Y. Takeda, 1995. Modeling stratified diffusion in biological invasions. *Amer. Natur.* 146: 229-251.
Skellam, J.G., 1951. Random dispersal in theoretical populations. *Biometrika* 38: 196-218
Snyder, J.P., 1987. *Map projections-a working manual.* U.S. Geol. Surv. Prof. Pap. 1395.
Talerico, R.L., 1981. Defoliation as an indirect means of population assessment. Pp. 38-49. In: C.C. Doane and M.L. McManus (eds.) *The gypsy moth: Research toward integrated pest management.* USDA Tech. Bull. 1584.

Population and Community Ecology for Insect Management and Conservation, Baumgärtner et al. (eds)
© 1998 Balkema, Rotterdam, ISBN 90 5410 930 0

Testing for correlation in the presence of spatial autocorrelation in insect count data

A.M.Liebhold
Northeastern Forest Experiment Station, USDA Forest Service, Morgantown, W.Va., USA

A.A.Sharov
Department of Entomology, Virginia Polytechnic Institute and State University Blacksburg, Va., USA

ABSTRACT: Spatial autocorrelation occurs when values of a variable sampled at nearby locations are more similar than those sampled at locations more distant from each other. Spatial autocorrelation can occur at multiple spatial scales or vary with spatial orientation (cardinal direction). It is common in ecological data. In landscape ecology research, scientists often are interested in testing the statistical association between two variables that have been sampled at multiple locations in space. The presence of spatial autocorrelation in one or both variables may violate the assumption of independence among samples and thereby inflate the degrees of freedom in the traditional test of significance of a Pearson correlation coeeficient. In this paper we used geostatistical simulation to illustrate this phenomenon and suggest strategies for overcoming this problem.

INTRODUCTION

In ecology, few generalizations are universal, yet nearly all data fulfill the generalization that values from samples taken near each other tend to be more similar than those taken farther apart. This tendency is termed spatial autocorrelation or spatial dependence (Cliff and Ord 1973; Rossi et al. 1992; Liebhold et al. 1993).

Despite this universal tendency, ecologists often assume that spatial autocorrelation does not exist. In many if not all studies, samples are replicated through space. The implicit assumption in this design is that separation of samples in space yields independent observations. This assumption is applied widely in ecological studies in which the association between two or more variables is tested. This assumption of independence is true in some cases, but in others, samples may not be separated by adequate distance such that spatial autocorrelation is negligible. Although this problem has been recognized by several authors (Student 1914; Bivand 1980; Clif and Ord 1981; Clifford et al. 1989), most ecologists apparently are unaware of it.

The study of landscape ecology has grown considerably over the last 20 years. Using a landscape approach, ecologists often address traditional ecological problems but focus on patterns and processes over large spatial scales. The adoption of this technique has been facilitated greatly by the advent of geographical information systems (GIS). A GIS is a computer program designed to input, output, and manipulate data that are referenced spatially. Often, landscape ecologists use a GIS to manipulate large data sets that form a matrix of cells (e.g., digital elevation models, remotely sensed vegetation data, gridded climatalogical data. While these exhaustive data are useful for examining landscape-level relationships, statistical problems may arise due to extensive autocorrelation in these and other data. The presence of spatial autocorrelation often violates the assumption of independence that is implicit in many statistical analyses.

In this paper we introduce geostatistical methods for quantifying spatial autocorrelation and examine how the presence of spatial autocorrelation affects error probabilities for tests of association/correlation. We also present an approach for testing for correlation between two spatially autocorrelated variables.

Much effort has been invested in characterizing spatial patterns of insect densities. Most earlier studies have attempted to described spatial patterns with the use of dispersion indices such as $s^2/\bar{x}$ (David and Moore 1954), coefficients of Taylor's Power Law (Taylor 1984), I_d (Morista 1959), Lloyd's Patchiness Index (Lloyd 1967), and Iwao's patchiness regression coefficients (Iwao 1972). These indices focus on the frequency distribution of samples (most quantify the relationship of the sample variance to the mean) but ignore the spatial location of samples. This property produces certain undesirable effects: (*a*) these indices often fail to differentiate among different spatial patterns (Jumars et al. 1977), and (*b*) their descriptions of spatial pattern may be highly dependent on the size of sample units (Sawyer 1989).

By contrast, geostatistics is a branch of applied statistics in which both the values and locations of samples are used to describe and model spatial patterns (Isaaks and Srivastava 1989; Liebhold et al. 1996). These methods originally were developed for geological applications but recently there has been considerable interest in their application to ecological problems.

Let $z(x)$ represent the value of a variable at location x and let $z(x+h)$ represent the value of the same variable some h distance, or lag, away. Consider the set of all possible combinations of samples that are separated by h. One way to express the similarity or dissimilarity between the paired values is to plot them in a scattergram of $z(x)$ vs. $z(x+h)$, known as plot is known as an h-scattergram. If the difference between all the $z(x)$ and $z(x+h)$ is small, then the scatter of points will be close to the 45° line and the variable is autocorrelated. Alternatively, the larger the difference between the pairs, the more diffuse the scatter of points around the 45° or $z(x) = z(x+h)$ line. When h is small, the scatter of points will, on average, be "tighter"; when h is large, autocorrelation is less and the scatter typically is more diffuse.

The h-scattergrams can be useful models of the degree of similarity or dissimilarity between samples separated by a common distance, but they are not practical. Because too many h-scattergrams would be required to adequately characterize the spatial similarity for all samples and for all h, a meaningful summary of these h-scattergrams is required. The most familiar tool in geostatistics is the "semi-variogram" or simply "variogram." A variogram summarizes all h-scattergrams for all possible pairings of data for all significant h:

$$\hat{\gamma}(h) = \frac{\sum_{i=1}^{N(h)} [z(x_i) - z(x_i + h)]^2}{2N(h)}, \qquad (1)$$

where $\hat{\gamma}(h)$ is the estimated variogram value for lag h and $N(h)$ is the number of pairs of points separated by h. Variograms plot the $\hat{\gamma}(h)$ as a function of distance h (Fig. 1). The variogram values can be

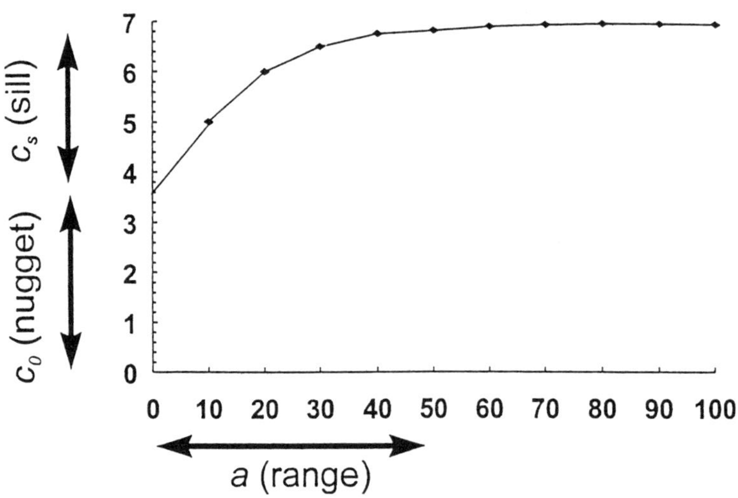

Figure 1. Typical variogram.

computed as averages over all directions, in which case the lag measure is scalar, or specific to a particular direction, in which case the lag measure is a vector.

Typically, when variogram values are plotted for all h, the values are small for low values of h; they increase with increasing distance, and usually level off or become constant after some distance (Fig. 1). Constant variogram values imply that the variance between values does not change with distance. Small values at short lags are indicative of data that are autocorrelated or spatially continuous. Large values, indicate that the paired samples are dissimilar and more spatially discontinuous.

If a variogram displays a leveling-off behavior, then the variogram value at which the plotted points level off is known as the "sill" (Fig. 1). The value of the sill usually is equivalent to the traditional sample variance. The distance at which the variogram values level off is known as the "range." The range designates the average distance within which the samples remain correlated spatially. Variograms that do not demonstrate a leveling off imply that the range is beyond the maximum lag distance analyzed.

Notice in (1) that, strictly speaking, $\hat{\gamma}(0) = 0$ since there is no variability between a sample and itself. In practice, however, when a variogram's scatter of points are extrapolated to lag zero, they often appear to intercept the ordinate at a value that is greater than zero. The variogram value at which the model appears to intercept the ordinate is known as the "nugget." A nugget represents two often co-occurring sources of variability. One source derives from spatial variability at a scale smaller than the minimum lag distance, so it cannot be modeled with the present sampling scheme. The other genesis of a nugget is experimental error, sometimes referred to as the "human nugget." Interpretations from variograms depend on the size of the nugget because the difference between the nugget and the sill (if there is one) represents the proportion of the total sample variance that can be modeled as spatial variability.

Variograms sometimes appear totally flat. These models have a complete lack of structure and their values are nearly identical to the sample variance over all h. In geostatistics these models are known as "pure nugget" variograms. Such variograms represent an absence of spatial dependence at the scale sampled. In this instance, the sample variance adequately summarizes the data's variability.

For entomological and environmental variables the similarity or dissimilarity between locations is rarely uniform with direction. For instance, one might expect, a priori, that a natural population of some insect would more often display different densities with different directions due to migration patterns or some environmental cue. When variograms computed for specific directions show different behaviors, the data are said to be "anisotropic."

Sample variograms often are not as regular as that shown in Figure 1. However, for other statistical procedures (e.g., kriging), it is necessary to fit a model to $\gamma(h)$ as a function of h to obtain values of $\gamma(h)$ for all values of h. Several forms of models used for variogram modeling; we describe only the exponential model of the form:

$$\gamma(h) = c_0 + c_s\left(1 - e^{-3h/a}\right) \tag{2}$$

where c_0 is the nugget effect, c_s is the sill, and a is the range.

HOW DOES SPATIAL AUTOCORRELATION AFFECT TESTS FOR SIGNIFICANCE OF CORRELATION?

The most widely used method for testing the correlation between two variables (a, b) is the computation of the Pearson product-moment correlation coefficient:

$$r_{ab} = \frac{s_{ab}}{s_a s_b} \tag{3}$$

where s_{ab} is the covariance between a and b and s_a is the standard deviation of a. We then test the null hypothesis that a and b are not correlated, $r_{ab} = 0$, by assuming that n independent samples of a and b are sampled from a bivariate normal distribution. It follows that the standard deviation of the correlation coefficient is:

$$s_r = \sqrt{\frac{1 - r^2}{n - 2}} \tag{4}$$

The null hypothesis is then tested as a t-test with n-2 degrees of freedom,

$$t_s = \frac{|r|}{s_r} \tag{5}$$

$$= |r|\sqrt{\frac{n-2}{1-r^2}}$$

Critical values of r for a given sample size (n) and error probability (α) can be found in most books of statistical tables.

Spatial autocorrelation in a and b violates the assumption of independence among samples. How does this affect the distribution of r? To answer this question we simulated samples of a and b where a and b were independent of each other but spatially correlated to varying degrees. Sample values were simulated using a technique called "sequential unconditional Gaussian simulation", which generates spatially distributed values that correspond to a specified variogram. Details of this technique are found in Borgman et al. (1984),and Deutsch and Journel (1992). With this procedure, the first step is to define a random path that visits each node of the grid to be simulated. The value at the first node is simply drawn from a normal distribution. At each subsequent node on the path, the variogram model is used to determine the mean and variance of the conditional cumulative distribution function (CCDF) at that location. A value is then drawn randomly from a Gaussian distribution with this mean and variance and added to the data set. The algorithm then proceeds to the next node where the process is repeated until all nodes are simulated.

We generated a 50 x 50 grid (2,500 nodes) of values for all simulations. Simulations were computed using the SGSIM procedure in the GSLIB software package (Deutsch and Journel 1992). This procedure simulates data in Gaussian space (mean = 0, standard deviation = 1) that honors a specific variogram model of sill = 1.0. We assumed an exponential variogram model for all simulations. One thousand simulated grids were generated for each of the following five combinations of variogram parameters ① c_0 = 0, a = 0.2, ② c_0 = 0, a = 2.0, ③ c_0 = 0.5, a = 0.2, ④ c_0 = 0.5, a = 2.0, and , ⑤ c_0 = 1.0 (Fig. 2), where c_0 is the nugget effect, expressed as a fraction of the sill, c_s is 1 - c_0, and a is the range, expressed as a fraction of the simulation domain length (50 cells). When c_0 = 1.0, the variogram was pure nugget effect and simulated values had no autocorrelation structure.

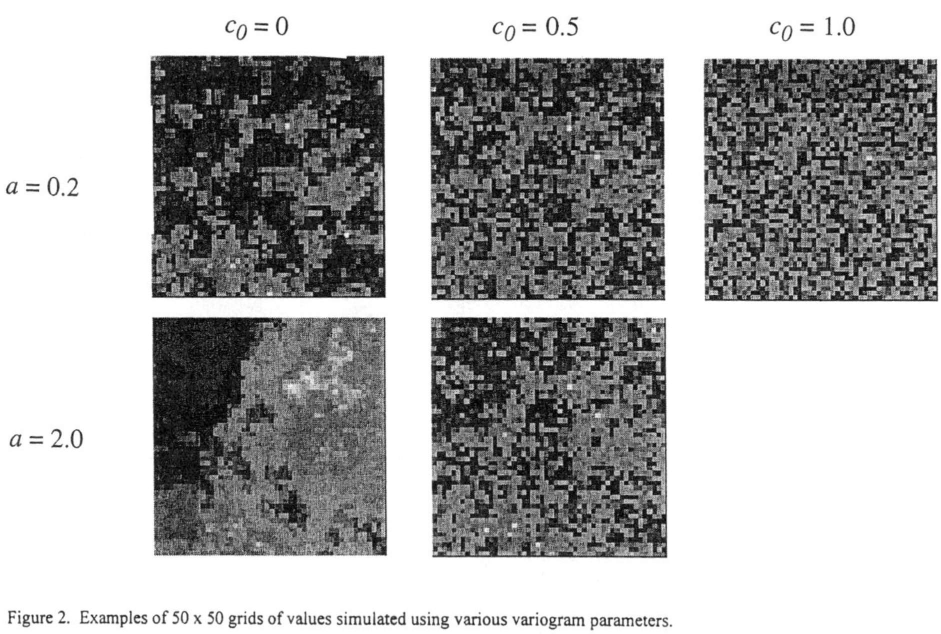

Figure 2. Examples of 50 x 50 grids of values simulated using various variogram parameters.

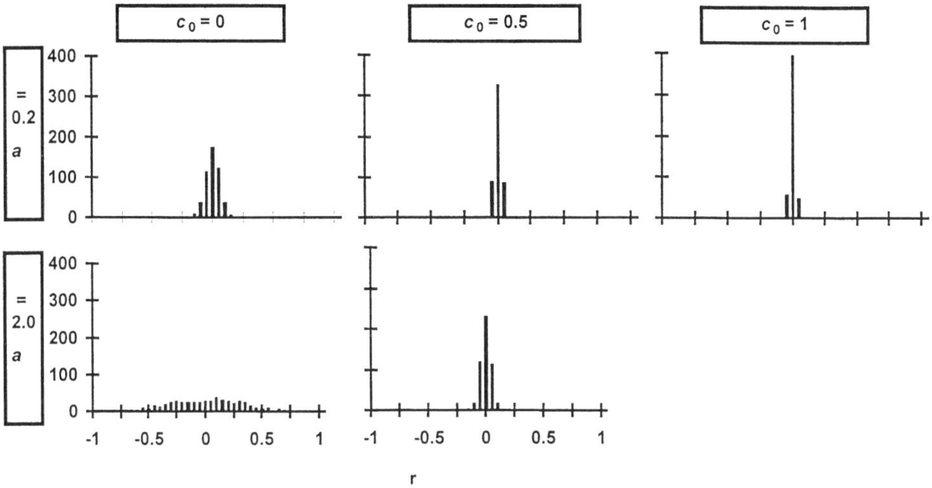

Figure 3. Frequency distribution histograms of correlation coefficients between simulated variables with identical variograms.

Table 1. Standard deviation of simulated correlation coefficients (all 2,500 points.)

Variogram Model Parameters	$c_0 = 0$ a = 0.2	$c_0 = 0$ a = 2.0	$c_0 = 0.5$ a = 0.2	$c_0 = 0.5$ a = 2.0	$c_0 = 1.0$
$c_0 = 0$ a = 0.2	0.0560	0.0913	0.0354	0.0463	0.0199
$c_0 = 0$ a = 2.0		0.2933	0.0480	0.1006	0.0210
$c_0 = 0.5$ a = 0.2			0.0267	0.0296	0.0200
$c_0 = 0.5$ a = 2.0				0.0438	0.0197
$c_0 = 1.0$					0.0199

Correlation coefficients of values were then computed between all combinations of these data sets. For example, 500 simulated grids for $c_0 = 0$, $a = 0.2$ were paired with 500 grids for $c_0 = 0.5$, $a = 0.2$. For each pair of simulations, values at each of the 2,500 grid nodes were paired and a correlation coefficient was calculated. Thus, obtained a frequency distribution of 500 correlation coefficients for each combination (Fig. 3). The standard deviations for simulated r are given in Table 1. When spatial autocorrelation was absent in either a or b, the frequency distribution of simulated correlation coefficients was the same as when autocorrelation was absent in both and a and b. However, when the nugget effect was less than 1 in both variables, the absolute value of r increased. The standard deviation of r was greatest when c_0 was small and a was large. Similar results were obtained when we used a random subset of 200 points rather than all 2,500 grid nodes (Table 2).

Table 2. Standard deviation of simulated correlation coefficients (200 randomly selected points.)

Variogram Model Parameters	c_0= 0 a = 0.2	c_0= 0 a = 2.0	c_0= 0.5 a = 0.2	c_0= 0.5 a = 2.0	c_0= 1.0
c_0= 0 a = 0.2	0.0834	0.1183	0.0813	0.0851	0.0729
c_0= 0 a = 2.0		0.3030	0.0853	0.1206	0.0739
c_0= 0.5 a = 0.2			0.0774	0.0768	0.0715
c_0= 0.5 a = 2.0				0.0854	0.0704
c_0= 1.0					0.0749

METHODS FOR TESTING FOR CORRELATION WHILE ADJUSTING FOR SPATIAL AUTOCORRELATION

As shown earlier, the use of conventional tests of the significance of correlation will yield incorrect results in the presence of spatial autocorrelation in both variables. The question then is: how do we modify these tests to account for this autocorrelation? One approach is to use unconditional simulation to simulate the spatial autocorrelation in each variable. By simulating uncorrelated variables with the same autocorrelation structure, we can estimate the probability that a greater *r* would occur by chance.

We illustrate this technique by examining the correlation between counts of gypsy moth, *Lymantria dispar* (L.), egg masses in 0.01 ha plots and elevation in the central Appalachian Mountains. In 1988, 4,822 egg-mass plots were sampled as part of a gypsy moth management program in this area (Reardon 1991; Liebhold et al. 1996). All egg-mass counts within the same 1-km^2 cell were averaged. Data on elevation for these same cells were obtained from U.S. Geological Survey digital elevation models (Elassal and Caruso 1983). The Pearson correlation coefficient between these two variables was calculated as r = 0.0402, n = 3,077. This value was greater than the critical value of 0.0355, so application of an ordinary test would indicate rejection of the null hypothesis of no correlation. Sample variograms were obtained for both the egg-mass and elevation data. The egg-mass variogram was modeled using an isotropic exponential model with c_0 = 0.26, c_s = 0.74, a = 7.0; the elevation data were fit using an anisotropic exponential model with c_0 = 0, c_s = 1.0, a_{max} = 33.0, a_{min} = 20.0, with the principal axis of anisotropy falling at a 45° angle (relative to the East - West line).

These variograms were used in the unconditional simulation procedure to generate five hundred 149 x 158 grids of simulated egg-mass and elevation data. Because egg mass data were lacking in many of the grid cells, those cells were deleted from both the simulated egg mass and elevation data. The data were then paired and a correlation coefficient was calculated for each of the 500 simulations. The frequency distribution of these correlation coefficients was examined. The observed value of r = 0.0402 was not greater than the 95 percentile or less than the 5 percentile of this distribution (r = 0.1782 for 1 tail test, r = 0.2177 for two-tailed test), so the null hypothesis of no correlation was not rejected.

One problem with this technique is that it is computationally intensive and requires that the user have access to software for conducting unconditional simulation. Clifford et al. (1989) developed a modified test of association that uses a Pearson correlation coefficient in which the degrees of freedom are adjusted by the spatial autocovariance in both variables. Our preliminary comparison of this method with our simulation technique indicates that they yield similar results. Ultimately, we believe that analytical methods such as these will be applied more widely in these situations.

ACKNOWLEDGMENTS

The authors thank B. Manly, J. Gurevitch, M. Hohn, and R.M. Muzika for conversations that led to this research. We also thank F. Curriero and T. Jacob for reviewing an earlier draft of this manuscript. This research was funded in part by grants 95-37313-1892 and 95-37302-1905 from the USDA NRICGP.

REFERENCES

Bivand, R. 1980. A Monte Carlo study of correlation coefficient estimation with spatially autocorrelated observations. *Quaestiones Geographicae*. 6: 5-10.

Borgman, L.,M. Taheri, and R. Hagan. 1984. Three-dimensional frequency-domain simulations of geological variables. P. 517-541 in G. Verly et al. (eds.), *Geostatistics for Natural Resources Characterization*. Dordrecht, The Netherlands: Reidel.

Cliff, A.D., and J.K. Ord. 1973. *Spatial Autocorrelation*. London: Pion Press.

Cliff, A.D. and J.K. Ord. 1981. *Spatial Processes*. London: Pion Press.

Clifford, P. S. Richardson, and D. Hémon. 1989. Assessing the significance of the correlation between two spatial processes. *Biometrics* 45: 123-134.

David, F.N. and P.G. Moore. 1954. Notes on contagious distributions in plant populations. *Ann. Bot. Lond. N.S.* 18: 47-53.

Deutsch, C.V., and A.G. Journel. 1992. *GSLIB Geostatistical Software Library and User's Guide*. New York: Oxford University Press.

Elassal, A.A. and V.M. Caruso. 1983. Digital elevation models. *U.S. Geol. Surv. Circ.* 895-B.

Isaaks, E.H., and R.M. Srivastava. 1989. *An introduction to applied geostatistics*. New York: Oxford University Press.

Iwao, S. 1972. Application of the $\overset{*}{m}$-m method to the analysis of spatial patterns by changing the quadrat size. *Res. Popul. Ecol.* 14: 97-128.

Jumars, P.A., D. Thistle, and M.L. Jones. 1977. Detecting two-dimensional spatial structure in biological data. *Oecologia* 28: 109-123.

Liebhold, A., E. Luzader, R. Reardon, A. Bullard, A. Roberts, F.W. Ravlin, S. DeLost, and B. Spears. 1996. Use of a geographical information system to evaluate regional treatment effects in a gypsy moth (Lepidoptera: Lymantriidae) management program. *J. Econ. Entomol.* 89: 1192-1203.

Liebhold, A.M., R.E. Rossi, and W.P. Kemp. 1993. Geostatistics and geographic information systems in applied insect ecology. *Annu. Rev. Entomol.* 38: 303-327.

Lloyd, M. 1967. Mean crowding. *J. Anim. Ecol.* 35: 1-30.

Morista, M. 1959. Measuring of the dispersion of individuals and analysis of the distributional patterns. *Mem. Fac. Sci. Kyushu Univ. E(Biol.)* 2: 215-235.

Reardon, R.C. 1991. Appalachian gypsy-moth integrated pest-management project. *For. Ecol. Manage.* 39: 107--112.

Rossi, R.E., D.J. Mulla, A.G. Journel and E.H. Franz. 1992. Geostatistical tools for modeling and interpreting ecological spatial dependence. *Ecol. Monogr.* 62: 277-314.

Sawyer, A.J. 1989. Inconstancy of Taylor's b: simulated sampling with different quadrat sizes and spatial distributions. *Res. Popul. Ecol.* 31:11-24.

Student. 1914. The elimination of spurious correlation due to position in time or space. *Biometrika* 10: 179-181.

Taylor, L.R. 1984. Assessing and interpreting the spatial distributions of insect populations. *Annu. Rev. Entomol.* 29: 321-357.

Population and Community Ecology for Insect Management and Conservation, Baumgärtner et al. (eds)
© 1998 Balkema, Rotterdam, ISBN 90 5410 930 0

Does host-plant exploitation influence the population dynamics of a rare weevil?

K. Schöps, R. M. Emberson & S. D. Wratten
Department of Entomology and Animal Ecology, Lincoln University, New Zealand

ABSTRACT: This paper provides evidence that an endangered monophagous weevil overexploits its host plant and can cause its local extinction. Over three consecutive summers a mark-recapture study was carried out on a geographically isolated weevil population on Mangere Island (New Zealand) to obtain population estimates and recruitment and survival rates. The host plants in the study area were mapped each summer. A release experiment was carried out to assess the host finding ability of the weevils and to simulate a local extinction of their host plants. At least 68% of the released weevils reached the host plants 100m away from the release site. During the three summers the weevil population more than quadrupled, while the area covered by the host plants halved. We predict the local extinction of the host plant population in the study area for the next summer and the migration of the weevils to other host plant patches.

INTRODUCTION

Most herbivorous insects are rare relative to the abundance of their food plant and usually have little or no impact upon the dynamics of their hosts. Only a small proportion of phytophagous insect species may become temporarily very abundant and inflict serious damage upon their hosts (Caughley & Lawton 1981). Outbreaks of phytophagous insects occur mostly in agricultural or forestry monocultures and in other modified landscapes. The simplification of agroecosystems compared with natural ecosystems is the most important contributing factor (Dent 1991).

However, many cases are known where the introduction of herbivores into areas outside their natural distribution has caused outbreaks (DeBach 1974) even in "near natural" habitats. In the absence of their normal natural enemy complex or of environmental constraints these introduced species are able to cause extensive damage to their host plants (Dent 1991). Even some monophagous native insects overexploit their host plants to such an extent that they deplete themselves of food and die of starvation. The impact of natural enemies does not prevent these outbreaks (Tscharntke 1990; Dempster 1971). Yet, only a few cases are known where overexploitation of host plants by monophagous herbivores causes the death of a considerable part of the host plant population (e.g. *Lochmea suturalis* (Thompson), heather beetle; Berdowski & Zeilinga 1987). In this paper we will give an example of an endangered monophagous weevil species that evidently causes local extinctions of its host plant which is also regarded as "threatened".

The weevil studied in this work, *Hadramphus spinipennis* Broun, belongs to a small endemic genus of large, flightless weevils (family Curculionidae; subfamily Molytinae; tribe Molitini). Its distribution is restricted to two outlying islands of the Chatham Islands group located 850 km east of New Zealand that are free from introduced vertebrate predators. The weevil also occurred on other islands of this archipelago before rodents and cats were introduced late last century. *H. spinipennis* is currently listed as an endangered species with highest priority for conservation by the New Zealand Department of Conservation (Molloy & Davis, 1992). The adults live for one to three years. They are up to 27mm long and feed on the flowers and leaves of the Dieffenbach's speargrass (*Aciphylla dieffenbachii* (F. Muell.) Kirk; Apiaceae) while the larvae feed on the roots of the plant. *A. dieffenbachii* is a dioecious perennial plant the status of which is regarded as threatened in conservation terms. Dieffenbach's speargrass is confined to grassland and open areas, where the plants form more or less distinct patches. Over the last decade large *Aciphylla* patches have suddenly disappeared from one year to the next and sometimes reappeared years later (Schöps, unpublished; E.C. Young pers. comm.). Our evidence strongly suggests that these local extinctions may be caused by the feeding damage of *H. spinipennis*.

METHODS

This study was carried out on Mangere Island in the Chatham Islands archipelago, from November 1993- December 1995. Once covered in low native forest, the island (113 ha) is now dominated by grassland and shrubs, as a result of 90% of the forest being burnt early this century to clear the island for farming. Today Mangere Island is a Nature Reserve and free of introduced vertebrate predators. One discrete medium sized patch of *A. dieffenbachii* was chosen as a study site and divided into thirty 5m x 5m quadrats. The weevil population in the area was estimated by a mark-recapture study over three seasons.

The island was visited in three consecutive summers and a cumulative total of over 7500 adult weevils was marked individually by gluing plastic bee tags on their prothoraces (glue: Zap-a-Gap, Pacer Technology, Great Britain; bee disks: Opalitplättchen, Striewski Bienenbedarf, Jevenstedt, Germany). As a contingency for the discs' being lost during the study, a combination of enamel paint colour and positions on their elytra was also used. On each visit to the island all *Aciphylla* plants in the study area were searched for weevils from five to seven times. All weevils found were marked, or if they had been captured before, their recapture was recorded. Then they were released back in the quadrat in which they were caught. The study site was geographically isolated and the next nearest *Aciphylla* patches were situated 600-700m away from it.

Pollock's robust design (Pollock 1982) was used to estimate population parameters for each mark-recapture trial. Therefore population sizes and their standard errors were estimated with CAPTURE (Otis *et al.* 1978) while survival and recruitment rates and their standard errors from one trial to the next were obtained from JOLLY (Pollock *et al.* 1990).

To follow the changes in the *Aciphylla* population, the study area was mapped once a year and the size and location of each plant was recorded.

The ability of weevils to migrate to another *Aciphylla* patch, if their initial host plant patch died off, was also investigated. To simulate this and to assess the host finding ability of the weevils, 150 were marked and released 100m away from the nearest *Aciphylla* plants in September 1995. The nearest host plants were situated in the study area, so if any of the weevils reached their host plants, a proportion of them were likely to be recorded during the mark-recapture trial in December 1995. The second-nearest *Aciphylla* plants were more than 600m away from the release site. The survival and the recapture rate from the mark-recapture trial in the study area were used to calculate the minimum number of released weevils that found their way to these nearest host plants. To obtain this estimate it was assumed that weevils from the release experiment were captured in the study area in the same proportion as the general population of the study area and that all weevils had the same chance to survive from September to December 1995.

RESULTS

The weevil population in the study area more than quadrupled during the time of the investigations (Figure 1). An increase in the population was observed between the summer of 93/94 and the summer of 94/95. However, a dramatic increase took place between the summers of 94/95 and 95/96. 29% of the weevils survived from 93/94 to 94/95 and 39% of

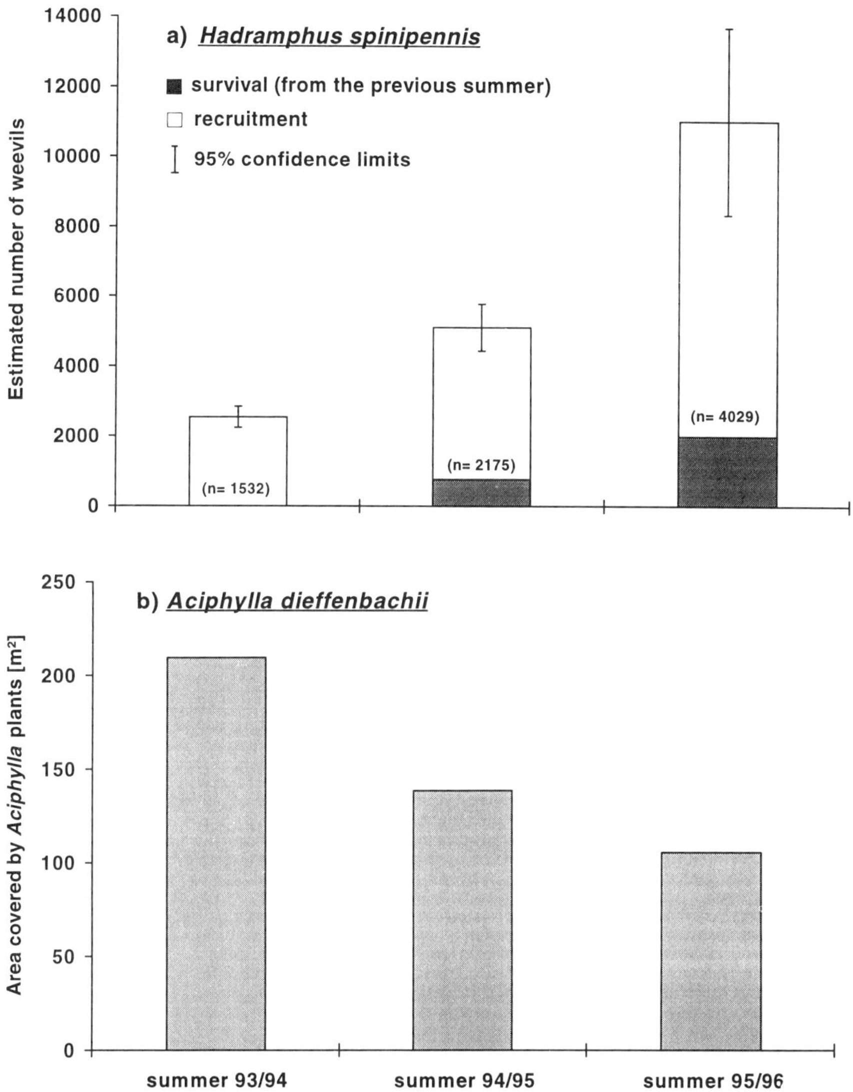

Figure 1. a) Estimated numbers of *Hadramphus spinipennis* from a mark-recapture study in an area of 750m² on Mangere Island (Chatham Islands). Population estimates and 95% confidence limits were calculated with CAPTURE (Otis *et al.* 1978) survival estimates were obtained with JOLLY (Pollock *et al.* 1990). b) Area covered by *Aciphylla dieffenbachii* plants

those which were alive in 94/95 survived until 95/96. The size of the study area was 750 m² and over the period of the investigations the proportion that was covered by *A. dieffenbachii* decreased steadily (Figure 1). From 93/94 to 95/96 the area covered by the plants had halved. Field observations also revealed that weevil feeding damage on the plants increased over the study period. In the summer of 93/94 most of the *Aciphylla* plants experienced less than 1% weevil feeding damage to their foliage. The plants were green and appeared healthy and the feeding activity of the larvae on the roots of the plants did not seem to have had a noticeable impact. In the summer of 95/96 a large number of plants were yellow and wilting. Some of them had roots that were heavily damaged while on others the roots had completely disappeared leaving only the wilting leaf-rosettes. This damage was caused by numerous larvae feeding on the roots of the plants. The browsing of the adult weevils generally had a less severe impact, although it caused some leaves to die. Ten percent of the plants had more than 5% feeding damage and a few of these plants were so heavily damaged that only a small green stump remained.

In the release experiment, at least 68% of the weevils reached the host plants 100m away from the release site after ten weeks.

DISCUSSION

Most herbivores are non-interactive, while only a few species are interactive and influence the rate of renewal of their host plants, which in turn influences the rate of increase of the herbivore (Caughley & Lawton 1981). In the case of *H. spinipennis* and *A. dieffenbachii*, the weevil has a considerable and sometimes catastrophic impact on its host plant population. As the weevil population of the study area grew, the area of it covered by *Aciphylla* decreased (Figrure 1). Over the three seasons the weevil population in the study area increased from about 12 weevils * m^{-2} of *Aciphylla* to almost 100 weevils * m^{-2} plants. The adult weevils damaged some of the plants severely and larval feeding frequently led to the deaths of plants. However the decreasing amount of available food appeared to have no negative influence on the rate of increase of the weevil population nor did it lead to reduced survival. The fact that approximately the same percentage of weevils survived from the summer of 93/94 to the summer of 94/95 and from the summer of 94/95 to the summer of 95/96 illustrates that at least at that stage, density dependence acting through intra-specific competition for food, did not appear to regulating the weevil numbers. If density dependence later played a role in regulating the weevil population, it set in after the carrying capacity of the habitat was reached. There are currently no known predators or parasitoids that significantly contribute to regulating the weevil population. The insect and spider fauna on Mangere Island is very species poor and none of the resident birds or lizards were ever seen to attack the weevils. However, subfossil material suggests that two large ground dwelling birds (family Rallidae) once occurred on Mangere Island and they could have been important weevil predators.

Local extinctions of *Aciphylla* patches, associated with high weevil densities that caused severe plant damage, have been observed in previous years (Schöps unpubl.; E.C. Young pers. comm.) and we are convinced that the key factor (Varley & Gradwell 1960) that contributes most to mortality and perturbation of densities in *H. spinipennis* populations is host plant death and consequent starvation of the herbivore. The state of the host plants and the high number of weevils in the study area in the summer of 95/96 strongly suggest that in the summer of 96/97 the plant population will collapse under the feeding damage inflicted on it by the weevils. However, competition for food followed by starvation does not seem to impact on the weevil population until extreme densities are reached. A similar case has been reported by Dempster (1971) where intraspecific competition for food and starvation in the cinnabar moth (*Tyria jacobeae* (L.)) occurred only in years when the larvae caused severe defoliation of ragwort (*Senecio jacobea* L.). Another example is a stem boring moth (*Archanara geminipunctata* (Haworth)) that in some years overexploits its host plant, the common reed (*Phragmites australis* (Cav.) Trin. ex Streudel) and reaches densities beyond the carrying capacity of the habitat. These outbreaks cause scramble competition between larvae and consequent starvation and death (Tscharntke 1990), but unlike *Hadramphus*, *Archanara* and *Tyria* do not kill their host plants. Their food source regenerates the following season from root stocks or seedlings and for both these insects emigration seems to be of minor importance as a response to high insect densities. *Hadramphus spinipennis* completely destroys its food source and there is no regeneration of the *Aciphylla* patch in the following year (Schöps unpubl.; E.C. Young pers. comm.). It often takes several years before the *Aciphylla* patch regenerates from seed banks and possibly from root fragments.

As a consequence of the local extinction of an *Aciphylla* patch the resident weevils either die out with their plant patch or migrate to a new one. From studies on the heather beetle it is known that after it destroys a patch of its host plant (*Calluna vulgaris* (L.) Hull) it disperses by flight to another patch. In spring and autumn seasonal migratory flights take place (Brunsting 1982), but if a patch dies outside the migration seasons the beetles, which at that time have reduced flight muscles, disperse by walking (Cameron *et al.* 1944). *Hadramphus spinipennis* is flightless, but we believe that the weevils are able to walk to new host plant patches if theirs becomes extinct. This is supported by the release experiment carried out in 1995 which showed that the majority of the weevils released were able to find their host plants over a distance of 100m. Some of the *Aciphylla* patches on Mangere Island are only a few hundred metres apart, but the patch nearest to the study area is 600m away. If our prediction is correct and the *Aciphylla* population in the study area collapses in the summer of 96/97 the weevils will have a long distance to migrate before they reach new host plants. However, we expect that at least some weevils will arrive at the new *Aciphylla* patch, although mortality is likely to be very high.

ACKNOWLEDGEMENTS

The senior author wishes to thank the Department of Conservation, New Zealand for funding this project and the German Academic Exchange Service (DAAD) for supporting her with a scholarship.

REFERENCES

Berdowski, J.J. M. and Zeilinga, R. (1987) Transition from heathland to grassland: damaging effect of the heather beetle. *Journal of Ecology* 75, 159-175.

Brunstig, A.M.H. (1982) The influence of a population of.herbivorous beetles on the development of vegetational patterns in a heathland system. pp. 215-223. In: Visser,J.H. and Minks, A.K. (eds.) *Proceedings of the 5th International Symposium on Insect-Plant Relationships,* Wageningen.

Cameron, A.E.; Hardy, J.W. and Bennett, A.H. (1944) *The Heather Beetle (Lochmea suturalis): Its Biology and Control.* Petworth, Sussex, British Field Sports Society.

Caughley,G. and Lawton,J.H. (1981) Plant-herbivore systems. pp. 133-166. In: May, R.M. (ed.) *Theoretical Ecology.* 2nd edn., pp. 133-166. Oxford, Blackwell Scientific Publications.

DeBach, P. (1974) *Biological Control by Natural Enemies.* Cambridge, Cambridge University Press.

Dempster, J.P. (1971) The population ecology of the cinnabar moth, *Tyria jacobaeae* (L.) (Lepidoptera, Arctiidae). *Oecologia* 7, 26-67.

Dent, D. (1991) *Insect Pest Management.* Wallingford, C.A.B. International.

Molloy, J. and Davis, A. (1992) *Setting Priorities for the Conservation of New Zealand's Threatened Plants and Animals.* Wellington, Department of Conservation.

Otis, D.L.; Burnham, K.P.; White, G.C. and Anderson, D.R. (1978) Statistical inference from capture data on closed animal populations. *Wildlife Monographs* 62, 135.

Pollock, K.H. (1982) A capture-recapture design robust to unequal probability of capture. *Journal of Wildlife Management* 46(3), 752-757.

Pollock, K.H.; Nichols, J.D.; Brownie, C.; Hines, J.E. (1990) Statistical inference for capture-recapture experiments. *Wildlife Monograph* 107, 1-97.

Tscharntke,T. (1990) Fluctuation in abundance of a stem-boring moth damaging shots of *Phragmites australis*: causes and effects of overexploitation of food in a late-successional grass monocuture. *Journal of Applied Ecology* 27, 679-692.

Varley, G.C. and Gradwell, G.R. (1960) Key factors in population studies. *Journal of Ecology* 29, 399-401.

Population and Community Ecology for Insect Management and Conservation, Baumgärtner et al. (eds)
© 1998 Balkema, Rotterdam, ISBN 90 5410 930 0

Dynamics of *Clavigralla tomentosicollis* in cowpea and trap influence on adult infestation

H. Dreyer
Plant Health Management Division, International Institute of Tropical Agriculture, Cotonou, Republic of Benin (Presently: Swiss College of Agriculture, Zollikofen, Switzerland)

J. Baumgärtner
International Centre of Insect Physiology and Ecology, Nairobi, Kenya

ABSTRACT: In southern Benin's cowpea fields, the temporal dynamics and adult movement of *Clavigralla tomentosicollis* Stål were investigated during 1991 and 1992 by visual counting and mark-recapture methods. The pest's phenology and adult movement were related to the pod and seed formation period. A rapid colonization by immigrating adults closely coincided with the pod formation phase of the plants. Oviposition occurred at the beginning of pod maturation and was succeeded by an increase of the nymphal density until pod harvest. The analysis of adult movement yielded daily emigration rates of 7 to 20% of the resident adult densities, while the daily percentage of in-situ adult mortality was below 5%. The decline in the number of adults towards harvest time is therefore mainly due to emigration rather than to field mortality. The high mobility of *C. Tomentosicollis* and the rapid field colonization suggest the use of trap crops as a control method. The use of pigeonpea as a trap border stripe in second season cowpea crops in 1991 and 1992 reduced the adult infestation level of *C. tomentosicollis*.

1 INTRODUCTION

Cowpea is grown throughout the subtropics and tropics and is one of the main sources of plant protein for the African smallholder (Okigbo, 1978). In Africa, grain yields are usually very low and range between 240 and 300 kg/ha (Rachie, 1985). Several arthropod pests seriously reduce the production potential of cowpea, particularly post-flowering insect pests which are reported to cause economically significant grain yield loss (Singh and Jackai, 1985). This pest complex can be roughly divided into three principal groups: the heteropteran pod-sucking bugs (PSB), lepidopteran larvae and coleopteran larvae (Dreyer et al. 1994). Their distribution and pest status in the tropics has been summarized by Singh et al. (1990).

The adults are highly mobile and appear to be economically relevant life stage. In previous work we focussed on the methodology to assess the movement of *C. tomentosicollis* adults into and out of the cowpea field (Dreyer and Baumgärtner, 1997).

In the present paper, we complete the study on adult dynamics and evaluate a method to influence *C. tomentosicollis* infestation pattern in cowpea fields.

2 MATERIALS AND METHODS

2.1 *Temporal dynamics*

Data on egg-, nymphal- and adult densities were obtained from three cowpea fields in monocrop row plantings. Details of the sampling procedures are given by Dreyer et al. (1994). Briefly, in each field and for each sampling data, the mean number of adults per 2m row and egg batches per plant on 20 randomly selected plants were estimated by visual counts. In addition, drop cloth samples for estimating nymphal densities per 2m row were taken at 3-day intervals as soon as oviposition by *C. tomentosicollis* was observed. The abundance of the individuals in the different life stages (adults, eggs and nymphs) was visually compared with the period of seed formation of the crop.

2.2 *Adult movement*

Adult movement was studied by means of mark-release-recapture methods. Three experiments (A, B and C) were carried out during the second cropping season of 1992 in southern Benin. In each experiment, a large cowpea plot, between 0.05 and 0.16 ha in size, was divided into two small plots of equal size (S1 and S2). The experiments started at pod setting, around 40 days after sowing (DAS), as soon as a large number of immigrants could be observed on the plants and ended with pod maturity, about 63 DAS, when the adult density was sharply declining. In each small plot, a sweep net (30 cm diameter) was used to capture marked and mark-free individuals of *C. tomentosicollis*. All individuals were given date- and plot-specific marks and released immediately thereafter in the same small plot.

2.2.1 *Immigration*

To estimate increases in the number of adults during pod formation, the first three sampling occasions in a field were used. Emigration was assumed to be negligible during this period. The 'dilution rate' λ'_t (Southwood 1978, p.108; Krebs 1989, p.40), given here by additions to the number of adults in the field, equals the immigration rate, being

$$\lambda'_t = N'_{t+1} / [\psi'_t (N'_t - (n_t - s_t))] \quad , \tag{1}$$

where λ'_t was calculated and averaged for the first three samplings. N'_{t+1} is the estimated population size at sample time $t+1$, and ψ'_t is the estimated probability that an animal present and alive at time t will be present and alive at time $t+1$. The term N'_t has to be corrected by the number of *C. tomentosicollis* that died during the markings (n_t-s_t). Etimates of the adult population size at capture time t, N'_t, and the survival rate between the captures, ψ'_t, were obtained by the Jolly-Seber method (Jolly 1965, Seber 1965), as explained below.

2.2.2 *Emigration & death*

Reduction in adult numbers may be due either to death in-situ or emigration. The probability, ϕ, of an insect remaining alive is the same in both small and large plots. However, the probability, ε, of emigrating from a small plot is higher than from a large plot, ε'. Adopting the notation of Manly (1985) the following relationships holds:

$$\psi_s = \phi (1 - \varepsilon) \quad , \tag{2}$$

and

$$\psi_l = \phi (1 - \varepsilon'), \qquad \text{while } \varepsilon' = \delta \cdot \varepsilon, \quad \delta = 2/3 \tag{3}$$

where ψ_s and ψ_l are the expected survival rates between sample times, i.e., the probability of an individual surviving in the small and the large plot, respectively. δ is the perimeter influence: the proportion of edge to area of the large plot in relation to that of the small plot. In experiment (A), small plots have perimeters of 120 m, while the large plot has a perimeter of

160 m, hence δ = 160m/240m = 2/3. ϕ and ε' are obtained by rearranging equations (2) and (3), using estimates of ψ_s and ψ_l. The latter were obtained by analyzing the mark-release-recapture data of the complete data set (large plot) and the subsets of the two small plots over the pod maturation phase, i.e., from samples taken after 50 DAS.
Manly (1970) found that the Fisher & Ford (1947) method provides accurate estimates of ψ_s and ψ_l if sampling intensity and survival rates are low (around 0.5), and/or population sizes are small (less than 1000). These conditions, however, were not always met in the present experiments, hence we prefer in this work the Jolly (1965)-Seber (1965) method which may be more adequate (Dreyer & Baumgärtner, 1997).
Computation of the survival rates ψ_l and ψ_s was thus carried out using the Jolly-Seber (1965) method, as described in Krebs (1989, pp.37-43). The various survival estimates between two sample times, ψ_t,were averaged by calculating the geometric mean ψ_l and ψ_s, during both, the immigration and emigration phase (see Conroy & Bishop, 1980).

For the analyses it was assumed that both immigration and emigration are constant because the sampling period was relatively short.

2.3 *Trap crops*

Trap crops are grown to attract insects or other organisms to protect target crops from pest attack. The principle of trap cropping rests on the fact that most pests show a distinct preference to certain plant species, cultivars, or a certain crop stage (Hokkanen, 1991). Accordingly, cowpea fields 0.05-0.15 ha in size, with and without a pigeonpea trap crop, were planted at the Cotonou station of the International Institute of Tropical Agriculture during the second cropping season of 1991 and 1992. Cowpea was planted in the same way as mentioned earlier. The trap crop was grown as a border stripe along one side of the cowpea field and made up 10-15% of the surface of cowpea. The fields had to be separated by large distances (100-300 m), to avoid the interactions between the experimental areas. To lure *C. tomentosicollis* into the trap crop, the latter had to be in an attractive phenological stage, i.e. formation of new flowers and pods, at the critical time in the target crop's phenology, i.e., during cowpea pod formation. Thus, the trap was sown prior to cowpea in such a way, that pod formation of the two crops coincided with each other. In our experiments a dwarf cultivar of pigeonpea was used.

As indicated above, in each field and for each sampling date, the mean number of *C. tomentosicollis* adults per 2 m cowpea row were estimated by visual counts. The mean pest load *Ls* for the interval Δt between the samples *s* and *s+1* being

$$Ls = [(N_s + N_{s+1}) /2] \Delta t \qquad \text{for } s = 1,..., (S-1), \qquad (4)$$

where N_S is the mean number of insects in the *s*-th sample of a total number of *S* samples and Δt is the duration in days between two sampling dates (DREYER et al., 1994). Infestation level of *C. tomentosicollis* adults in trapped and untrapped cowpea fields is expressed by the cumulative load [CL] per 2 m row, i.e., the number of adults times their presence in the field. The increase of the CL along pod formation in trapped and untrapped cowpea at a given location were visually compared.

3 RESULTS AND DISCUSSION

3.1 *Temporal dynamics*

Infestation by immigrating *C. tomentosicollis* adults from neighbouring areas usually started

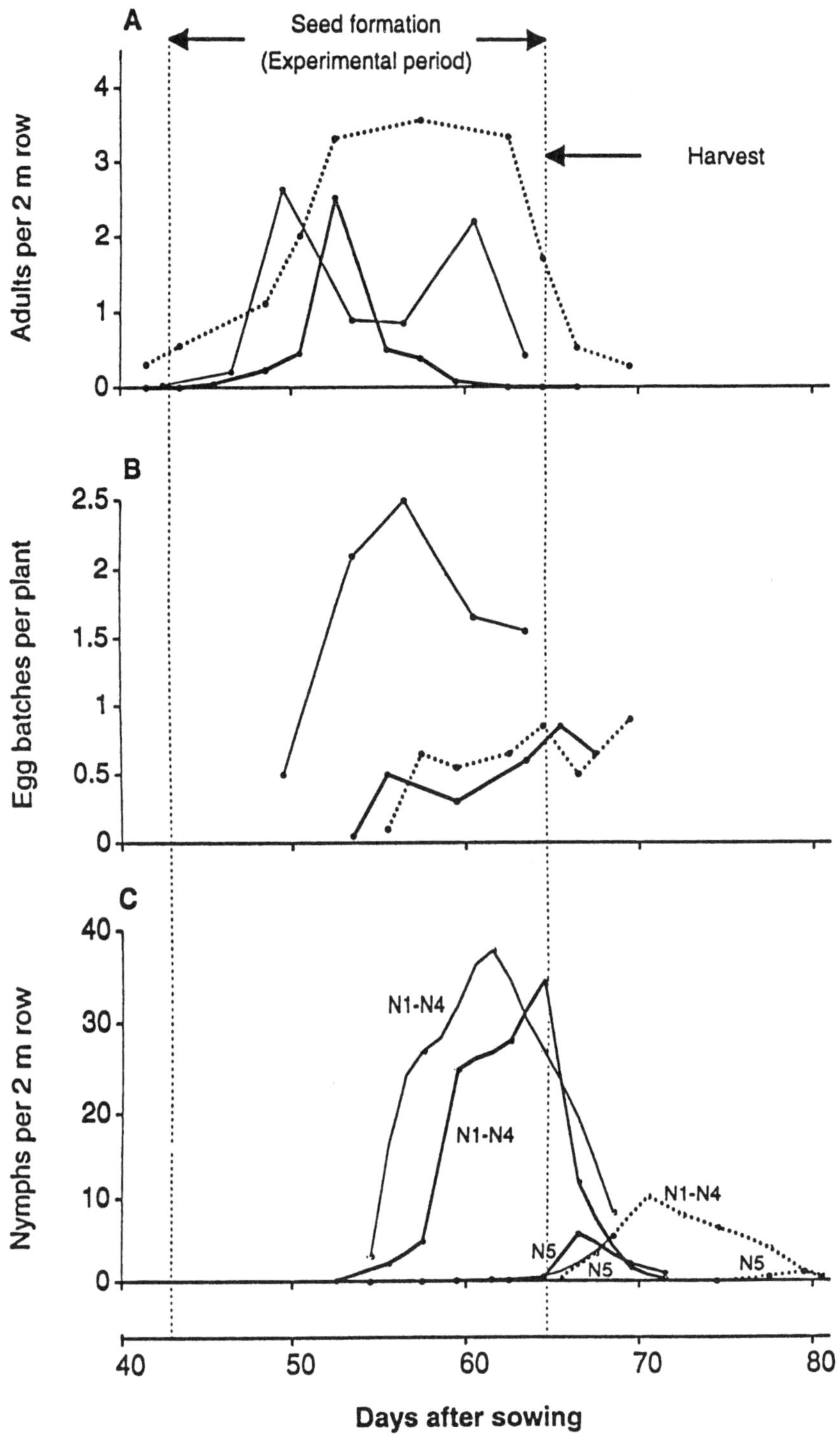

Figure 1. Mean density of *Clavigralla tomentosicollis* adults [A], egg batches [B]and number of nymphal instars N1-N4 and N5 [C] in relation to cowpea crop phenology in second season fields of 1991 (▬) and 1992 (┄) at IITA Station and of 1992 on-farm at Paouignan (—), harvest = regular harvest time of dry pods.

soon after the first pods appeared on the plants (40-45 DAS, Fig. 1A). In 1991 the maximum number was observed when the pods started to mature. In 1992, peak densities ocurred over a period of 10 days, when pods were maturing, supporting the findings of Singh & Jackai (1985).

Oviposition started between 50-55 DAS in both years, soon after the adult density had reached its maximum (Fig. 1B). The recorded appearance of the nymphs was delayed by about one week in relation to the first record of egg masses. This reflects the egg incubation period. Peak numbers of fifth instar nymphs were observed when the pods became dry.

A time gap of 10 to 14 days was observed between the peak of immigrating adults and fifth nymphal instars (Fig. 1A,C). In the system under study, adults emerging from the first generation cannot be expected before pod harvest and thus do not contribute to an increase in the number of adults during pod formation. Consequently, any change in adult density during pod growth and development is due to movement and death. Since damage by nymphs is negligible the observations illustrate that the seed damage is mainly caused by immigrating adults (Dreyer & Baumgärtner, 1997)

3.2 *Adult movement*

The above observations indicate that movement and mortality were solely responsible for the changes in number of adults.

The overall adult survival rates are given in Table 1. The procedure assumes that the probability of capture is the same in the two small plots, S1 and S2, and that emigration is not in any particular direction (Manly, 1985). The partitioning of the total population loss into field mortality (in-situ death) and emigration is presented in Table 1 for the large field plots in each experiment. The estimated overall survival rate, ψ'_l, in experiment B was particularly high (ψ'_l= 0.735). It is interesting to note that, in this field, fresh pods were available during a longer period of time than in the other two fields. This might have retained *C. tomentosicollis* in this field longer.

In all three experiments, the decrease in number of adults was mainly due to emigration rather than to death (Table 1). The daily emigration rates in experiment C (ε'= 0.197), and in

Table 1. Per day probabilities of field emigration ε' and of in situ death *d* for adult *C. tomentosicollis* in cowpea fields planted during the 2nd cropping season 1992 in southern Benin.

Field	IITA Station (A)	IITA Station (B)	On-Farm (C)
i (days)	7	2.5	3.5
ψ'_l	0.364	0.735	0.392
ψ'_{s1}	0.155	0.680	0.317
ψ'_{s2}	0.123	0.662	0.169
ε'	0.205	0.073	0.197
d	0.000	0.046	0.048
λ	1.567	1.227	2.617

i, mean interval between 2 sampling occasions; $\psi'_{l/s}$, survival rates between sample times, *i*, in the large (*l*) and the 2 small plots (S_1, S_2) in an experiment; λ, dilution (immigration) rate, calculated for the first 3 marking occasions.

experiment A (ε'= 0.205), were high. Population loss due to death was estimated to be zero in one field and less than 5% per day for the remaining fields. The difference between emigration (ε'=0.073) and death (d= 0.046) was less pronounced in experiment B.

Clavigralla tomentosicollis appears to be very mobile and may remain in a cowpea field for less than two weeks. The decrease in adult density, usually observed when the pods are drying, seems chiefly caused by emigration of the adults from the field. However, at the beginning of colonization, emigration can be disregarded and, hence, the dilution rate λ_t (eq. 1) represents population gains by immigrating adults. Between two marking sessions at early pod formation, generally between 42 and 52 DAS, λ_t was roughly the same in experiment A and B (Table 1). In experiment C, the number of adults more than doubled (λ_t = 2.62) between two markings, i.e., during 3-4 days. The figures illustrate that *C. tomentosicollis* is able to colonize cowpea rapidly.

During the experimental period, a daily rate of adult loss (emigration and mortality) of 0.2, 0.12 and 0.24 has been calculated for the three fields (Table 1). This indicates that the immigration rate is small before 45 DAS and after 55 DAS, while at the beginning of pod formation about 0.6 adults per day per 2 m row entered the field.

In summary, the dispersal of *C. tomentosicollis* adults was characterized by both high immigration and emigration rates into and from the fields, respectively. Mortality was of less importance in explaining decreases in adult density. Although reproduction in the field occurred, the emerging first generation adults join the population only around pod harvest and are suspected to immediately emigrate from the field. Immigration and emigration were assumed to be constant in this study, although immigration may prevail over emigration at early podding, while emigration increases with the crop senescing. This is clearly reflected in the characteristic abundance curves recorded at the two locations during two seasons (Fig. 1A). The emigration pattern is influenced by a number of factors, such as the presence of other nearby cowpea fields in an attractive phenological stage or alternative host plants.

The results indicate that both movement and mortality of the adults vary according to the field site and experimental conditions. In the system under study, the adults have been identified as the economically relevant life stage of *C. tomentosicollis*. Our detailed analysis of the spatio-temporal dynamics suggests that the high mobility of adults is a salient feature of *C. tomentosicollis* population dynamics. Consequently, a method controlling adult movement may be a promising approach in managing *C. tomentosicollis* populations. Among the methods available, we considered the use of trap crops appropriate for the system under study.

3.3 *Trap crops*

The infestation level of adults in trapped and untrapped cowpea is presented in Fig. 2. The adult infestation was consistently lower in the fields with a trap crop. The number of insect days accumulated at harvest differred by a factor 5 in both years. In 1991 the number of adults started to increase at the same time. In the experiment in 1992, the colonization of the two trapped cowpea fields was delayed by about 1 week if compared to the control without trap. Additional studies in 1993 yielded conflicting results (Dreyer, 1994), possibly because of lower *C. tomentosicollis* population levels. A trap impact may become obvious only under higher population densities, as found in 1991 and 1992. While we observed an influence on adult densities by pigeonpea trap border stripes, little effect on parasitism between trapped and untrapped cowpea could be recognized. Egg parasitism was similar in a trapped and an untrapped first and second season cowpea field and the proportions parasitized by *Gryon fulviventre* (Crawford) and *Ooencyrtus utetheisae* (Risbec) were comparable during this period. In summary on the basis of detailed studies the potential of using trap crops as a control method could be assessed. Our understanding of the importance of adult movement of *C. tomentosicollis* is clearly reflected by the results of the trap crop experiments. In general, trap

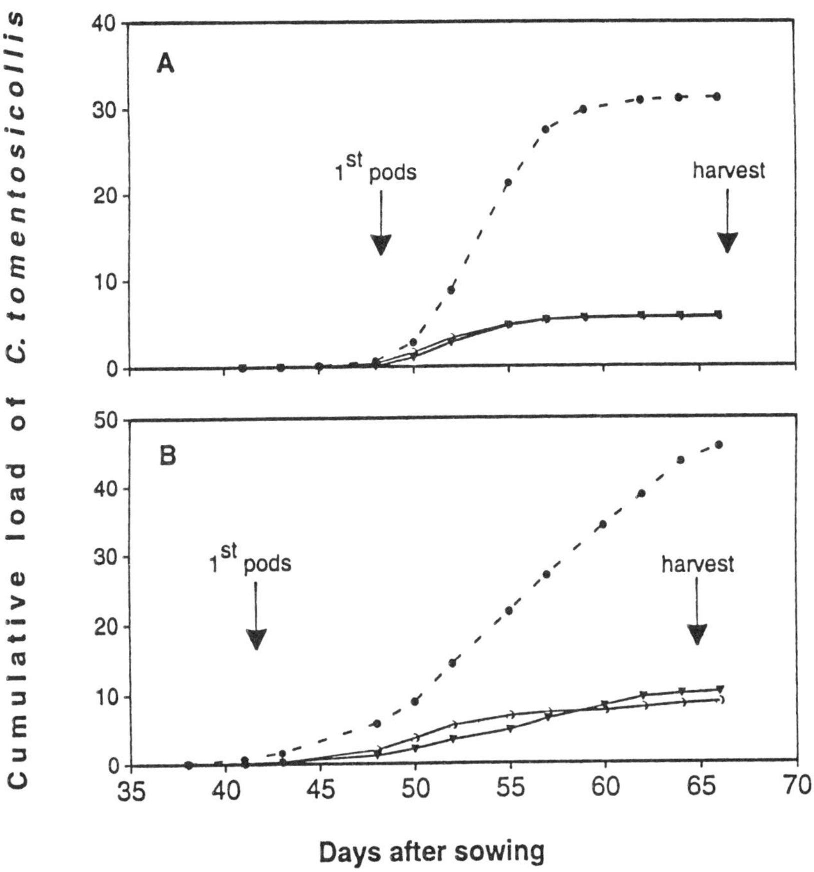

Figure 2. Infestation pattern [CL] of *Clavigralla tomentosicollis* adults in cowpea fields with a pigeonpea (*Cajanus cajan*) trap crop, planted as a border strip [solid lines] and without pigeonpea trap crop [dashed lines] during the 2nd cropping season on-station [A: 1991 and B: 1992]. Each line represents a field; CL = cumulative load: load = number of bugs per 2m cowpea row times their presence during the interval between two sample times; harvest = regular harvest time of dry pods.

crops may offer significant economic and environmental benefits in comparison with other pest management strategies (Hokkanen, 1991). This also because they can successfully be integrated in a pest managment system comprizing cultural, biological and chemical control methods.

REFERENCES

Conroy, B. A. & J. A. Bishop 1980. Maintenance of the polymorphism for melanism in the moth *Phigalia pilosaria* in rural north Wales. *Proc. R. Soc. Lond.* 210: 285-298.

Dreyer, H. 1994. Seed damaging field pests of cowpea (*Vigna unguiculata*) in southern Benin, with special reference to *Clavigralla tomentosicollis* Stål (Het., Coreidae). *Ph.D. dissertation*. Zurich: Swiss Federal Institute of Technology.

Dreyer, H. & J. Baumgärtner 1995. The influence of post-flowering pests on cowpea seed yield with particular reference to damage by Heteroptera in southern Benin. *Agric. Ecos. Environ.* 53: 137-149.

Dreyer, H. & J. Baumgärtner 1997. Adult movement and dynamics of *Clavigralla tomentosicollis* (Heteroptera: Coreidae) populations in cowpea fields of Benin, West Africa. *J. Econ. Entomol.* (submitted).

Dreyer, H., J. Baumgärtner & M. Tamò 1994. Seed damaging field pests of cowpea (*Vigna unguiculata* L. Walp.) in Benin: occurrence and pest status. *Internat. J. Pest Manag.* 40(3): 252-260.

Fisher, R.A. & E.B. Ford 1947. The spread of a gene in natural conditions in a colony of the moth *Panaxia dominula* L. *Heredity* 1: 143-174.

Hokkanen, H.M.T. 1991. Trap cropping in pest management. *Annu. Rev. Entomol.* 36: 119-138.

Jolly, G. M. 1965. Explicit estimates from capture-recapture data with both death and immigration-stochastic model. *Biometrika* 52: 225-247.

Krebs, C. J. 1989. *Ecological Methodology*. New York: Harper Collins.

Manly, B. F. J. 1970. A simulation study of animal population estimation using the capture-recapture method. *J. Appl. Ecol.* 7: 13-39.

Manly, B. F. J. 1985. A test of Jackson's method for separating death and emigration with mark-recapture data. *Res. Popul. Ecol.* 27: 99-109.

Okigbo, B.N. 1978. Grain legumes in the agriculture of the tropics. In S.R. Singh, H.F. Van Emden & T.A. Taylor (eds.), *Pests of Grain Legumes: Ecology and Control*: 1-14. London: Academic Press.

Rachie, K.O. 1985. Introduction. In S.R. Singh & K.O. Rachie (eds.), *Cowpea Research, Production and Utilization*: xxi-xxviii. Chichester: Wiley & Sons.

Singh, S. R. & L. E. N. Jackai 1985. Insect pests of cowpeas in Africa: Their life cycle, economic importance and potential for control. In S.R. Singh & K.O. Rachie (eds.), *Cowpea Research, Production and Utilization:* 217-231. Chichester: Wiley & Sons.

Singh, S.R., L.E.N. Jackai, J.H.R. Dos Santos & C.B. Adalla 1990. Insect pests of cowpea. In S.R. Singh (ed.), *Insect Pests of Tropical Food Legumes:* 43-89. Chichester: Wiley & Sons.

Seber, G. A. F. 1965. A note an the multiple-recapture census. *Biometrika* 52: 249-259.

Southwood, T.R.E. 1978. *Ecological Methods*. 2nd edition. London: Chapmann & Hall.

Population and Community Ecology for Insect Management and Conservation, Baumgärtner et al. (eds)
© 1998 Balkema, Rotterdam, ISBN 90 5410 930 0

Metapopulation structure of *Carabus problematicus* in a fragmented landscape, significance of simulation results for nature conservation

H.Gruttke & H.Engels
Bundesamt für Naturschutz, Institut für Tierökologie, Bonn, Germany

ABSTRACT: Within an agricultural landscape in Germany in which only a few isolated patches of woodland have remained, the spatial distribution of the eurytopic forest species *Carabus problematicus* was studied for three years using pitfall traps. In more remote forest patches the species could not be found every year, which is interpreted as an indication of the existence of metapopulation dynamics of this species in this area.
On the basis of GIS-generated maps, stochastic metapopulation models are developed using the standard modelling program RAMAS-GIS. Simulation results obtained correspond well with field data. According to the models created, the spatial arrangement of habitats and distance between suitable neighbouring habitats belong to the most important parameters regulating habitat colonization. Dispersal is a key factor in most metapopulation models, but good data on migration are rare and hard to obtain. Applicability and limitations of simulation models as well as consequences for nature conservation planning are critically discussed.

1. INTRODUCTION

The metapopulation concept - first proposed by Levins (1970) but modified and differentiated many times since (e.g. Pulliam 1988, Harrison 1991, Hanski 1994, Poethke et al. 1996) -, plays an increasingly important role in ecology and conservation biology (Brunzel & Reich 1996, Veith & Klein 1996). In particular, studies dealing with the effects of habitat fragmentation in order to assess the decline or extinction risk of endangered species often apply the metapopulation concept (Lahaye et al. 1994, Hanski et al. 1995, Bascompte & Solé 1996, Reich & Grimm 1996). In such studies simulation models are developed and population viability is analysed to determine conservation and management priorities (e.g. Shaffer 1990, Liu et al. 1995).

For the present study the species *Carabus problematicus* HBST. (*COLEOPTERA, CARABIDAE*) was chosen because it shows some characteristics which make it a suitable candidate for finding some evidence for metapopulation dynamics in the field. It is a monomorphic brachypterous but very mobile species (Gruschwitz 1992) which mainly lives in forests under the climatic conditions of western central Europe, but also exhibits a certain degree of dispersal activity outside wooded habitats (Den Boer 1970, Rijnsdorp 1980, Assmann 1994, Mossakowski 1996).

It is less likely to find typical metapopulation dynamics for more stenotopic species which are much more confined to their preferred habitat because, for such species, only very low exchange rates between neighbouring populations, if at all, can be expected, when their habitats are isolated from each other.

On the other hand, eurytopic species, possibly also with a high colonization capacity, do not show metapopulation dynamics because they are able to colonize every suitable habitat and thus are present nearly everywhere or wherever habitats are available.

Another problem when creating models is that the empirical database on the ecology, population biology and dispersal characteristics is often very weak (Doak & Mills 1994, Reich & Grimm 1996). In the case of *Carabus problematicus* the amount and quality of available information is relatively high (e.g. Den Boer 1970, 1977, Neumann 1971, Rijnsdorp 1980, Loreau 1984, Glück & Kreisel 1986).

The species is an autumn breeder with winter active larvae (Betz 1992) and two activity peaks of the adults: a smaller one in May and a higher one, the reproduction period, between the end of August and the beginning of October. Young adults appear first in July (Rijnsdorp 1980). The maximum age of *C. problematcus* registered by Rijnsdorp (1980) and Glück & Kreisel (1986) was three years. According to detailed results of mark-recapture experiments in enclosures carried out by Rijnsdorp (1980), the mean population density is 2.5 individuals per 100 m^2 in an oak forest in the Netherlands. Loreau (1984) measured a value of 2 individuals per 100 m^2 in a similar investigation in a beech forest in Belgium.

The maximum migration distance of *C. problematcus* per year outside woods was estimated by Den Boer (1970) at about 1100 m, by extrapolating a regression-line of catch numbers from sample sites at different distances from the nearest wood. Rijnsdorp (1980) recaptured 19 marked beetles at a distance of 300 m away from their place of release. The proportion of beetles leaving the forest habitat during the breeding season is calculated by Rijnsdorp (1980) at 4-8 %.

The main aims of the present study are:

1. to develop a realistic metapopulation model which corresponds well with the three years' field results on habitat occupancy of *Carabus problematicus* in the fragmented landscape under investigation, by using a standard modelling package (RAMAS-GIS).
2. to evaluate and discuss analytical and predictive potentialities, limitations and problems of ecological models.
3. to consider the possible significance of ecological metapopulation models for nature conservation planning.

2. MATERIAL AND METHODS

Within an intensively cultivated agricultural region in Germany (about 30 km west of Bonn) the spatial distribution pattern of *Carabus problematicus* was studied in a landscape section of 100 km^2 for three years - 1992, 1994 and 1995 (May to October) - using circular pitfall traps with a metal roof (1992: 5 traps of 4.5 cm in diameter per site, half filled with ethyleneglycol; 1994, 1995: 8 unbaited life traps of 7.1 cm in diameter per site).

Samplings carried out before (1982-1991) in a different context but in the same region (see Gruttke & Kornacker 1995, Gruttke in press), were analysed as well, in order to get a picture of the spatial distribution pattern of this species at the landscape level as precise as possible. The standardized field program comprised altogether samplings at 26 wooded sites of different age and size and at 10 open habitats (including the sites studied in 1990). At 6 old wooded habitats (forests Bb, Wa, Wb, Wc, Wd and woodlot Gi) samples were taken in all 3 years. Another 7 sites were studied for only two years (forests Mh, Ba, We, Wf, and woodlots W, Ah, Rb), the isolated forest patch Bc only in 1995 (cf. Figure 1 and 2). Additionally, collection by hand was carried out at all forest sites.

The natural vegetation in this area had been beech- or oak-hornbeam forest. Most of today's forest remnants belong to the latter vegetation type. The climate is mild and relatively dry in summer with an average precipitation of 570 mm/year and a mean temperature of 9,6 °C.

All relevant landscape patterns - the wooded landscape features being potentially particularly suitable habitats or migration corridors of *Carabus problematicus* - have been digitized and analysed using the GIS-system Arc-Info. Additionally, vegetation cover and soil parameters were registered in the field. Statistical calculations were carried out with the program NCSS 5 (Number Cruncher Statistical System, by Dr. Jerry L. Hintze, Kaysville, Utah USA). Stochastic metapopulation models were developed using the program RAMAS-GIS (Akçakaya 1994).

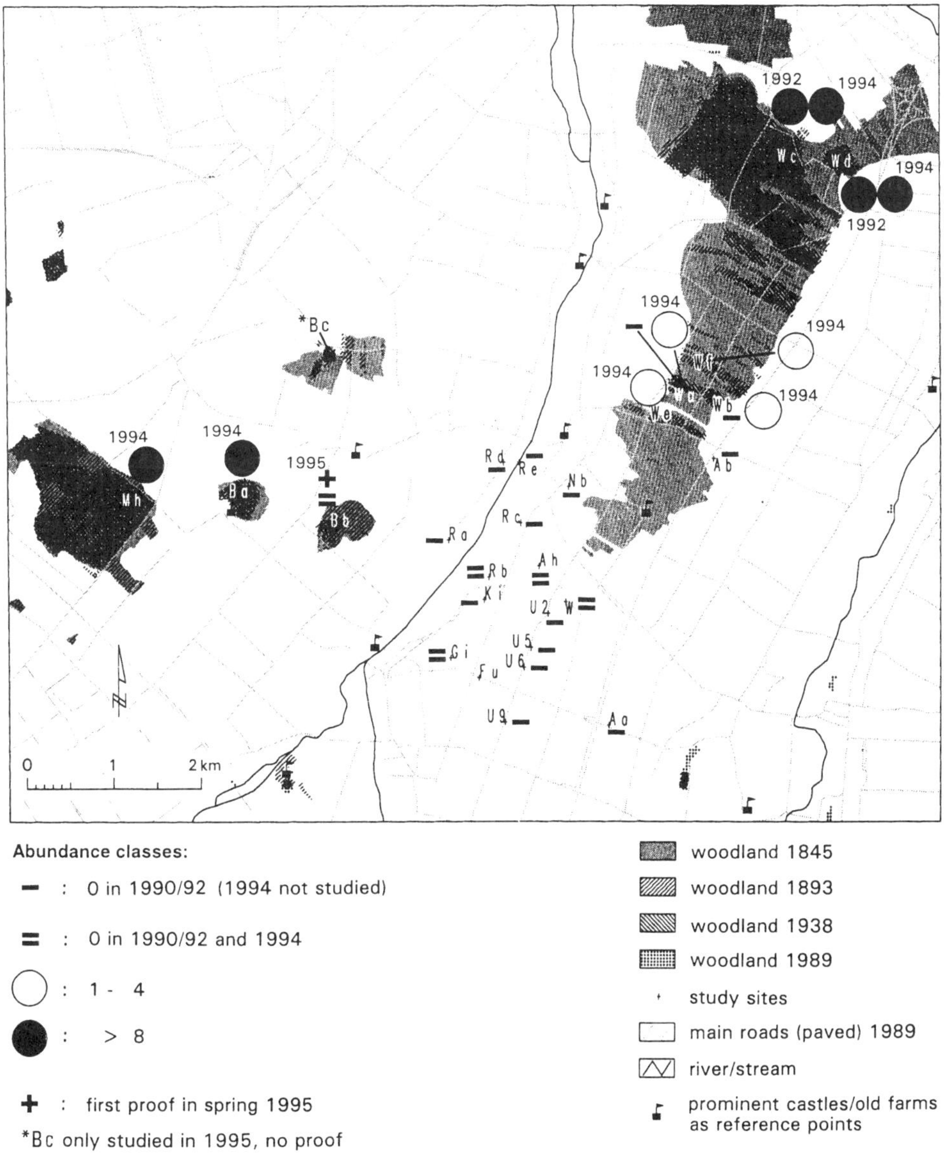

Figure 1. Distribution and abundance (scale stepwise in classes of 4 specimens/ yearcatch) of *Carabus problematicus* in the study area in 1990/92 and 1994 (1995 only one new find in Bb) as well as decline of woodland area between 1845 and 1989. The darkest hatched areas represent mainly the forest distribution today, because nearly all forest areas had been reduced in size during that period. Small wooded sites such as hedges are not particularly marked, but all study sites in the map (sign = +) are wooded habitats of at least 400 m^2 in size.

3. RESULTS

3.1 Field results on the spatial distribution of Carabus problematicus

In the landscape studied, today only a few, mostly isolated patches of woodland have remained (Fig. 1, darkest hatched areas). *Carabus problematicus* was caught only in large forest remnants or in forest islands not too distant from these. However, in some of the smaller and more remote forest patches the species could not be found every year (Fig. 1 and 2).

The forest patch Wa in the eastern part of the area was occupied in 1994 and 1995, but not in 1992. In the highly isolated forest island Bb, in the western part of the area, *C. problematicus* had not been found before 1995, in spite of high catch numbers in the neighbouring larger forests Ba and Mh in 1994. Catch numbers in the latter forests were high in 1995, too. In the very remote and small wooded site Gi this species was never detected over three years of sampling. Nor was it found in the larger and old but very isolated forest island Bc which was studied only in 1995.

Figure 2 demonstrates that catch numbers and local occupancies (here taken as number or percentage of years, in which the species has been positively proven to be present in a habitat) are clearly decreasing with growing isolation measured as distance to the next large forest

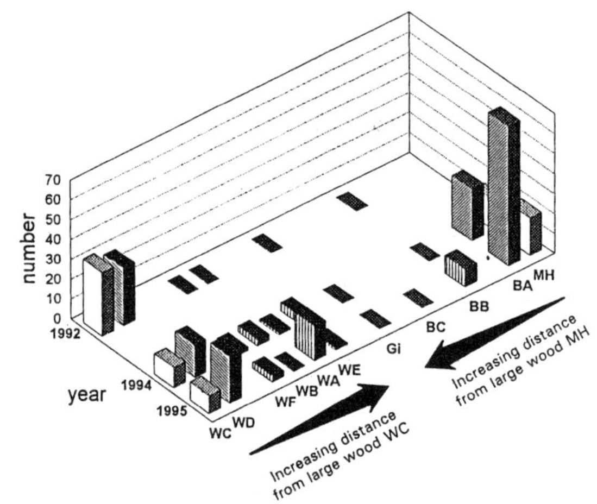

Figure 2. Scheme of catch numbers (standardized to 400 traps × catching days) of *Carabus problematicus* at old wooded sites between 1992 and 1995 in relation to the distance to next large forest. Only study sites marked by hatched squares (= zero catch) or columns had been studied in the respective year. For instance, in 1992 samples had been taken at only 6 sites. With increasing distance or isolation from the large forests Wc and Mh habitat occupancy is decreasing (for geographical location of sites see Fig. 1).

Table 1: Correlations (r) between different landscape parameters and local occupancy of *Carabus problematicus* in the field (I: 6 wooded sites, 3 years occupancy, II: 13 wooded sites, 2 years occupancy) n.s: not significant; +: $p \leq 0.05$; +++: $p \leq 0.005$.
Comparison between field data and simulation results: correlation of data on local occupancies, calculated for sites Wa, Wb, Wc, Wd, Bb, Gi, all sampled for 3 years: 1992/1994/1995: Spearman r_s: 0.8956, $p < 0.02$; Pearson r: 0.8402, $p < 0.04$.

Landscape parameters	r	I (n= 6 / 3 y) Significance	r	II (n= 13 / 2 y) Significance
size of forest patch	0.5658	n.s	0.3923	n.s
$\log_{10}$ size of forest patch	0.7789	n.s	0.5943	+
area of forest in 200 m radius	0.8150	+	0.6379	+
area of forest in 500 m radius	0.6551	n.s	0.5347	n.s
area of forest in 1000 m radius	0.7431	n.s	0.4874	n.s
distance to next forest > 1 ha	-0.7587	n.s	-0.8931	+++
distance to next forest > 100 ha	-0.8890	+	-0.9044	+++

(>100 ha, site Wc or Mh). This is also confirmed by statistical analysis (Table 1). Local occupancies are highest negatively correlating with distance to or isolation from the next forests.

The result, that *C. problematicus* has only occasionally been found in some isolated patches, is interpreted as an indication for the existence of a kind of metapopulation dynamic of this species in the landscape under study, with local extinctions and recolonizations occurring from time to time.

To test whether such a dynamic can be supported by a model and thus gain plausibility, stochastic metapopulation models were developed in which all relevant information available from literature on the different ecological and biological parameters was utilized (mainly Den Boer 1970, 1977, Rijnsdorp 1980, Loreau 1984).

3.2 Metapopulation simulation

On the basis of a GIS generated map, containing the spatial distribution of suitable habitats at landscape scale, stochastic metapopulation models were created using the standard program RAMAS-GIS.

RAMAS-GIS is a program which allows the testing of both the influence of different external factors, such as habitat suitability, geographic configuration or environmental variability, and intrinsic species or population-specific factors, for instance reproduction or migration parameters (Fig. 3).

Several important parameters had rarely or never been examined by empirical field studies, e.g. the Allee-effect, which describes the phenomenon that population growth is negatively affected as population size reaches very low levels, drawing small populations away from carrying capacity (c.f. Allee et al. 1949). Other, little-known parameters are density-dependent migration or variability in reproduction and survival rates. This is true for *C. problematicus* but also for the majority of other species.Thus the impact of such parameters could be tested only by modelling.

By varying several input factors and testing exponential population growth as well as different types of density dependence (cf. Akçakaya 1994), a high number of simulation results can be

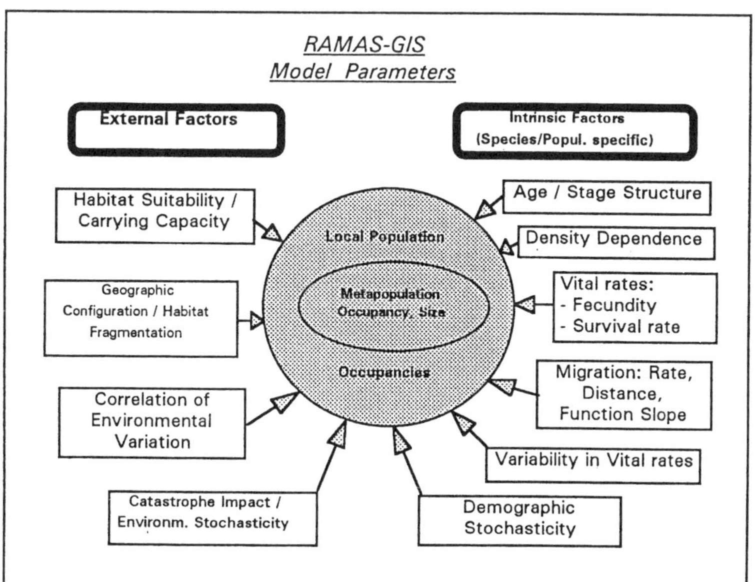

Figure 3. Overview of the main input variables of the modelling program package RAMAS-GIS, by which the simulated metapopulation can be influenced.

obtained. Some of the simulations we produced corresponded well to the field results, others did not. In the following, only one specific model approach will be described in detail. Of course, this model explains the distribution pattern found in the field very well, but it also affords the possibility of demonstrating some critical aspects.

The metapopulation map in Figure 4 is the result of the first step of the modelling procedure, in which the most reliable data from literature have been put in the model. The map shows 115 potentially suitable habitats, the circle size representing carrying capacities, and the connecting lines reflecting possible dispersal routes or potential population exchange between the habitat patches.

Parameter values used in the negative exponential migration function (1) of this model are: D_{max}= 1.1, a= 0.1, b= 0.1 and c=2,4.

$$m_{ij} = a \times \exp\left(\frac{-D_{ij}^{c}}{b}\right) \quad , \text{ if } D \leq D_{max} \text{ ; } \quad 0, \text{ if } D > D_{max} \qquad (1)$$

where m_{ij} is the migration rate between population i and j and D_{ij} the distance between the two populations. D_{max} (maximum migration distance), a (maximum emigration rate), b and c are function parameters influencing the slope of the function curve.

The picture obtained is very similar in all model-configurations tested, because it simply shows the spatial arrangement of all suitable habitats, those which are close enough to each other being connected, so that the distance can be potentially bridged by dispersing beetles.

In the next step we tested whether - respectively to what extent - certain input parameters affect the recolonization of forest islands, if only the two biggest populations of *C. problematicus* were left. Figure 5 shows the development of metapopulation occupancy (the total sum of all occupied habitats) over a timespan of 100 years. After no more than about 30 to 40 years the mean number stabilizes at a level of 26 populations. At that time a stable equilibrium of extinction and colonization is reached in the study area under the basic assumptions made (e.g. that there is no habitat loss or reduction in habitat quality during the timespan considered).

Some simulation results achieved with the model corresponded especially well with the field data, which was also confirmed by statistics. The Spearman correlation coefficient r_s between

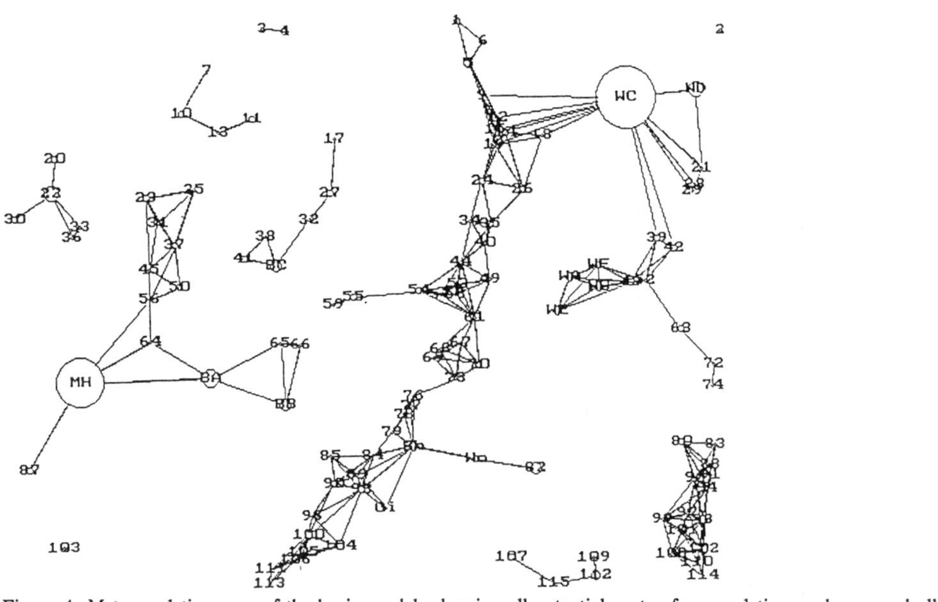

Figure 4. Metapopulation map of the basic model, showing all potential routes for population exchange and all suitable habitats of *Carabus problematicus* in the study area.

field data and model results for the local occupancy in those 6 habitats which had been studied for 3 years is 0.8956, $p < 0.02$ (see Table 1).

Another interesting simulation result is shown in Figure 6. Inserting relatively high standard deviations of survival and reproduction rates into the model (100% higher than the standard

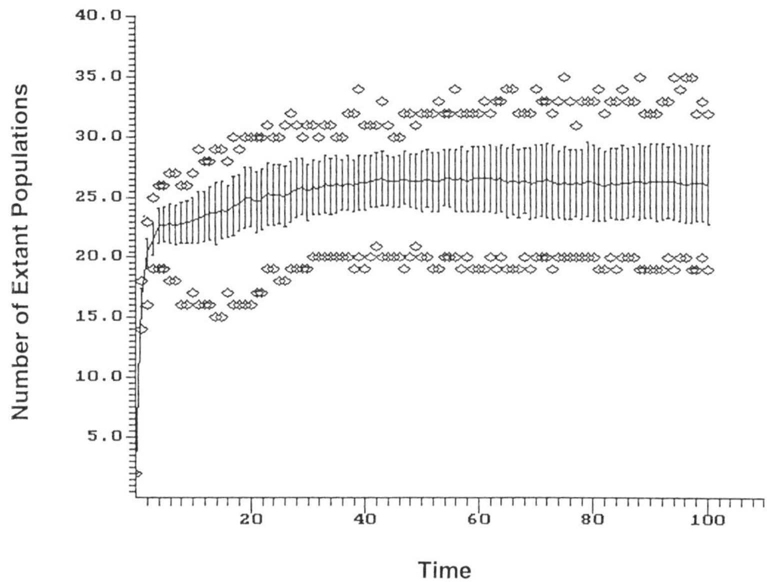

Figure 5: Metapopulation occupancy (number of occupied habitats, with standard deviations, maxima and minima per 100 replications) in the study area after a simulation time of 100 years. Model configuration used in the basic standard model: 3 age classes with fecundities (± standard deviations): 0.42 ± 0.21; 0.74 ± 0.25; 0,1 ± 0.05; survival rate 0.58 ± 0.15; initial stage abundances (%): 1. 49.5; 2. 31. 0 ; 3. 19.5; max. reproduction rate (R_{max}): 6.1; type of density dependence: scramble allee; Allee parameter: 75; carrying capacity: 2.5 Individuals/ 100 m^2.

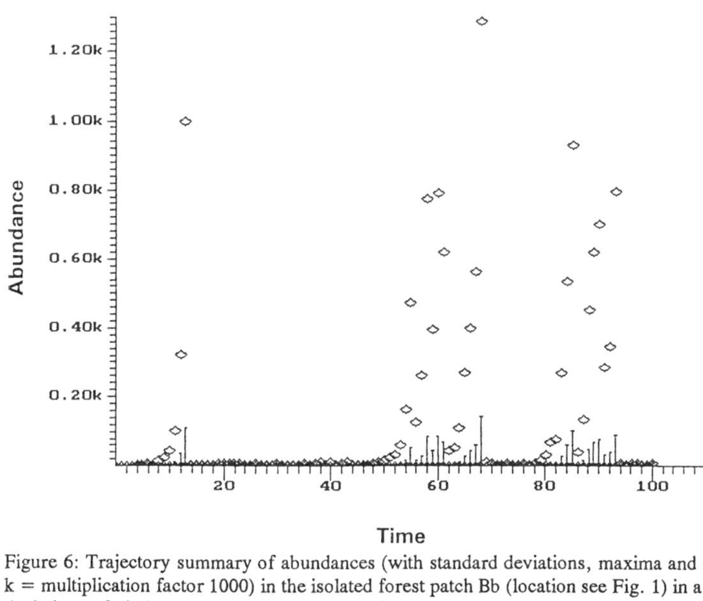

Figure 6: Trajectory summary of abundances (with standard deviations, maxima and minima per 100 replications; k = multiplication factor 1000) in the isolated forest patch Bb (location see Fig. 1) in a simulation in which standard deviations of vital rates have been increased for 100 % in comparison to the standard model (see Fig. 5).

used), the isolated forest patch Bb is colonized only from time to time, and not continuously, which is typical in metapopulation dynamics and is also in line with the field results obtained for this forest patch (see Fig. 1 and 2). Such fluctuations are not produced by the model, when the variability of these population parameters is smaller.

According to the models created, the spatial arrangement of habitats and distance between neighbouring suitable habitats are among the most important parameters regulating colonization of habitat patches and stabilizing metapopulations of *C. problematicus* at landscape level. Nevertheless, a number of specific population parameters, which have never been studied in the field but are easily modelled with RAMAS-GIS, also have a strong impact on the colonization success of *C. problematicus*.

One example of such a critical factor is the Allee-parameter (Allee et al. 1949). Figure 7 demonstrates that increasing the value of this parameter leads to reduced local occupancies in two forest islands - Bb and Wa - but not in the third, Wd. The latter habitat is the biggest of the three sites considered, so the Allee-parameter may have no effect, because of the large population size. However, the forest patch Bb is larger than Wa but affected at a lower level by the Allee-parameter. In this case isolation plays a more important role than habitat size, because Bb is more separated from other woods than Wa.

The complex of dispersal parameters will be regarded more closely in the next example (Fig. 8). There are 4 different input parameters in the migration function used by RAMAS-GIS (see formula (1)). For the estimation of two of them - the proportion of the population leaving a habitat (parameter *a*) and maximum migration distance (D_{max}) - data are available in the literature (Den Boer 1970, Rijnsdorp 1980). The other two factors (*b* and *c*) influencing the shape of the migration curve have to be put in more or less intuitively, because no adequate data are available in literature.

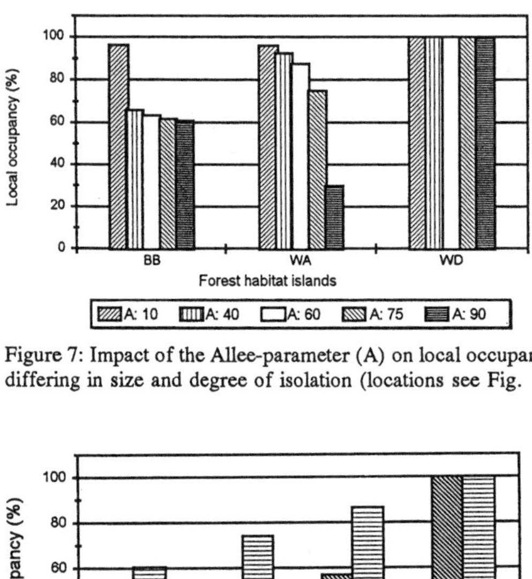

Figure 7: Impact of the Allee-parameter (A) on local occupancies of *Carabus problematicus* in three forests islands differing in size and degree of isolation (locations see Fig. 1).

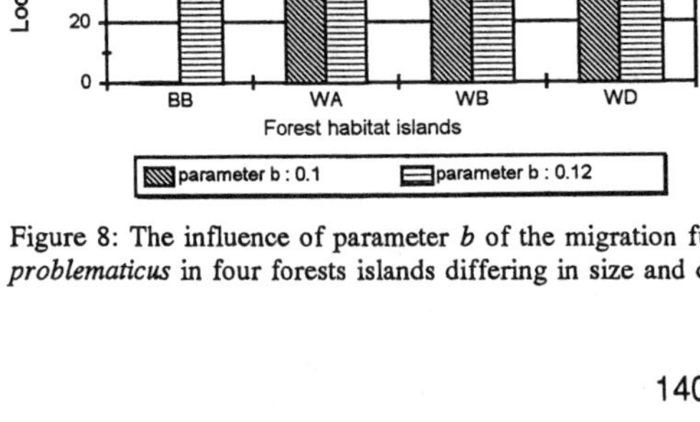

Figure 8: The influence of parameter *b* of the migration function (1) (see text) on local occupancies of *Carabus problematicus* in four forests islands differing in size and degree of isolation (locations see Fig. 1).

However, by changing the value of the above-mentioned *b*-parameter in the migration function only from 0.1 to 0.12, the local occupancy in habitat Bb is increased from zero to about 60%. According to Akçakaya (1994) this parameter affects the death rate of migrants per unit distance. A small change in this factor influences the slope of the migration function only to a minor degree, but it has a significant impact on habitat occupancy which can have consequences for the evaluation of the importance of a habitat, if such models are used in nature conservation planning.

4. DISCUSSION AND CONCLUSIONS

With RAMAS-GIS it is possible to create realistic metapopulation models for *Carabus problematicus* which correspond well with field results. Both model and field results show that habitat isolation has a significantly negative impact on local occupancies of *C. problematicus* in the landscape studied. However, the survival of this species in the study area is not threatened. Even moderate habitat loss would not bring the species close to extinction. On the other hand, the species is absent in large parts of the area - even in old forest patches -, a fact which is confirmed by simulation results as well.

Forest carabid species of a more stenotopic character and with a much more limited distribution in the study area are, of course, considerably more affected by habitat loss and fragmentation than *Carabus problematicus* (e.g. *Abax parallelus* see Assmann 1994 or Gruttke in press).

Of course, model results cannot be taken as a proof of the existence of metapopulation dynamics of *C. problematicus* in the fragmented landscape studied, however plausibilty can be strengthened thereby. Reich & Grimm (1996) set up four basic criteria which have to be fulfilled by a real metapopulation system for being accepted as such: 1. the discernibility of local boundaries of populations, 2. extinction of at least some local populations, 3. migration between local populations and 4. (re-) colonization of suitable habitats. All of these criteria seemed to be fulfilled by the population complex of *C. problematicus* studied, although migration between habitat patches has not been proved experimentally in our study, but has often been shown by others in geographical regions not very far away (Den Boer 1970, Rijnsdorp 1980, Glück & Kreisel 1986).

The best possible knowledge of the dispersal behaviour of a species (e.g. average and maximum migration distance, mortality during migration) and of the factors influencing it, is necessary in order to develop realistic metapopulation models, because dispersal is the major parameter determining stability of metapopulations and functioning of metapopulation dynamics (Pagel & Paye 1996, Reich & Grimm 1996).

Models are good tools to identify important factors and to test the potential significance of population and landscape parameters or to compare different scenarios in order to determine priorities for landscape planning and nature conservation (Akçakaya 1994, Poethke et al. 1996).

However, the quality of a model depends considerably on the quality of the empirical biological and ecological information available (Doak & Mills 1994, Reich & Grimm 1996). For most species - not only beetles - this is exactly the main problem, because knowledge is usually poor (e.g. on conditions for successful reproduction, density dependence including the Allee-effect). Hence, there is a strong need for further field research.

Moreover, modelling offers many opportunities for manipulation, particularly if empirical data are poor. This reduces its usefulness for nature conservation planning, for instance when determining the best location and size of nature reserves. Even for a good expert it can be very difficult to assess the quality of simulation results when not all model parameters are verifiable. It is nearly impossible for any non-expert, e.g. for a planner or administrator, to evaluate the predictive reliability of a simulation model.

That abuse through the application of spatially explicit models is possible has been demonstrated impressively by Harrison et al. (1993) for the classic example of the spotted owl in North America. With the help of a model developed for a forestry department, the minimum size necessary for the maintenance of the species over a certain period of time was calculated and used as an argument to cut down most of the rest of the large ancient forest, which is the habitat of the spotted owl. Critical aspects and problems arising from uncritical applications of metapopulation models are also discussed by Reich & Grimm (1996) and Veith & Klein (1996). It may be stated as a conclusion that no model can be a substitute for empirical data, and that the results of each simulation model have to be evaluated carefully. If wide-ranging decisions concerning nature conservation issues are to be made, safety factors should be implemented and respected because, once a species is locally extinct, it is very difficult to get it reestablished in that area.

5. REFERENCES

Akçakaya H.R. 1994. RAMAS/GIS: *Linking Landscape Data with Population Viability analysis* (version 1.0). Applied Biomathematics. New York: Setauket .

Allee, W.C., Emerson, A.E., Park, O., Park, T. & Schmidt, K.P. 1949. *Principles of Animal Ecology*. Philadelphia: Saunders.

Assmann, Th. 1994. Epigäische Coleopteren als Indikatoren für historisch alte Wälder der Norddeutschen Tiefebene. *NNA-Berichte* 7(3): 142-151.

Baskompte, J. & v. Solé, R. 1996. Habitat fragmentation and extinction thresholds in spatially explicit models. *Journal of Animal Ecology* 65: 465-473.

Betz, J.O. 1992. Studies on winter-active larvae of the ground beetle *Carabus problematicus (Coleoptera, Carabidae)*. *Pedobiologia* 36: 159-167

Brunzel, S. & Reich, M. 1996. Zur Metapopulationsstruktur des Roten Scheckenfalters (*Melitaea didyma* Esper 1779) auf der Schwäbischen Alb. *Z. Ökologie u. Naturschutz* 5: 243-253.

Den Boer, P.J. 1970. On the significance of dispersal power for populations of Carabid beetles (*Coleoptera,Carabidae*). *Oecologia* 4: 1-28.

Den Boer, P.J. 1977. Dispersal power and survival. Carabids in a cultivated countryside. *Missellaneous Papers L.H. Wageningen* 14: 1-190.

Doak, D.F. & Mills, L.S. 1994. A useful role for theory in conservation. *Ecology* 75(3): 615-626.

Glück, E. & Kreisel, A. 1986. Die Hecke als Lebensraum, Refugium und Vernetzungsstruktur und ihre Bedeutung für die Dispersion von Waldcarabiden. *Laufener Seminarbeiträge* 10/86: 64-83.

Gruschwitz, M. 1992. Zur Verbreitung und Ökologie von Laufkäfern der Gattung *Carabus* im Rheinland (*Coleoptera: Carabidae*). *Tier und Museum* 3(1): 12-19.

Gruttke, H. & Kornacker, P. 1995. The development of epigeic fauna in new hedges - a comparison of spatial and temporal trends. *Landscape and Urban Planning* 31: 217-231.

Gruttke, H. (in press): Impact of landscape changes on the ground beetle fauna (*Carabidae*) of an agricultural countryside. - *Proceed. of the Intern. Conf. on Habitat Fragmenation & Infrastructure*.

Hanski, I. 1994. A practical model of metapopulation dynamics. *J. Anim. Ecol.* 63: 151-162.

Hanski, I., Pakkala, T., Kuussaari, M. & Lei, G. 1995. Metapopulation persistence of endangerd butterfly in a fragmented landscape. *Oikos* 72: 21-28.

Harrison, S. 1991. Local extinction in a metapopulation context: an empirical evaluation. *Biol. J. of the Linnean Society* 42: 73-88.

Harrison, S., Stahl, A. & Doak, D. 1993. Spatial models and spotted owls: exploring some biological issues behind recent events. *Conservation Biology* 7: 950-953.

Lahaye, W.S., Gutiérrez, R.J. & Akçakaya, H.R. 1994. Spotted owl metapopulation dynamics in Southern California. *Journal of Animal Ecology* 63: 775-785.

Levins, R. 1970. Extinction. - In Gerstenhaber, M. (ed.): *Some mathematical problems in biology*: 77-107. Providence, R.I.: American Mathematical Society.

Liu, J., Dunning, J.B. jr. & Pulliam, H.R. 1995. Potential Effects of a Forest Management Plan on Bachman's Sparrows (*Aimophila aestivalis*): Linking a Spatially Explicit Model with GIS. *Conservation Biology* 9(1): 62-75.

Loreau, M. 1984. Population density and biomass of *Carabidae (Coleoptera)* in a forest community.*Pedobiologia* 27: 269-278.
Mossakowsky, D. 1996. Regionaler Biotopwechsel und Verbreitungsareal bei Arten der Gattung *Carabus* und *Cychrus* (Coleoptera: Carabidae). *Abh. Naturw. Verein Bremen* 43 (2): 609-615.
Neumann, U. 1971. Die Ausbreitungsfähigkeit von Carabiden in den forstlichen Rekultivierungen des Rheinischen Braunkohlenreviers. In P. J. Den Boer (ed.): Dispersal and Dispersal Power of Carabid Beetles. - *Miscellaneous papers* 8: 89-100. Wageningen: H. Veenman & Zonen N.V.
Pagel, M. & Payne R.J.H. 1996. How migration affects estimation of the extinction threshold.*Oikos* 76: 323-329.
Poethke, H.J., Gottschalk, E. & Seitz, A. 1996. Gefährdungsgradanalyse einer räumlich strukturierten Population der Westlichen Beißschrecke (*Platycleis albopunctata*): Ein Beispiel für den Einsatz des Metapopulationskonzeptes im Artenschutz. *Z. Ökologie u. Naturschutz* 5: 229-242.
Pulliam, H.R. 1988. Sources, sinks, and population regulation. *The American Naturalist* 132: 652-661.
Reich, M. & Grimm, V. 1996. Das Metapopulationskonzept in Ökologie und Naturschutz: Eine kritische Bestandsaufnahme. *Z. Ökologie u. Naturschutz* 5: 123-139.
Rijnsdorp, A.D. 1980. Pattern of movement in the dispersal from a dutch forest of *Carabus problematicus* Hbst. (*Coleoptera, Carabidae*). *Oecologia* 45: 274-281.
Shaffer, M.L. 1990. Population Viability Analysis. *Conservation Biology* 4(1): 39-40.
Veith, M. & Klein, M. 1996. Zur Anwendung des Metapopulationskonzeptes auf Amphibienpopulationen. - *Z. Ökologie u. Naturschutz* 5: 217-228.

Population and Community Ecology for Insect Management and Conservation, Baumgärtner et al. (eds)
 ISBN 90 5410 930 0

The age structure in a local population of *Carabus auronitens* under the influence of an abiotic key factor for reproduction

P. Hockmann, B. Horstmann, A. van den Boom, M. Landwehr, A. Bechtel, V. Kliewe, R. Baumgärtner, S. Mond, U. Fennemann, D. Kurz & F. Weber
University of Münster, Institute of General Zoology and Genetics, Germany

ABSTRACT: On an area of 0.2 ha in a forest of the Westphalian Lowlands abundance and net reproduction of the carabid beetle *Carabus auronitens* have been measured since 1982. In summer 1992 this area was surrounded by a beetle-proof fence, after which the interseasonal mortality and age structure of the population could be measured. Beetles were caught by pitfall trapping and were individually marked. *C. auronitens* is a spring breeder with summer larvae. The young are obligatorily active after hatching in late summer-autumn, during which season surviving old beetles are dormant. Net reproduction is calculated as the quotient "number of new beetles in the late summer-autumn season divided by number of females active in the spring of the same year". 73.58 % of the variability of net reproduction was found to depend on meteorological mean temperature in the 2nd ten-day period of May. The maximum net reproduction rate measured in 1993 was 3.11; in 1987 and 1996 net reproduction was near zero. The average probability of surviving from the season of hatching until the 1st spring season was around 75 % for females and males of the generations autumn 1992 - autumn 1994; for both females and males of the generation autumn 1995 the survival probability was considerably smaller (30.4 % / 17.5 %). The survival probabilities of the adults that have reached the first spring season were nearly invariable in the years of investigation (about 65 % from the 1st until the 2nd spring season; about 60 % from the 2nd until the 3rd spring; about 55 % from the 3rd until the 4th spring and about 40 % from the 4th until the 5th spring). The extreme dependence of reproduction on temperature in the 2nd ten days of May is evidently tolerable only because of these conspicuously high survival rates of the adults.

1 INTRODUCTION

Viable populations of a given species will be founded and maintained only in places (habitats) where particular requirements, essential for reproduction and survival of the individuals of this species, are fulfilled. Amongst the total of essential habitat factors, a more or less limited number are used as factors of habitat choice ("habitat selection") (Evans, 1983). These are "cues" by which individuals recognize places that may be suitable for reproduction and survival. Habitat cues comprise non-directional stimuli that are used by the individuals to remain within the boundaries of a suitable habitat, as well as directional stimuli, which guide migrating individuals. A cue factor is not necessarily an essential habitat factor; it can be an "auxiliary" factor signalizing a suitable habitat.

In a given habitat occupied by a given species some essential habitat factors will have a more or less constant influence on the parameters of the population (at least during the time of observation). The effects of other factors on rates of reproduction and mortality may change in time, triggering fluctuations of abundance that can, occasionally, be so great as to extinguish the population. Such environmental factors constitute a third category, called key factors (Morris, 1959). Key factors are powerful agents in selection, and the evolution of local adaptations that attenuate their negative influence on a population's abundance can be expected.

In carabids, much information has been gathered on esssential habitat factors as well as on factors of habitat choice. In many species illumination, parameters of the microclimate (temperature, humidity) and the soil (texture and chemistry) are essential habitat factors as well as non-directional habitat cues (as revealed by preference experiments: Lindroth, 1949; Thiele, 1977; Paje & Mossakowski, 1984). Dispersing individuals are attracted by directional cues, for example by volatile metabolites (Evans, 1984) or visual stimuli (Neumann, 1971). On the other hand, little is known in carabids about key factors influencing reproduction and mortality. In this article we show that in a Westphalian lowland population of *Carabus auronitens* a temperature factor influences the rate of net reproduction to such an extent that from time to time the population effectively does not reproduce. Thus, the viability of this population is guaranteed essentially by the longevity of the adults. Adult survivorship and iteroparous reproduction have been demonstrated for several carabid species and are interpreted as a strategy against reproductive failures (Murdoch, 1966; Schjøtz-Christensen, 1968; van Dijk, 1972, 1979 a; Thiele, 1977; Houston, 1981; Butterfield, 1986; Sota, 1987; Rusdea, 1992). Our data confirm this hypothesis.

The *C. auronitens* range predominantly embraces mountain and high mountain countries from South France to Poland and Romania (Breuning, 1932-1936); it is only at the northern border of the range that woods and forests in lowlands are occupied (Niehues et al., 1996). Life history phenomona of the *C. auronitens* populations in the Westphalian Lowlands have been described by Hemmer et al. (1986), Hockmann et al. (1989) and Althoff et al. (1992). *C. auronitens* is a spring breeder with summer larvae. Young beetles emerge in late summer-early autumn ("autumn season").

2 MATERIALS AND METHODS

2.1 *Site of investigation*

The site of investigation is a semi-natural forest near Münster in the Westphalian Lowlands ("Forst Tinnen"; 525 ha). The soil is a water-damming sandy loam (pseudogley). Natural vegetation is the *Querco-Carpinetum* in the form of a species-poor *Stellario-Carpinetum* (Burrichter, 1973). The forest contains a relatively high proportion of beech trees, the growth of which is enabled by drainage ditches. The ground is covered by a layer of rotten leaves and a layer of rough mould (each 3 - 5 cm thick). The present forest developed during the 19th century from a brushwood. *C. auronitens* has probably colonized the forest from a south and a northwest refugium shortly after 1900 (Niehues et al., 1996).

The climate of the Westphalian Lowlands is influenced by the Atlantic Ocean. Summers are frequently relatively cool with much precipitation, winters mild with little snow. However, occasionally summers are hot and dry and winters rather cold.

2.2 *Trapping and marking*

We have studied the population dynamics of *C. auronitens* on an area of 0.2 ha since 1982 (Althoff et al., 1992). In spring 1982 114 pitfall traps were set in this area, in autumn 1988 the number was increased to 152 traps, and in spring 1989 again to 171 traps (Ø of a trap: 9.5 cm). During the activity seasons of the nocturnal beetles the traps were usually opened twice a week overnight (in autumn season 1993 there were some gaps in the capture program). A piece of cork was placed in the traps as an "island" in case rainwater flowed in. From spring 1982 until spring 1992 the area was surrounded by a fence which was not beetle-proof but considerably reduced the exchange of individuals with neighbouring areas (Althoff et al., 1992). In summer 1992 the area was closed by a beetle-proof fence, the tightness of which is routinely monitored by trapping with 64 pitfalls along the outside of the fence (Niehues et al., 1996). Since spring 1982 beetles were individually marked, usually by polishing an individual pattern of small marks on the ribs of the elytra. In the autumn season of 1993 teneral beetles were marked at first capture by colour spots; they were marked individually by polishing after recapture in the hardened state in autumn 1993 or spring 1994. In autumn 1995 newly hatched beetles were marked individually by cauterization (Schjøtz-Christensen, 1965).

2.3 *Temperature*

For calculating the influence of temperature on reproduction we used the daily means of air temperature measured by the meteorological stations in Münster and nearby in Greven. Both are open-field stations.

2.4 *Estimation of abundances*

Abundances were estimated by the method of Jolly & Seber (Southwood, 1978). Using tables which contain the individual capture-recapture data (cf. Table 1) we calculated Jolly-Seber estimates for every catching day. Because of asynchronies of individual seasonality we used the data from two or three successive seasons for estimating the abundances in a given autumn season (A_i) and in a given spring season (S_i), respectively. For the spring season S_i usually S_{i-1}-A_{i-1}-S_i tables were constructed (Table 1 shows an excerpt of an S_{i-1}-A_{i-1}-S_i table); S_i-A_i-S_{i+1} tables were used for the first spring season (1982), and S_{i-1}-S_i-S_{i+1} tables for the spring season 1996, when many females were so much less active that they were not caught in this season (Baumgartner et al., in press). A_i-S_{i+1} tables were constructed for estimating the abundance in the autumn season A_i. During spring the single-day estimates were fairly constant from the middle of April until usually the middle of May; during autumn the single-day estimates were usually constant over several catching days at the end of the season. A "season estimate" ($\tilde{N}$) was calculated by averaging the single-day estimates for these "plateau phases" (the single-day abundances were weighted by their standard deviation; Althoff et al., 1992). As expected, in the beetle-proof enclosure the number of individuals caught in a spring season was equal or somewhat larger than the Jolly-Seber estimate for the "plateau-phase". Jolly-Seber estimates were calculated separately for females and males as well as for the pooled data of females and males ("total estimates"). In the open as well as in the beetle-proof enclosure the sum of the Jolly-Seber estimates for females and males was similar to the respective total Jolly-Seber estimate (Althoff et al., 1992).

2.5 *Calculation of net reproduction*

Net reproduction rate (B_i) is defined as the quotient "number of beetles hatched in autumn season i divided by number of females active in spring season i". It is probable that both parameters were to some extent mis-estimated for the population living in the open enclosure. (1) In the autumn seasons the females' Jolly-Seber estimates and the number of females caught exceeded the respective values for the males in the beetle-proof as well as in the open enclosure; in the beetle-proof enclosure the proportions of both the females' Jolly-Seber estimates and of the number of the females caught were also larger than 0.5 in the spring seasons; on the other hand, in the open enclosure the proportions of females according to both measures were mostly smaller than 0.5 in spring (Althoff et al., 1994). A mis-estimation in the open enclosure in spring was probably due to two effects: (i) some males that emigrated from the area returned later; (ii) in spite of a high number of traps and a dense catching program the capture probability of some females was rather low during spring (Hockmann, 1997). In order to correct the mis-estimation of the females' spring abundance in the open enclosure, we used the half of the total Jolly-Seber estimates for the spring seasons to calculate the net reproduction rate for the years the enclosure was open. (2) During autumn the individual season is rather short, and some young beetles were caught for the first time in the following spring season (see below). This means that the beetles' autumn abundance was to some extent underestimated in the years the enclosure was open. Despite these restrictions the Jolly-Seber estimates are consistent with each other insofar as the year-to-year differences of the total spring estimates ($\tilde{N}_{i+1}$ - $\tilde{N}_i$) are positively

Table 1. An excerpt from the trapping table of females caught in spring 1994, autumn season 1994 and spring 1995. Left column: decoded individual number (arranged in ascending order). The symbol "1" represents a single capture of that individual (point: no capture). Spring 1994: March 3 - July 7 (31 trapping days); autumn season 1994: July 12 - Oct. 26 (25 days); spring 1995: March 15 - July 4 (31 days). The total table (519 females) is an S_{i-1}-A_{i-1}-S_i table from which the parameters of the Jolly-Seber formula were evaluated for the catching days in spring 1995.

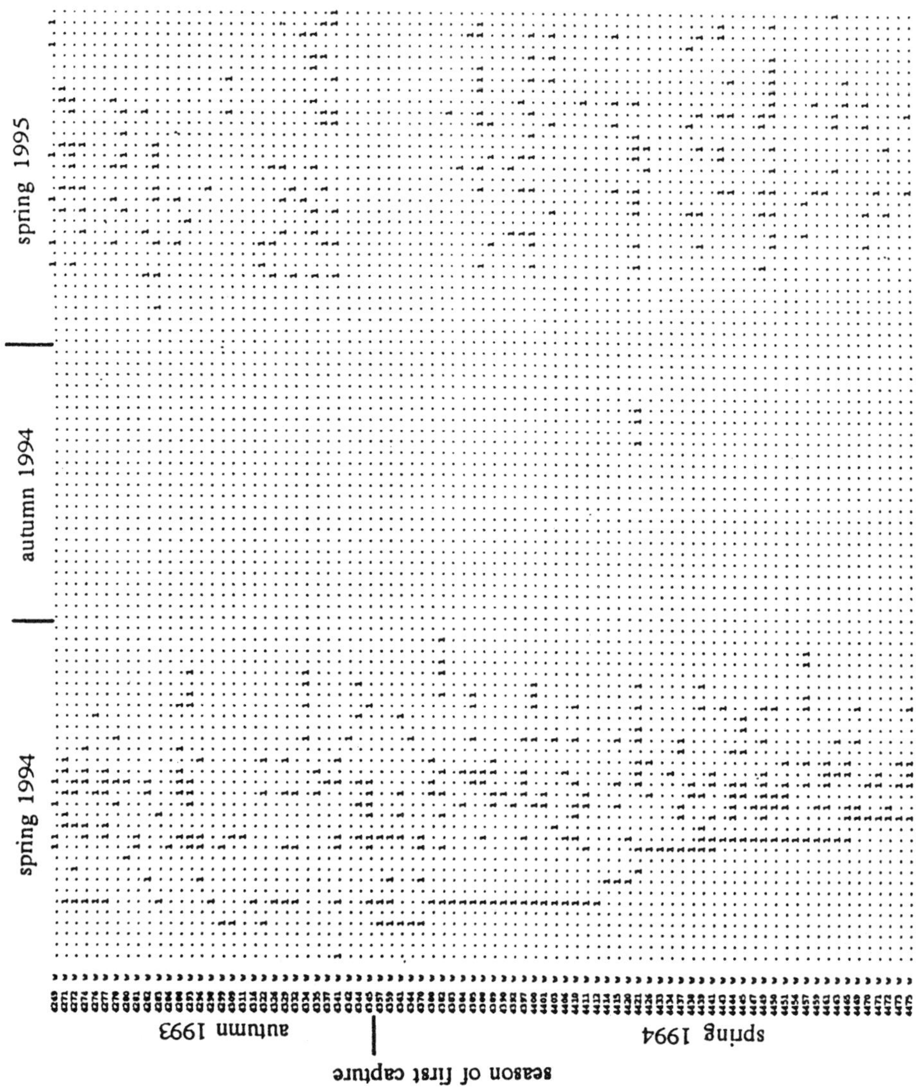

autumn 1994

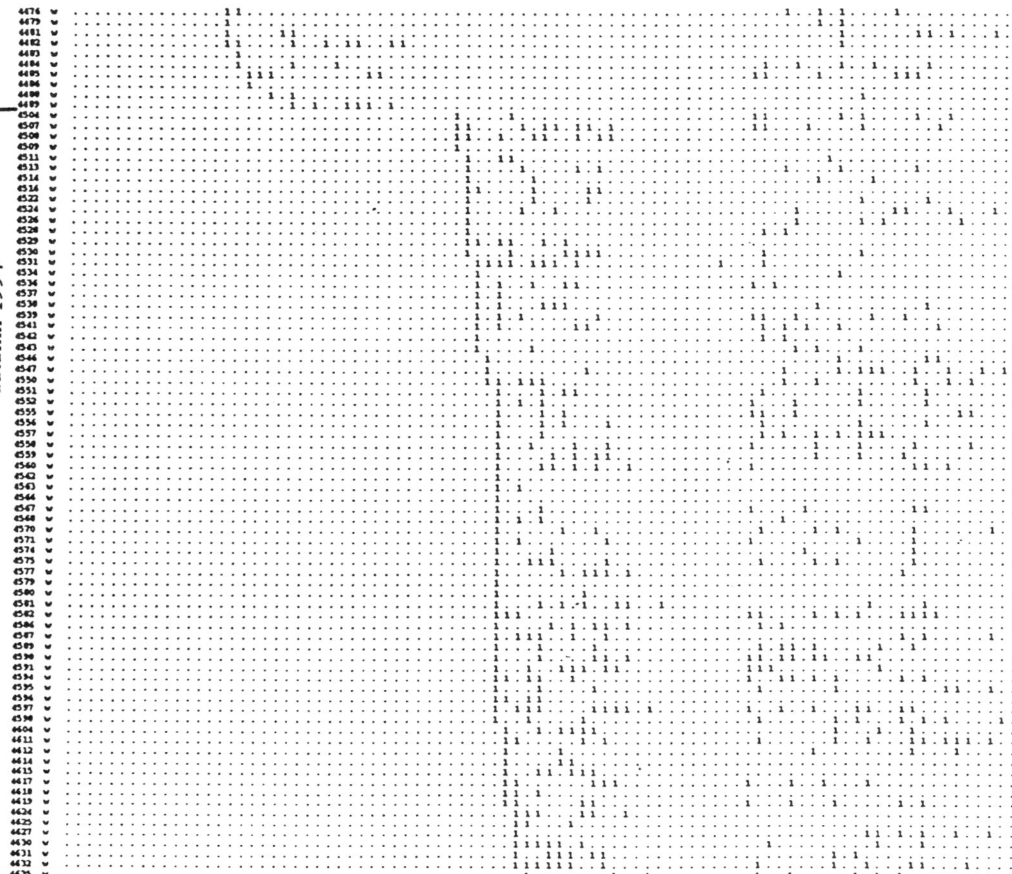

Table 1. spring 1994 | autumn 1994 | spring 1995

Table 2. Age structure of the enclosed *C. auronitens* population under observation in spring seasons 1996 and 1997. Area of the enclosure: 0.2 ha. For each generation the number of individuals caught in successive seasons is given. For spring seasons the percentages of surviving individuals are indicated. - A: autumn season, S: spring season.

	A '92	S '93	A '93	S '94	A '94	S '95	A '95	S '96	A '96	S '97	age structure S '97
females	147	113 (76.9%)		66 (58.4%)		29 (43.9%)		17 (58.6%)		7 (41.2%)	5th generation
males	112	85 (75.9%)		50 (58.8%)		22 (44.0%)		5 (22.7%)		3 (60.0%)	10 (3.8 %)
females		(13)[1]		(7) (53.8%)		(3) (42.9%)		(1) (33.3%)		0	
males		(15)[1]		(10) (66.7%)		(4) (40.0%)		(2) (50.0%)		0	
females			201	157 (78.1%)		105 (66.9%)		66 (62.9%)		45 (68.2%)	4th generation
males			163	124 (76.1%)		90 (72.6%)		44 (49.4%)		22 (50.0%)	95 (36.5%)
females				61[2]		36 (59.0%)		26 (72.2%)		17 (65.4%)	
males				60[2]		39 (65.0%)		20 (51.3%)		11 (55.0%)	
females					182	140 (76.9%)		96 (68.6%)		70 (72.9%)	3rd generation
males					135	102 (75.6%)		71 (69.6%)		47 (66.2%)	136 (52.3%)
females						21[2]		14 (66.7%)		11 (78.6%)	
males						22[2]		15 (68.2%)		8 (53.3%)	
females							46	14 (30.4%)		11 (78.6%)	2nd generation
males							40	7 (17.5%)		4 (57.1%)	16 (6.1%)
females								1[2]		1 (100%)	
males								0			
females									0		1st generation
males									0		3 (1.1%)
females										1[2]	
males										2[2]	
							Σ females:	235 (+3)[3]		163	
							Σ males:	164		97	

[1] Amongst these beetles there are possibly immigrants from the end of the spring season 1992; [2] probably hatched in the previous autumn season; [3] hatched before autumn 1992.

correlated with the autumn estimates $\tilde{N}_i$ (for all years: $r = 0.84$, $DF = 12$, $p < 0.001$; for 1982 - 1992: $r = 0.76$, $DF = 8$, $p < 0.05$).

Concretely, we used as parameters of the females' spring abundance: for spring 1982 until spring 1992, half of the total season estimates; for spring 1993 until spring 1996, the season estimates for females. As parameter of autumn abundances we used: for autumn 1982 until autumn 1991, the total season estimates for newly hatched beetles; for autumn 1992 until autumn 1996, the total number of newly hatched individuals (= number of unmarked individuals caught in autumn i and spring i+1 under consideration of winter mortality; as the area of investigation was closed in summer 1992, we considered half of the unmarked beetles found in spring 1993 as hatched during autumn 1992; cf. Table 2).

2.6 Statistics

In order to analyse the dependence of net reproduction rate B on abiotic parameters the coefficient r of the product-moment correlation and the coefficient R of multiple correlations were calculated (Sachs, 1991).

3 RESULTS

3.1 Seasonality

During the spring season (from March until June) adults of several generations are active (Table 2). Preimaginal stages develop during 3 - 4 months in the summer (Hemmer et al., 1986; Althoff et al., 1989). Young beetles emerge asynchronously in late summer-early autumn; they are active in foraging during 3-5 weeks (Table 1). Some of the young beetles were not caught during the season of hatching, but during the following spring season (especially when - as in autumn 1993 - there were gaps in the trapping program; cf. Table 2). Most surviving old beetles are dormant from the end of the spring season until the beginning of the next spring season (Table 1; 0 - 3 % of a spring stock was caught during the autumn seasons).

3.2. Adult longevity

After the beetle-proof fence had been erected in summer 1992, the interseasonal survival rate of the adults was measured. The average probability of surviving from the season of hatching until the 1st spring season was around 75 % for females and males of the generations A' 92, A' 93 and A' 94 (Table 2). For both females and males of the generation A' 95, however, the survival probability was rather small (30.4 % / 17.5 %). It might be suspected that the greater mortality was due to individual marking by cauterization in autumn 1995 (see above). However, this is improbable as the individuals hatched in autumn 1995 were unusual in some more respects: (i) they were on average of small body size and of relatively low weight; (ii) many freshly hatched beetles were found to be injured; (iii) even those beetles caught three times or more during the autumn season were recaught with relatively low probability in spring 1996 (only 10 of such 25 females and 3 of such 11 males were recaught).

On the other hand, the probabilities of surviving the spring seasons were nearly invariable: the average survival rate from the 1st until the 2nd spring season was about 65 %; from the 2nd until the 3rd spring about 60 %, from the 3rd until the 4th spring about 55 % and from 4th until 5th spring about 40 % (Table 2). Iteroparous reproduction seems to be common in species of the genus *Carabus*. However, little is known on adult survival probabilities. Schøtz-Christensen (1968) reported that about 25 % of the stock of a *Carabus hortensis* population survived from a reproduction season to the next one. In *Carabus yaconinus*, a spring breeder, the survival rate of new adults from emergence to the 1st breeding season was about 48 %, in *Leptocarabus kumagaii*, an autumn breeder, about 55 %. In *C.*

yaconicus about 8 % and in *L. kumagaii* about 20 % survived until the 2nd reproduction season. In the latter species a small proportion still lived in the 3rd season (Sota, 1987).

3.3. *Age pyramid*

In *C. auronitens* usually beetles of 4 - 5 generations seem to be active during a spring season (Table 2). In spring 1996 three females (0.7 % of the total population) that had hatched before autumn season 1992 were still active in the enclosure; 29 beetles (7.2 %) had hatched during autumn season 1992; 155 beetles (38.6 %) during autumn season 1993 and 196 beetles (48.9 %) during autumn season 1994. Remarkably, the generation A '95 was small in 1996: in spring 1995 reproduction had been relatively low (see below), and winter mortality of the young beetles had been rather high (Table 2). In spring 1997 the oldest generation was that from autumn 1992 (10 individuals: 3.8 %); 95 beetles (36.5 %) originated from autumn 1993 and 136 (52.3 %) from autumn 1994; as expected, the generation autumn 1995 was small (16 individuals: 6.1 %); generation autumn 1996 was negligible (probably 3 individuals: 1.1 %), as in spring 1996 reproduction had failed nearly completely (see below).

3.4. *Net reproduction rate*

Between spring 1982 and spring 1997 the total Jolly-Seber estimates of the number of beetles present during the "plateau phase of abundance" in spring fluctuated between 252 in 1984 and 576 in 1995 (Table 3). The number of newly hatched beetles in the autumn season varied much more widely (between less than 5 in 1996 and 523 in 1993). Obviously, net reproduction rate fluctuated dramatically in the years between 1982 and 1996. The maximum value measured in 1993 was 3.11; in 1987 and 1996 B was near zero (Table 4). There are several investigations on egg production in carabid species; however, net reproduction rates have seldom been measured. From data reported by Sota (1987) we calculated a value of about 4 for *C. yaconinus* and of about 3 for *L. kumagaii*.

It seems that the critical period for reproduction is rather restricted in the *C. auronitens* population under study: from 1982, the start of our investigation, until 1996, 100 x r^2 = 73.58 % of the variability of net reproduction rate was found to depend on meteorological mean temperature in the 2nd ten-day period of May (Table 4, Fig. 1). When in this period the temperature was ≤ 10.6°C (1987, 1991, 1995, 1996) net reproduction was small or even zero. Temperatures in the 1st ten days of May had little, but possibly a significant effect on net reproduction rate: 100 x R^2 = 79.65 % of the net reproduction rate can be explained when the mean temperature during the 1st ten days of May is considered as a second

Table 3. Abundances in the 0.2 ha enclosure. Spring 1982 - 1996 and autumn 1982 - 1991: Jolly-Seber estimates (in brackets: females) (weighted mean for the "plateau-phase" of the season). - Autumn 1992 - 1996: number of individuals (corrected by winter mortality, cf. MATERIALS AND METHODS).

year	'82	'83	'84	'85	'86	'87	'88	'89	'90	'91	'92	'93	'94	'95	'96
spring	421	375	252	256	315	403	271	316	532	452	307	290 (168)	501 (282)	576 (314)	381 (233)
autumn	289	84	166	267	300	< 20	223	370	325	41	277	523	373	89	<5

Table 4. Net reproduction rates (B_i) and meteorological mean temperature (°C) in the 1st, 2nd and 3rd ten days of May. Small deviations of B_i from the data reported in Niehues et al. (1996; Fig. 2) are due to a re-evaluation of spring and autumn abundances.

year	'82	'83	'84	'85	'86	'87	'88	'89	'90	'91	'92	'93	'94	'95	'96
1st (°C)	8.2	10.5	9.4	11.3	14.5	9.9	14.8	12.3	17.8	8.3	10.7	13.6	11.3	14.1	8.7
2nd (°C)	15.8	12.3	11.7	15.4	14.3	8.8	15.5	14.8	12.0	10.3	16.3	16.3	14.5	9.1	10.6
3rd (°C)	16.4	11.2	11.9	16.2	14.1	11.9	15.5	16.8	13.2	12.3	20.6	14.9	12.6	15.9	13.9
B_i	1.37	0.45	1.32	2.09	1.90	<0.10	1.65	2.34	1.22	0.18	1.80	3.11	1.32	0.28	<0.02

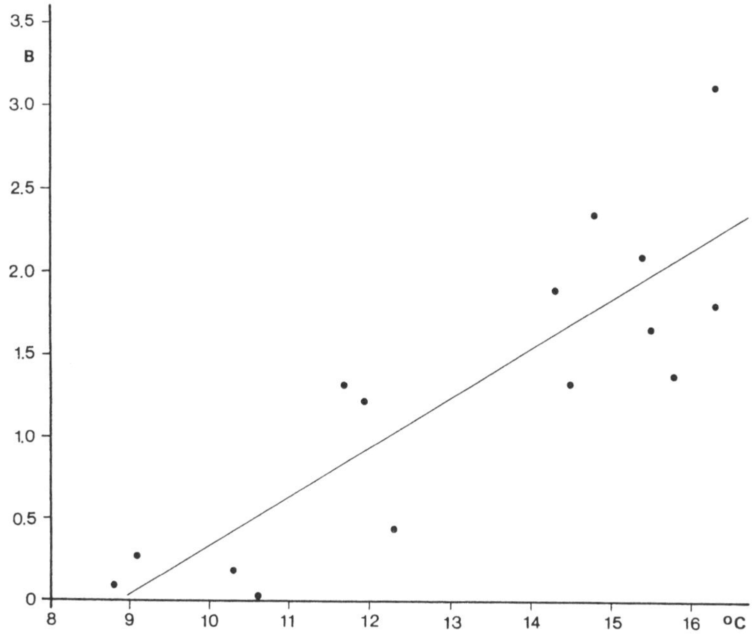

Figure 1. Dependence of net reproduction rate (B) on mean meteorological temperature (°C) in the 2nd ten days of May from 1982 - 1996. Product-moment correlation: $r = 0.86$; $p < 0.001$. Linear regression: $y = 0.30x - 2.66$.

independent variable ($p < 0.1$, F-test). Low temperatures in the 2nd ten days were not compensated by high temperatures in the 3rd ten days of May or at the beginning of June (considering the mean temperatures in the 2nd and 3rd ten days of May as independent variables, 100 x R^2 amounts to 73.64 %). The temperature in April also did not influence net reproduction rate.

In order to consider whether the net reproduction rates 1982 - 1992 were probably less accurately estimated than the rates 1993 - 1996 let us distinguish only three classes of net reproduction: namely the means (1) of the years 1987, 1991 and 1996; (2) of the years 1983 and 1995; (3) of all other years (Table 4). Then 68.25 % of the variability of net reproduction is still due to the variability of mean temperature in the 2nd ten days of May. The result is entirely the same if net reproduction in 1991 is averaged with the values of the years 1983 and 1995. This confirms that the high dependence of the net reproduction rate on this abiotic factor is not an artefact resulting from estimation errors.

4 DISCUSSION

4.1. Essential habitat factors and factors of habitat choice in C. auronitens

In *C. auronitens* temperature and humidity of the soil and adjacent air are essential habitat factors (Schmidt, 1955, 1956 a, b). Other environmental factors of this kind are unknown. The floor of the forests occupied by *C. auronitens* is usually covered by layers of rotten leaves and rough mould. These layers, however, do not seem to be essential factors: we observed *C. auronitens* reproducing, developing and surviving in enclosures erected on fresh fallow land with loamy soil (Niehues et al., 1996). By preference experiments it has been shown that in this species humidity and temperature are non-directional factors of habitat selection (Schmidt, 1956 b, 1957; Weber, 1965). *C. auronitens* uses visual stimuli as directional factors: specimens put out on fallow land orientated visually towards elevated horizon silhouettes (they were predominantly active during the daytime; Niehues et al.,

1996). Also, they seemed to be able to maintain constant directions of locomotion without local orientation stimuli ("directed movement" in the sense of Baars, 1984). Inside the forest, specimens were observed to orientate towards trees for climbing up (Hockmann et al., 1989).

4.2. Key factors in carabid population biology

In carabid species temperature and humidity have been found to be critical factors for reproduction and survival. Den Boer (1981) has observed marked fluctuations of abundance in local populations of the autumn breeder *Calathus melanocephalus* and the spring breeder *Pterostichus versicolor.* In *C. melanocephalus* the amount of winter precipitation seems to be a key factor for larval mortality (van Dijk & den Boer, 1992). Moreover, in this species low adult survival and recruitment were found to coincide with high temperatures during spring and the reproductive season (Baars & van Dijk, 1984). In *P. versicolor* also, temperature and rainfall influenced survival and reproduction; however, the effects depended on the structure of the habitat (Baars & van Dijk, 1984). In this species no eggs are produced below 8 - 9°C; moreover, eggs are only laid after a certain temperature-total has been reached (van Dijk, 1979 b). In *Carabus yaconinus* egg production is stopped at 25°C (Sota, 1986). In *Carabus creutzeri* the larvae do not hatch at temperatures over 24°C; moreover, in this species the embryonic development is extremely sensitive to dryness (Brandmayr & Zetto-Brandmayr, 1979).

4.3. A key factor in the population under study

From our observations we conclude that the meteorological temperature during the 2nd ten days of May determines a local (habitat-specific) temperature regime that operates as a key factor of reproduction in the *C. auronitens* population under study. As the survival rates of the adults that have reached the first spring season were found to be nearly invariable, the actual age structure of the population under study depends largely on the reproductive success in the previous years; thus, the age structure is predominantly influenced by the May temperatures.

In the population under study other factors influencing reproduction or mortality, apart from the temperature during May as key factor for reproduction, are unknown. For example, summer dryness seemed not to be a critical factor, because even in hot, dry summers the mould layer of the forest floor is kept humid by the loamy water-damming underground. Moreover, adults hibernating in rotten wood are insensitive to deep frost even when there is no snow cover (unpubl. observations). By feeding experiments analogous with those of Lenski (1984) we were able to show that during the spring seasons food for the adults seems not to be limited (beetles were kept in circular enclsosures {Ø 20 m}; the experiments were performed over 2 - 2.5 months during three spring seasons, twice with the maximum abundance observed up to now in nature, once with an abundance by 50 % higher; unpubl. results).

4.4. Problems and consequences of a the short critical period of reproduction in the population under study

Two questions arise from the observation that the mean temperature in the 2nd ten days of May is a key factor for net reproduction in the *C. auronitens* population under study: (1) how does cold weather within such a short period influence reproduction; (2) why do the beetles react so inflexibly to cold weather in the 2nd ten days of May? - Regarding 1: We know from dissections that in spring seasons with "normal" temperatures egg production was continued until June (Hemmer et al., 1986: Fig. 8); beetles hatching late (during October) probably developed from those eggs. Thus, it seems that cold weather did not merely interrupt egg ripening but terminated it. Moreover, we assume that the development of embryos and early larvae was negatively influenced by series of days with cold weather: the greatly reduced survival fitness of the adults of generation A' 95 could be the consequence of being exposed to pessimum low temperatures during early development. - Regarding 2: It could be that in the lowland *C. auronitens* population under study we are confronted with a regional trait, the origin and adaptive significance of which is still obscure (Baumgartner et al., in press.). Mountain and high mountain populations are very unlikely to react critically to low temperatures during May as has been observed in this lowland population. Adaptations of high mountain populations of *Carabus* species concerning the seasonality of reproduction and development have been described by Butterfield (1986) and Sota (1996).

Such an extreme dependence of reproduction on temperatures in the 2nd ten days of May is evidently tolerable by this lowland population only because of the conspicuously low, and nearly invariable, mortality rates of the adults that have reached their first reproductive season. After two spring seasons with little or even no reproduction, in spring 1997 the population was mainly composed of two generations, hatched in autumn 1993 and in autumn 1994. If in spring 1997 reproduction were to have failed once more, the population would still not be condemned to die out: the survivors (estimated as 73 females and 46 males on the area of 0.2 ha) would have a chance to reproduce. And even if this chance were also missed, in spring 1999 some old specimens will be active and could propagate the population (by then estimated as 22 females and 14 males). It may be speculated that the relatively low mortality rate of adults and the high dependence of reproduction on the temperature during a rather short critical period have evolved as mutually conditioning traits in the *C. auronitens* lowland population under study.

Acknowledgement. We thank the "Deutscher Wetterdienst Essen" for providing the meteorological data. F. W. wants to thank all his highly motivated students for productive teamwork.

REFERENCES

Althoff G.-H., M. Ewig, J. Hemmer, P. Hockmann, M. Klenner, F.-J. Niehues, R. Schulte & F. Weber 1992. Ergebnisse eines Zehn-Jahres-Zensus an einer *Carabus auronitens*-Subpopulation im Münsterland (Westf.). *Abhandl. Westf. Mus. Naturk.* 54 (4): 1-64.

Althoff G.-H., P. Hockmann, M. Klenner, F.-J. Niehues & F. Weber 1994. Dependence of running activity and net reproduction in *Carabus auronitens* on temperature. In K. Desender, M. Dufrene, M. Loreau, M.L. Luff & J.-P. Maelfait (eds.), *Carabid Beetles: Ecology and Evolution.*: 95-100. Dordrecht: Kluwer Academic Publishers.

Andrewartha H.G. & L.C. Birch 1954. *The distribution and abundance of animals.* Chicago: University of Chicago Press.

Baars, M.A. 1979. Patterns of movement of radioactive carabid beetles. *Oecologia* 44: 125-140.

Baars, M. & Th.S. van Dijk 1984. Population dynamics of two carabid beetles at a Dutch Heathland. I. Subpopulation fluctuations in relation to weather and dispersal. *J. Animal Ecol.* 53, 375-388.

Baumgartner R., A. Bechtel, A. van den Boom, P. Hockmann, B. Horstmann, V. Kliewe, M. Landwehr & F. Weber 1997. Age pyramid of a local population and viability fitness of phenotypical fractions in *Carabus auronitens* (Coleoptera, Carabidae). Italian J. Zool. in press

Boer, P. J. den 1981. On the survival of populations in a heterogeneous and variable environment. *Oecologia* 50, 39-53.

Brandmayr P. & T. Zetto-Brandmayr 1979. Contribution to ecology of an euryhypsic ground-beetle of eastern alps and dinaric karst, *Carabus creutzeri* Fabr. *Zool. Jb. Syst.* 106: 50-64 .

Breuning, S. 1932-1936. Monographie der Gattung carabus L. Troppau: Edmund Reitter`s Nachf. Emmerich Reitter.

Burrichter, E. 1973. Die potentielle natürliche Vegetation in der Westfälischen Bucht. *Landeskundliche Karten und Hefte der Geographischen Kommission für Westfalen. Reihe Siedlung und Landschaft in Westfalen* 8, 1-58.

Butterfield, J.E.L. 1986. Changes in the life-cycle strategies of *Carabus problematicus* over a range of altitudes in Northern England. *Ecol. Entomol.* 11: 17-26.

Dijk T. S. van 1972. The significance of the diversity in age composition of *Calathus melanocephalus* L. (Col., Carabidae) in space and time at Schiermonnikoog. *Oecologia* 10: 111-139.

Dijk T.S. van 1979 a. On the relationship between reproduction, age and survival in two carabid beetles: *Calathus melanocephalus* L. and *Pterostichus coerulesecens* L. (Coleoptera, Carabidae). *Oecologia* 40: 63-80.

Dijk T.S. van 1979 b. Reproduction of young and old females in two carabid beetles and the relationship between the number of eggs in the ovaries and the number of eggs laid. *Miscell. Pap. (Wageningen, Netherlands)* 18: 167-183.

Dijk, Th. S. van & P.J. den Boer: The life histories and population dynamics of two carabid species on a Dutch heathland. *Oecologia* 90, 340-352 (1992).

Evans, W.G. 1983. Habitat selection in the Carabidae. *The Coleopterists Bulletin* 37, 164-167.

Evans, W.G. 1984. Odor-mediated responses of *Bembidion obtusidens* (Coleoptera; Carabidae) in a wind tunnel. *Can. Ent.* 116: 1653-1658.

Hemmer J., Klenner M., Niesing H., Weber F., 1986 - Life history phenomena in a subpopulation of *Carabus auronitens* from the Westphalian Lowland. *In*: P.J. Den Boer, M.L. Luff, D. Mossakowski, F. Weber (eds.), Carabid Beetles - Their Adaptations and Dynamics. G. Fischer, Stuttgart, pp. 439-464.

Hockmann, P. 1997. Zur Problematik der exakten Messung populationsdynamischer Parameter: Analyse der Daten aus Freilanduntersuchungen an *Carabus auronitens* und Simulationsversuche auf der Grundlage dieser Daten. *PhD thesis at Department of Biology of the University of Münster.*

Hockmann P., P. Schlomberg, H. Wallin & F. Weber 1989. Bewegungsmuster und Orientierung des Laufkäfers *Carabus auronitens* in einem westfälischen Eichen-Hainbuchen-Wald (Radarbeobacchtungen und Rückfangexperimente). *Abhandl. Westf. Mus. Naturk.* 51(1): 1-71.

Houston, W.W.K. 1981. The life cycles and age of *Carabus glabratus* Paykull and *C. problematicus* Herbst (Col.: Carabidae) on moorland in northern England. Ecol. Entomol. 6: 263-271.

Lenski, R.E. 1984. Food limitation and competition: a field experiment with two *Carabus* species. J. Anim. Ecol. 53: 203-216.

Lindroth, C.H. 1949. Die Fennoskandischen Carabidae. Eine tiergeographische Studie. III. Allgemeiner Teil. *Göteborgs Kungl. Vetenskaps- och Vitterhets-Samhälles Handlingar.* S.F. Ser. B. Bd. 4, No. 3: 1--911.

Morris, R.F. 1959. Single-factor analysis in population dynamics. *Ecology* 40, 580-588.

Murdoch, W.W. 1996. Population stability and life history phenomena. *Am. Nat.* 100: 5-12.

Neumann, U. 1971. Die Sukzession der Bodenfauna (Carabidae {Coleoptera}, Diplopoda und Isopoda) in den forstlich rekultivierten Gebieten des Rheinischen Braunkohlenreviers. *Pedobiologia* 11: 193-226.

Niehues F.-J., P. Hockmann & F. Weber 1996. Genetics and dynamics of a *Carabus auronitens* metapopulation in the Westphalian Lowlands (Coleoptera, Carabidae). *Ann. Zool. Fennici* 33: 85-96.

Paje, F. & D. Mossakowski 1984. ph-prefences and habitat selection in carabid beetles. *Oecologia* 64: 41-46.

Rusdea E. 1992. Stabilisierende Selektion bei microphthalmen Höhlentieren: Untersuchungen zur tageszeitlichen Aktivitätsverteilung und Populationsdynamik von *Laemostenus schreibersi* (Küster) (Carabidae). *Mém. Biospéol.* XIX: 1-110.

Sachs L. 1992. *Angewandte Statistik.* 7. Aufl. 846 pp. Berlin: Springer.

Schjøtz-Christensen, B. 1965. Biology and population studies of carabidae of the *Corynephoretum.* Nat. Jut. 11: 1 - 173.

Schjøtz-Christensen, B. 1968. Some notes on the biology and ecology of *Carabus hortensis* L. (Col. Carabidae). *Nat. Jutl.* 14: 127-151.

Schmidt, G. 1955. Physiologische Untersuchungen zur Transpiration und zum Wassergehalt der Gattung *Carabus* (Ins. Coleopt.). *Zool. Jb. Physiol.* 65: 357-504.

Schmidt, G. 1956 a. Der Stoffwechsel der Caraben (Ins. Coleopt.) und seine Beziehungen zum Wassergehalt. *Zool. Jb. Physiol.* 66: 273-294.

Schmidt, G. 1956 b. Der Einfluß des Wasserhaushalts und Stoffwechsels auf die Vorzugstemperatur der Insekten. *Biol. Zentralbl.* 75: 178-205.

Schmidt, G. 1957. Die Bedeutung des Wasserhaushalts für das ökologische Verhalten der Caraben (Ins. Coleopt.). *Z. angew. Entomol.* 40: 390-399.

Sota, T. 1986. Effects of temperature and photoperiod on the larval development and gonad maturation of a carabid beetle, *Carabus yaconinus* (Coleoptera: Carabidae). *Appl. Ent. Zool.* 21: 89-94.

Sota, T. 1987. Mortality pattern and age structure in two carabid populations with different seasonal life cycles. *Res. Popul. Ecol.* 29: 237-254.

Sota, T. 1996. Altitudinal variation in life cycles of carabid beetles: life cycle strategy and colonization in alpine zones. Artic Alpine Res. 28: 441-447.

Southwood, T.R.E. 1978. *Ecological Methods with Particular Reference to the Study of Insect Populations.* 2nd ed. London: Chapman & Hall.

Thiele, H.-U. 1977. *Carabid Beetles in Their Environments.* 369 pp. Berlin: Springer.

Weber, F. 1965. Vergleichende Untersuchungen über das Verhalten von *Carabus*-Arten in Luftfeuchtigkeitsgefällen. Z. *Morph. Ökol. Tiere* 55: 233-249.

Population and Community Ecology for Insect Management and Conservation, Baumgärtner et al. (eds)
© 1998 Balkema, Rotterdam, ISBN 90 5410 930 0

Adult longevity – A factor stabilizing the abundance of a *Laemostenus schreibersi* population (Coleoptera, Carabidae)

Evelyn Rusdea
Institut für Landespflege, Universität Freiburg, Germany

ABSTRACT: A population of the microphthalmic carabid *Laemostenus schreibersi* (Küster 1846) was studied in a cave near Villach in Carinthia/Austria during 15 seasons of investigation between 1986 and 1995. The beetles were trapped in pitfalls, individually marked and released in the cave near the trapping site. The following aspects of population structure and dynamics are investigated: seasonal and spatial distribution of the beetles in the cave, patterns of movement between the different cave-compartments and between the cave and the surrounding fissure-system, seasonality of breeeding and development, reproduction rate, mortality, age-structure, sex-ratio and population size. Obviously, in some seasons all individuals living inside the explored cave-compartments were marked. The size of the population was small, but rather constant (50 - 100 individuals). Some beetles attained the remarkable age of 7.5 and >8 years. Longevity of the individuals is discussed as one factor stabilizing population size. Measurements of climatic parameters complemented the biological investigations.

1 INTRODUCTION

In the evolution to cavernicolous life, besides the morphological adaptations (for example depigmentation, reduction or even absence of eyes, elongated appendages), the adaptations of life history traits are of special interest. As caves represent rather stable environments, the question arises: how stable are cave populations in time, and - in this context - how large are the rates of reproduction and individual survival? Long adult life-spans of epigean carabid beetles as a cause for stability of population have already been discusssed by Van Dijk (1972), Den Boer (1979), (Althoff et al. 1992), Niehues et al. (1996) and others.

With respect to these questions, structure and dynamics of a cavernicolous population of the carabid beetle *Laemostenus schreibersi* Küster 1846 were studied in a natural habitat over a period of ten years. A part of the results of this long-term investigation have been already published (Rusdea, 1992, 1994, in press).

2 MATERIAL AND METHODS

Laemostenus (Antisphodrus) schreibersi (Coleoptera, Carabidae, Sphodrini) is a microphthalmic carabid beetle, considered as an "eucavernicolous" species (in the sense of Vigna-Taglianti 1982) - for a revision of the genus see Casale (1988). This carabid beetle inhabits caves and cave-like cavities (like mines for example), as well as superficial underground compartments and holes beneath deeply embedded stones (the "milieu souterrain superficiel"-habitat of Juberthie et al. 1980). The distribution area of this species covers the southern part of the Eastern Alps, from Northern Italy to Carinthia, Styria, Slovenia and Croatia.

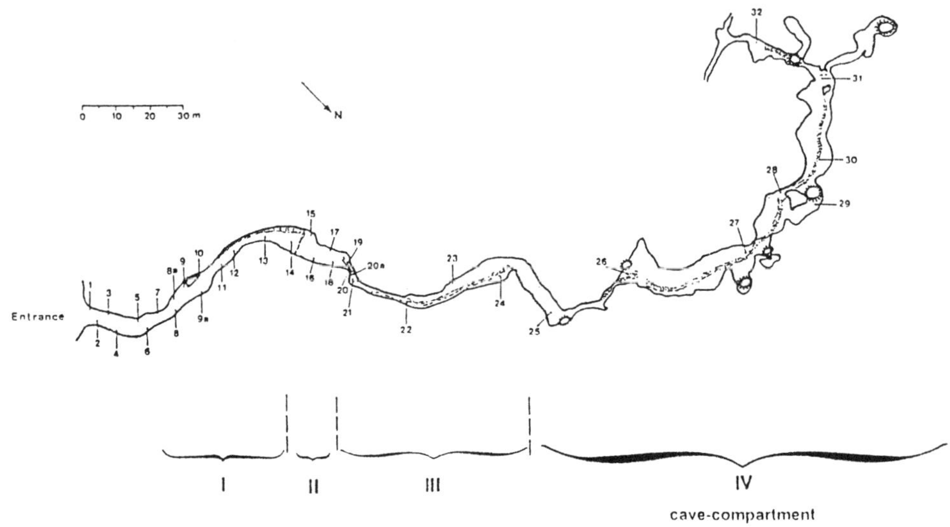

Figure 1: Map of the Eggerloch-cave with the positions of the pitfall-traps (compartments I - IV were distinguished).

The investigations were carried out in the cave called "Eggerloch" near Villach (Austria), located at 565 m above sea-level in the triassic carbonates of the Villacher Alpe-mountains. The total length of the whole cave-system is 650 m and the vertical height amounts to 119 m. According to the topography of the, we distinguish the cave mouth region and four cave-compartments (Figure 1). A number of 35 pitfall traps baited with meat were exposed along the cave gallery covering a distance of 331 m.

Table 1. Periods of field observations in the Eggerloch-cave.

Year	Study seasons	Number of days
1986	Spring (25.5.-12.7.)	49
	Autumn (24.9.-25.10.)	32
1987	Spring (6.4.-27.6.)	83
	Late-summer (3.8.-16.9.)	45
1988	Spring (30.5.-17.7.)	49
1990	Early-summer (30.6.-17.7.)	18
	Autumn (25.8.-13.9.)	20
1992	Spring (24.5.-10.6.)	18
	Autumn (5.9.-23.9.)	19
1993	Spring (23.5.-10.6.)	19
	Autumn (4.9.-20.9.)	17
1994	Spring (29.5.-16.6.)	19
	Autumn (20.8.-29.9.)	41
1995	Spring (1.6.-17.6.)	17
	Autumn (3.9.-19.9.)	17
Total	**15 seasons**	**463 days**

Field observations were carried out during 15 periods of investigation from 1986 to 1995 (Table 1). The first seasons of observation (1986 - 1988) lasted 1 - 2 months; beginning with 1990 each period of observation lasted for about 18 days. During the seasons of 1986 and 1987 the pitfalls were inspected twice a day (in the morning before sunrise and in the evening after sunset) and, beginning with 1988, every two days. The carabids were individually marked (outside the cave under binocular) with notches on the elytra by a small drill and released again inside the cave nearby the trap they were caught in. Measurements of the topoclimatic factors, which complemented the biological investiagations concern: air and soil temperature, relative humidity and cave-ventilation.

3 RESULTS AND DISCUSSION

Measurements of the climatic parameters inside the cave indicate average values in temperature of 8 - 9 °C and 97 % in relative humidity (Figure 2). Variations from outside are damped inside the cave up to the second cave-compartment. In a distance of 95 m from the entrance (beginning with compartment III) temperature is almost constant: 8.6 °C during summer-time and 8.0 °C in winter-time. The variations of relative humidity are damped within a shorter distance: at 20 m from the entrance humidity amounts already the value of 90 % in summer- and winter-seasons. Moreover, in a distance of 80 m from the entrance, the average values of relative humidity are constantly over 97 %. For a detailed description of the seasonality of topoclimate see Rusdea 1992.

Laemostenus schreibersi revealed a seasonality of locomotion activity. During winter months no animals were caught inside the cave. The first beetles were found in the pitfalls in spring at the end of May. They were present regularly during summer and autumn; at the end of October

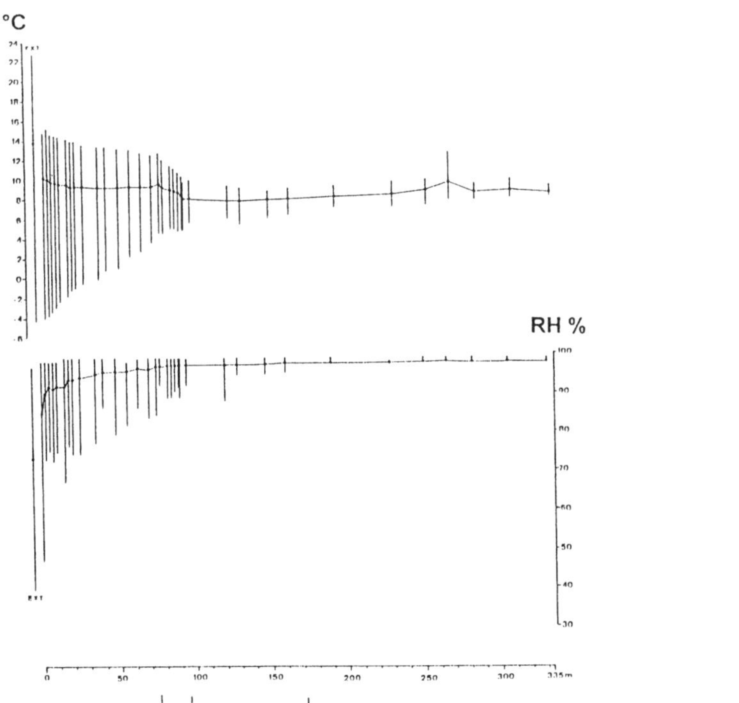

Figure 2: Average values and amplitudes of air-temperature (°C) and relative humidity (RH %) in the Eggerloch-cave (EXT = external values).

their number decreased. Then the beetles seem to leave the cave and to retire into the surrounding fissure-system. Obviously, this change is an adaptation to the seasonal changes of the topoclimatic factors: during winter-time, when cave-ventilation is ascending, temperature inside decreases and the beetles respond by retiring into the fissure-system. Beginning with spring/early-summer, when cave-ventilation is descending, the temperature increases and the beetles reappear inside the cave. Thus, appearance and disappearance of the beetles in the cave correspond very well to the changes of ventilation direction: in April/May and October/November.

Spatial distribution of *Laemostenus schreibersi* inside the cave is not uniform (Figure 3). In the cave mouth region (up to 20 m from the entrance) no specimens were caught (pitfalls 1-5). The beetles concentrate in the anterior cave-compartments I and II (20 - 90 m from the entrance). Less frequently they were found in the posterior compartments III and IV (95 - 331 m from the cave entrance).

In Figure 4 the detailed spatial distribution of *Laemostenus schreibersi* in the Eggerloch-cave is represented over the years of study. The number of captured beetles is shown for each pitfall and each season. The total number of captured beetles amounts to 1567 (including recaptures). The number of captures during the seasons of 1986 and 1987 was relatively high, because of the long investigation period of 1 - 2 months (30 - 50 days) comparing with the investigation-periods of about 18 days during the seasons from 1988 - 1995 (see Table 1). At the same time, the small number of beetles captured in autumn 1993 becomes evident. One possible explanation could be, that - because of the very scarce precipitations which occured during winter 1992/1993 and springtime 1993 - the cave was rather dry. But the captures from 1994 (spring and autumn) demonstrate, that the population recovered to its usual size.

During the investigation period from 1986 - 1995 a total of 597 individuals were marked. The number of individuals captured in each period of observation (without the recaptures) is presented in Figure 5. Their number was rather constant: population size varies between 50 and 110 individuals (with the exception of autumn 1993 - already mentioned). At the end of the spring/early-summer seasons 1986 and 1987 all beetles

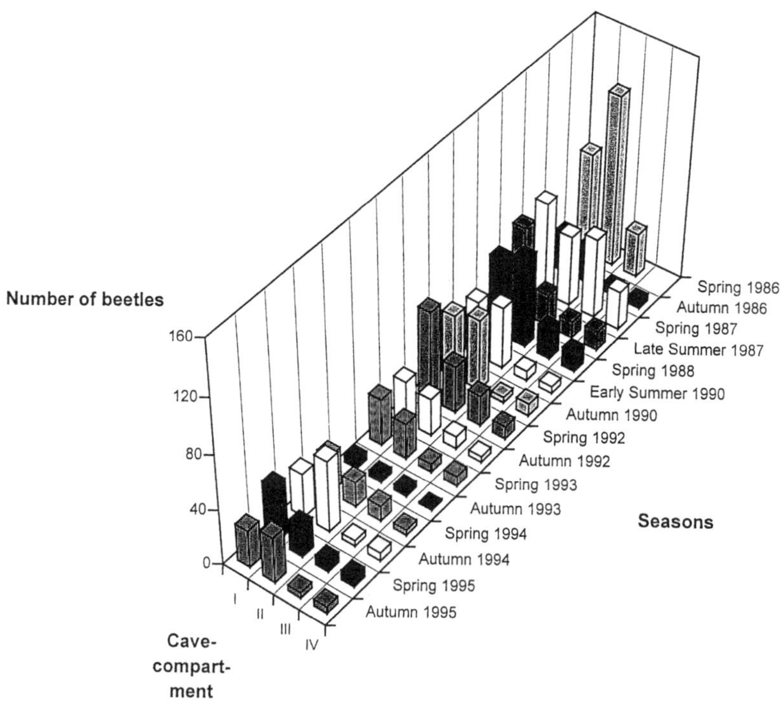

Figure 3: Captures of *Laemostenus schreibersi* in the different cave-compartments (recaptures included).

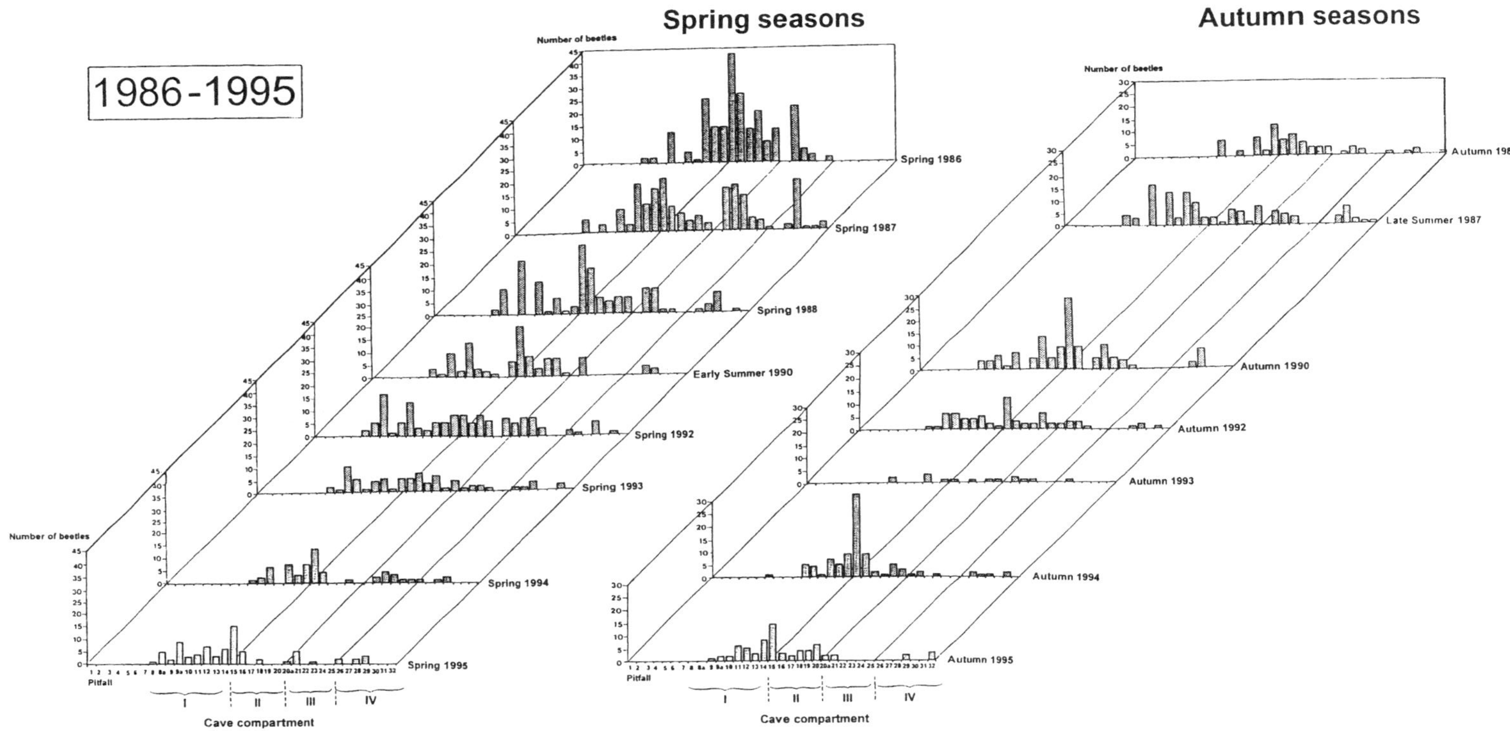

Figure 4: Spatial distribution of the captures (including recaptures) for each pitfall-trap throughout the series of investigated seasons.

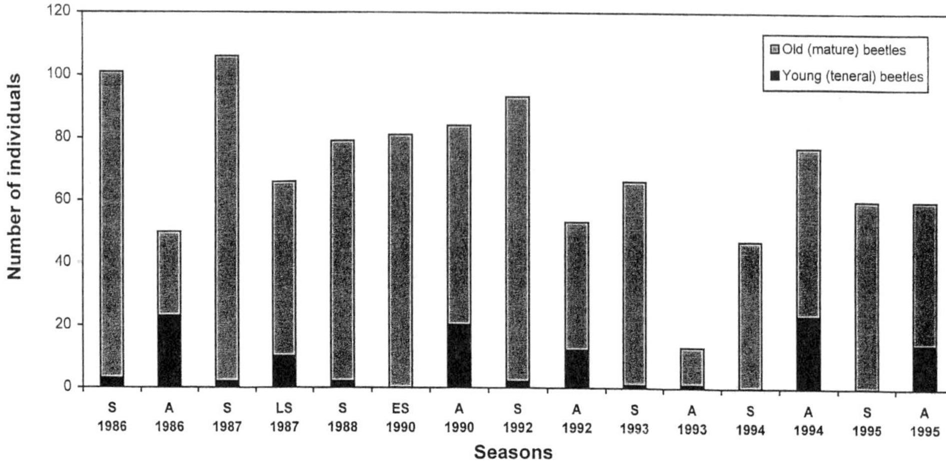

Figure 5: Population size (= number of individuals, recaptures excluded) caught per season and distinguished in teneral and mature beetles. S = Spring, ES = Early-Summer, LS = Late-Summer, A = Autumn.

living in the explored cave-compartments had been marked: new (unmarked) beetles were not found any more. Obviously, in these seasons the number of captured individuals corresponded to the size of the population living at that time in the explored parts of the Eggerloch-cave. In the late-summer/autumn seasons this "saturation" level was not attained: a few unmarked beetles were still captured at the end of these seasons. The population size was estimated using the Jolly-Seber-model for each season. Population size seems to be relatively small, however rather constant throughout the years: 100-300 individuals. For the spring-seasons of 1986 and 1987 (when all beetles living in the cave at that time were captured – that means no unmarked carabids were registered), the estimated data fit very well with the observed number of individuals (Rusdea 1992).

For each beetle the distance covered between two consecutive captures was calculated. The data demonstrate the high mobility of this carabid beetle, which is able to cover distances of 40 m and more during 10 hours (not shown, cf. Rusdea 1992). Concerning the captures and recaptures within a study-season, most of the beetles were recaptured in the same pitfall or in the pitfalls nearby. However the beetles of compartment IV did not migrate into the other compartments. Therefore, one might consider these beetles as a separated population. But regarding the positions of captures and recaptures from one season to the next it can be confirmed that the carabids from compartment IV migrated into the other cave-compartments and vice versa. Thus, we can conclude, that the adult individuals of *Laemostenus schreibersi* living in the "Eggerloch" belong to the same (meta)population (Rusdea 1994).

From the total number of 1567 captures 62 % represent recaptures. The probability of staying in the explored cave-area was rather high. 63 % of the beetles were captured more than once, 38 % more than twice; one beetle was caught for 18 times. Figure 6 represents the distribution of new-captured and recaptured beetles per season. The white sectors figure the recaptured beetles from previous seasons. Their portion varied between 28 % and 77 % (on the average 54 %). At the beginning of the investigation in spring 1986, of course, there are no recaptures and in the spring seasons 1990 and 1992 their portion was rather small, because the investigations were interrupted in 1989 and 1991. The black segments show the portion of teneral first-captured beetles, which were observed more often in the late-summer/autumn seasons than in spring/early-summer seasons. In the spotted segments the first-captured old animals (mature beetles) are figured. They represent the rate of exchange with other groups of *Laemostenus schreibersi* of the fissure-system inside the massif. In the spring/early-summer seasons their portion varied between 38 % and 60 % - on the average 45 % (taking on account the seasons with continuous sampling only). These "newcomers" represent the rate of exchange of mature beetles with other groups, which obviously occurred during winter time, when the beetles retire into the fissure-system. During summer and autumn-seasons, when the carabids were regularly captured in the cave, this exchange was lower (on the average 23 %).

The existence of more compartments nearby, inhabited by the beetles but inaccessible for man, is also proven by the fact that some beetles were recaught after they had been missing inside the cave during one or more seasons. There are different patterns of appearance to be observed (Table 2). Some individuals usually "skip" one or two seasons; two beetles were missed for 7 consecutive seasons. It is of high interest, that *Laemostenus schreibersi* survived for long periods of time.

On the basis of captures and recaptures during the series of investigation seasons, it was possible to estimate the age of the beetles (Figure 7). The individuals were distributed into a fictive "age-pyramid" according to their age at last capture (state: autumn 1995). The pyramid is divided in two parts. On the left side there are represented the beetles with "definite age": they were first captured as young tenerals and therefore their age can be exactly determined (as tenerals their age does not exceed 0.5 years, because postecdysial maturation lasts no longer). As soon as they were recaptured in a later season, they pass to the next age-categories. The age of beetles captured for the first time as hardened mature beetles (right part of the pyramid) can not be determined exactly, only estimated: they already have the minimum age of > 0.5 year. Remarkably, some beetles attained the age of 7-8 years. A survival time of several years has been reported for some carabid species (see 1 Introduction), however such an extreme longevity has not been reported from a beetle before.

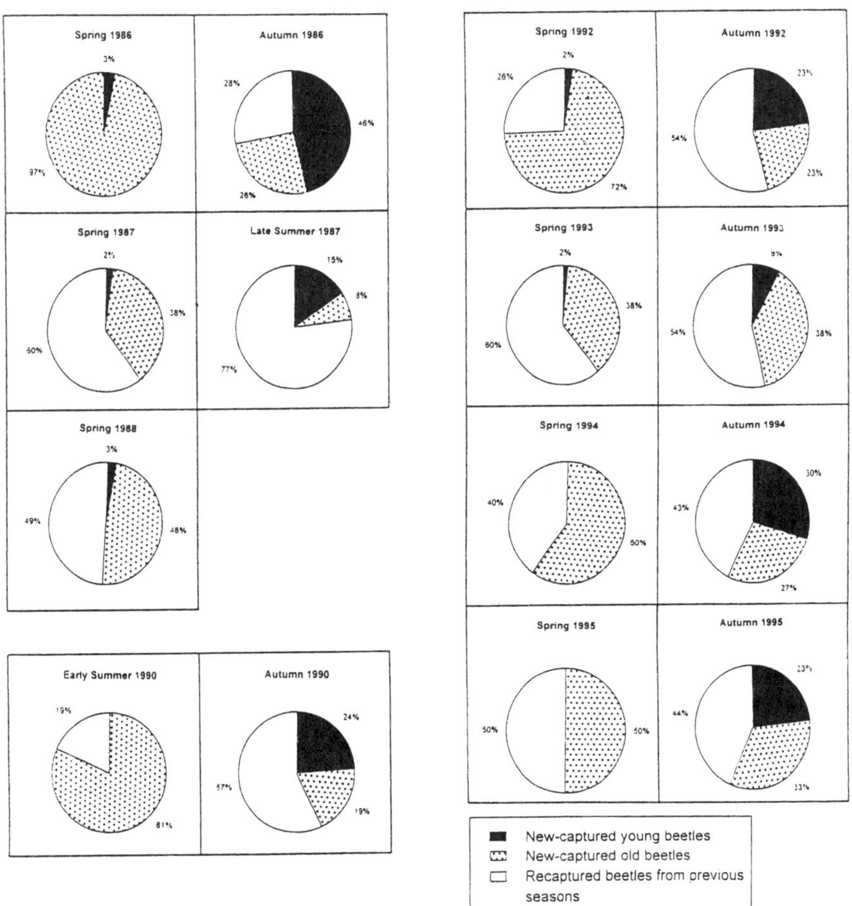

Figure 6: Distribution of new-captured and recaptured beetles in the Eggerloch cave.

Table 2: Patterns of appearance of *Laemostenus schreibersi* over the years of investigation (■ = the beetle was captured at least once during the season; □ = the beetle was not captured in that season). The sequence of boxes corresponds to the succession of the seasons under study; S = Spring, ES = Early-Summer, LS = Late-Summer, A = Autumn.

Example: in late-summer 1987 from the total number of 15 marked individuals, 13 were captured only in that season, 1 beetle was recaptured two seasons later (that means in early-summer 1990) and 1 beetle 3 seasons later (in autumn 1990 - consequently).

Patterns of appearance		S 1986	A 1986	S 1987	LS 1987	S 1988	ES 1990	A 1990	S 1992	A 1992	S 1993	A 1993	S 1994	A 1994	S 1995	A 1995	Total
1 season	■	45	13	12	13	29	28	22	34	11	11	1	10	26	25	34	314
2 seasons	■■	11	8	14	-	3	22	4	7	7	1	1	6	8	5		
	■□■	12		8	1	2	1	1	8	-	2	2	4	8			
	■□□■	-	1	1	1	-	-	1	-	1	2	1	-				
	■□□□■	1	-	-	-	-	-	-	4	-	2	-					
	■□□□□■	-	-	1	-	-	-	-	-	-	1						
	■□□□□□■	-	-	-	-	-	1	-	-	-							
	■□□□□□□□■	-	-	-	-	-	1	-									170
3 seasons	■■■	1	2	4	-	4	7	3	6	-	-	-	2	2			
	■□■■	13	1	1	-	-	-	-	-	-	2	-	-				
	■■□■	-	3	-	-	-	-	3	-	1	1	-	2				
	■□■□■	6	-	-	-	-	1	-	1	-	-	-					
	■□□■■	1	-	-	-	-	2	-	-	-	-	1					
	■■□□■	-	-	-	-	1	-	-	-	-	-	-					
	■□□□■■	-	-	-	-	-	-	-	-	1	-						
	■■□□□■	-	-	-	-	-	-	-	-	1	-						
	■□□□■□■	-	-	-	-	-	-	-	-	1							
	■□□□□■■	-	-	-	-	-	-	-	2	-							
	■□□□□□■□■	-	-	-	-	-	-	1									77
4 seasons	■■■■	1	5	-	-	-	1	1	-	-	-	-	2				
	■□■■■	5	-	-	-	-	-	-	-	-	-	-					
	■■□■■	-	-	1	-	-	1	-	-	-	-	-					
	■■■□■	1	-	-	-	-	-	-	2	-	-	-					
	■□■□■■	-	-	-	-	-	-	-	1	1	-						
	■□■■□■	-	-	-	-	-	-	-	-	-	1						
	■■□■□■	-	1	-	-	-	-	-	-	-	-						
	■■■□□■	-	1	-	-	-	-	-	-	-	-						
	■□■■□□■	2	-	-	-	-	-	-	-	-							
	■■■□□□□□□■	-	-	-	-	-	1										
	■■■□□□□□□□■		-	-	-	1											29
5 seasons	■■■□■■	-	-	-	-	-	-	-	2	-	-						
	■■□■□■■	-	-	-	-	-	-	-	1	-							
	■■■□□■■	-	-	-	-	-	-	-	1	-							
	■□■■□■■	1	-	-	-	-	-	-	-	-							
	■□■□□■■■	1	-	-	-	-	-	-	-								
	■■□■□■□□■	-	1	-	-	-	-	-									7
Number of beetles		101	36	42	15	40	66	36	69	24	26	6	28	44	30	34	597

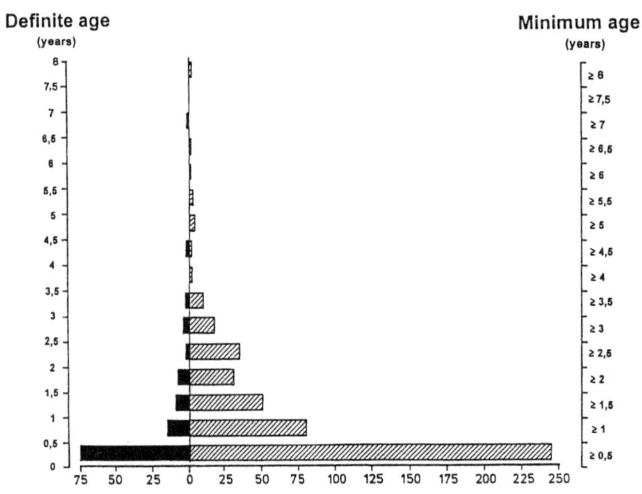

Figure 7: Distribution of maximum-age for the total number of marked individuals of *Laemostenus schreibersi* (cf. text).

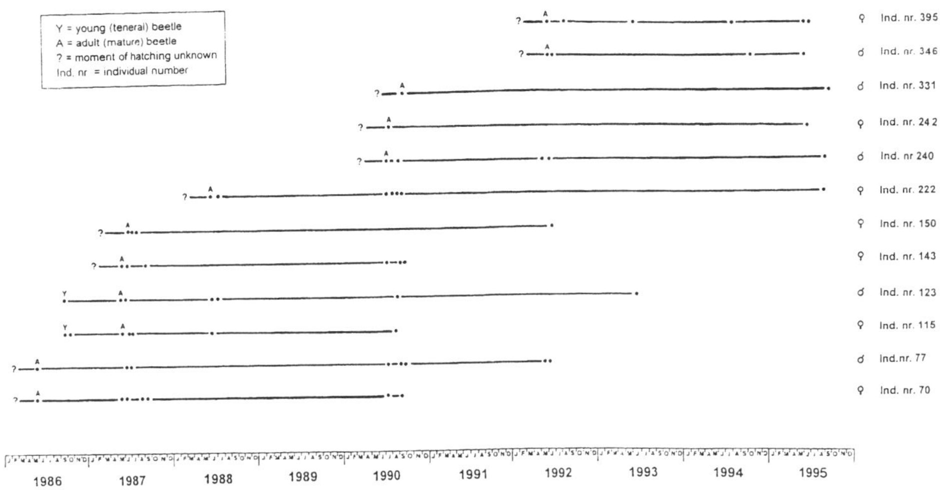

Figure 8: Examples of longevity for some individuals of *Laemostenus schreibersi* in the Eggerloch-cave.

Thus, life-span of *Laemostenus schreibersi* in its natural habitat can be very large. Figure 8 represents some examples of individual longevity. Beetle nr. 123 was first captured as teneral in autumn 1986 and recaptured until spring 1993. Thus, the age of this individual was 6.5 - 7 years at last capture. Beetle nr. 222 was first captured as mature beetle in summer 1988 and then recaptured in autumn 1990 and autumn 1995. Consequently this individual was more than 8 years old!

Concerning the age structure, not only adults, but also larvae were attracted by the pitfalls. From the Eggerloch-cave larvae of *Laemostenus schreibersi* were reported to occur in April and from July to November (Mais & Schmid, 1963). During my investigations larvae were captured only in late-summer/autumn: during August, September and October. Teneral beetles (which had not completed postecdysial ripening) were not equally distributed over the year (Figure 5). In the spring/early-summer seasons their portion was only 1% - 3%; during late-summer/autumn however, their portion varied between 15 % and 46 % (on the average 24 %). From laboratory breeding we know that hardening and colouring of the cuticle of *Laemostenus schreibersi* takes 3 - 4 months. Inside the cave, animals with soft and light-brown coloured elytra were observed from the beginning of August. From the seasonal occurrence of larvae and teneral adults, it can be concluded, that the population of *Laemostenus schreibersi* in the Eggerloch-cave exhibits a seasonality of reproduction and development. Brandmayr & Zetto Brandmayr (1988) reported that in the Dolomites and Carnic Prealps, populations of *Laemostenus schreibersi* seem to reproduce aperiodically. A seasonality of egg-production and development was also observed on the cavernicolous carabid *Typhlotrechus bilimeki tergestinus* living in the Triestine Karst (Colombetta et al. 1989). In the Grotta delle Torri di Slivia – the locus classicus of this beetle – it reproduces only during the summer, at temperatures of about 8°C. It is possible, that periodical microclimatic changes in this cave operate as an exogenous factor, which determines the seasonality of reproduction. The same could be true for the *Laemostenus schreibersi* populations of the Eggerloch and of the Dolomites and Carnic Prealps. After Casale (1988) all species of the genus *Laemostenus* are spring-breeders with the larval development during the summer.

Sex-ratio demonstrates a predominance of the females (59 % females comparing to 41 % males) - concerning the number of captures (including recaptures) as well as the number of new-marked beetles (Figure 9). The probability of being recaptured was identical for males and females; thus, it seems that sexes did not differ with respect to activity behaviour. From the number of females during the spring-season and the number of recent hatched tenerals in the autumn season of the same year, the individual reproduction rate could be calculated. This reproduction rate is very low: on the average 0.39. That means, that the rather constant population size throughout the years of investigation was mainly due to the low adult mortality. The survival of this population is mainly guaranteed by adult longevity.

4 CONCLUSIONS

The population of *Laemostenus schreibersi* living in the Eggerloch cave is part of a larger metapopulation, which comprises also groups living in the inaccessible compartments of the fissure-system of the massif. This population shows several characteristics of a K-strategy of survival. The size of thc population is rather constant. Rates of reproduction and adult mortality are low. Sex ratio is shifted in favour to the females. Individual longevity could be a factor which stabilises the population size in the "Eggerloch" and also the size of the larger metapopulation.

REFERENCES

Althoff, G.-H., Ewig, M., Hemmer, J., Hockmann, P., Klenner, M., Niehues, F.-J., Schulte, R. & F. Weber 1992. Ergebnisse eines Zehn-Jahres-Zensus an einer *Carabus auronitens*-Subpopulation im Münsterland (Westf.). *Abh. Landesm. Naturkde. Münster*, 54 (4): 1-64.

Brandmayr, P. & T. Zetto Brandmayr 1988. Communitá a coleotteri carabidi delle Dolomiti Sudorientali e delle Prealpi Carniche. *St. trentini Sc. Nat.*, Acta biologica 64: 125-250.

Casale, A. 1988. *Revisione degli Sphodrina (Coleoptera, Carabidae, Sphodrini).* Museo Reg. Sc. Nat., Monografie V., Torino, 1024 p.

Colombetta, G., Solda, R., Zetto, T. & P. Brandmayr 1989. Ritmo riproduttivo e habitat di un coleottero troglobio nella Grotta delle Torri di Slivia *Typhlotrechus bilimeki tergestinus* Müller (Coleoptera, Carabidae, Trechinae). *Atti Comm. "E. Boegan"*, Trieste, 28: 51-81.

Den Boer, P. J. 1979. The individual behaviour and population dynamics of some carabid beetles of forests. In: P.J. den Boer, H.U. Thiele & F. Weber (eds.*), On the evolution oj behaviour in carabid beetles. Miscellaneous Papers*, Landbouw Hogeschool Wageningen, 18: 151-166.

Juberthie, C., Delay, B. & M. Bouillion 1980. Extension du milieu souterrain en zone non-calcaire: description d'un nouveau milieu et de son peuplement par les coleoptères troglobies. *Mém. Biospéol.,* Moulis, VII: 19-52.

Mais, K. & M. Schmid 1963. Faunistische Untersuchungen im Eggerloch bei Warmbad Villach und in der Dachstein-Mammuthöhle. *Höhlenkundl. Mitt.*, Wien, 19 (4): 44.

Niehues, F.-J., Hockmann, P. & F. Weber 1996. Genetics and dynamics of a *Carabus auronitens* metapopulation in the Westphalian lowlands (Coleoptera, Carabidea). *Ann. Zool. Fennici*, 33: 85-96.

Rusdea, E. 1992. Stabilisierende Selektion bei microphthalmen Höhlentieren: Untersuchungen zur tageszeitlichen Aktivitätsverteilung und Populationsdynamik von *Laemostenus schreibersi* (Küster) (Carabidae). *Mém. Biospéol.,*Moulis, XIX: 1-110.

Rusdea, E. 1994. Population dynamics of *Laemostenus schreibersi* (Carabidae) in a cave in Carinthia (Austria). In: K. Desender et al. (eds.): *Carabid beetles: ecology and evolution*: 207-212, Dordrecht.: Kluwer Academic Publishers

Rusdea, E. (in press). Stabilization of a *Laemostenus schreibersi* population (Carabidae) as an effect of adult longevity. *Mém. Biospéol.*, Moulis, XXIV.

Van Dijk, Th. S. 1972. The significance of the diversity in age composition of *Calathus melanocephalus* L. (Col., Carabidae) in space and time at Schiermonnikoog. *Oecologia* (Berlin), 10: 111-136.

Vigna-Taglianti, A. 1982. Le attuali conoscenze sui Coleotteri Carabidi cavernicoli italiani. *Lav. Soc. ital. Biogeogr.* (n. s.), 7: 339-430.

2 Population interactions

Population and Community Ecology for Insect Management and Conservation, Baumgärtner et al. (eds)
© 1998 Balkema, Rotterdam, ISBN 90 5410 930 0

The population dynamic role of parasitoids: Some inroads to a synthesis

Michael E. Hochberg
Institut d'Ecologie, Paris, France

ABSTRACT: Insect parasitoids are often employed as theoretical models to better understand what drives population dynamics and the factors responsible for success in biological control. This is, in part, because parasitoids have more restricted host ranges than other classes of natural enemies, leading to putative 'tightly-coupled' interactions with their host(s). This paper presents how a number of recent studies support an integrated view of the role of monophagous parasitoids in host population dynamics and biological control.

1 INTRODUCTION

Over the past three decades considerable theoretical work has been dedicated to the mechanisms responsible for the persistence of associations between monophagous insect parasitoids and their herbivorous insect hosts. Perhaps the most persuasive reason for this focussed attention is that parasitoids are the most frequently used taxa in classical biological control programs, indicating that they are capable of limiting the population densities of their hosts. In more natural, undisturbed situations however, the importance of parasitoids in driving host dynamics remains equivocal.

If these cursory observations do indeed reflect real differences between communities created by the intentional introduction of a parasitoid and more natural ones, which are honed by ecological and evolutionary forces, then two interesting questions follow. When parasitoids are important limitors of host densities, are they indeed regulating their hosts in time and space? May it be that strong top-down host limitation is evanescent in evolutionary time, and that monophagous parasitoid-host systems inevitably evolve towards bottom-up control?

The purpose of this paper is three-fold. First, I present the basic elements of an integrated conceptual model which is emerging of the role of monophagous parasitoid in biological control programmes. Second, recent studies are discussed which address important features of this model. And third, I present possible ways forward in the empirical and experimental testing of the model.

2 TOWARDS A UNIFIED CONCEPT OF THE IMPACT OF MONOPHAGOUS PARASITOIDS

Virtually all theoretical studies on monophagous parasitoid-host dynamics model in one way or another unequal exploitation of hosts by the parasitoid. Many have considered, sometimes in considerable detail, how mechanisms generating unequal host exploitation can affect host population dynamics. Depending on the assumptions of the model and the particular mechanism investigated, dynamics may either be stabilised or destabilised by parasitoids, or other forms of host density dependence may be brought into dynamic play (review in Taylor 1993).

A recent survey of host-parasitoid models (Hochberg & Holt, submitted) showed that two major notions determine the impact of monophagous parasitoids on their host populations. These are the *uniformity* and *density* of host exploitation by the parasitoid.

Exploitation uniformity is the host to host variation in risk of parasitism. The distribution of exploitation is influenced by a number of factors including space, phenology, behaviour, morphology, physiology, or associations with plants. Tendencies away from uniformity have been referred to in the literature as 'heterogeneity in parasitism' and 'refuges from parasitism'.

Exploitation density is the power of population increase of the parasitoid; that is, the potential number of hosts an average fecund parasitoid female parasitizes when host numbers are not limiting. Exploitation density will be a function of a number of parasitoid life-history and demographic parameters, including egg load and maturation rate, life-expectancy, and sex ratio. Mathematical models have shown that sufficient exploitation density is a necessary condition for the distribution of exploitation to be operational in the dynamics of host-monophagous parasitoid systems.

Figure 1 represents how hypothetical combinations of exploitation uniformity and density lead to dynamic outcomes. Under a range of circumstances certain members of the community surrounding the tightly-coupled host-parasitoid association will impinge on host dynamics. Consider two main possibilities.

First, when the impact of the parasitoid is sufficiently extreme (combining high density and uniformity of exploitation), it will destablize the dynamics (Area A, Fig. 1). Unstable dynamics will mean two things. First, populations can from time to time reach high enough levels for other density dependent forces acting on the host to come into play. Second, it will mean that the parasitoid and host are vulnerable to local extinction in spatially extended systems. If the

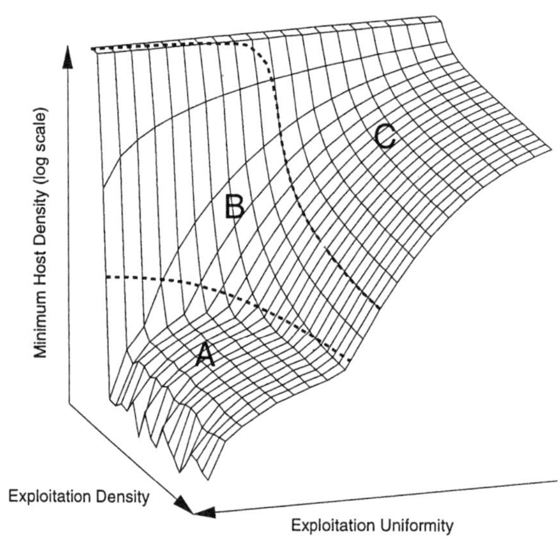

Figure 1. Hypothetical plot of minimum host population levels observed over the initial generations following the introduction of a monophagous parasitoid. Area A = host goes extinct or populations exhibit persistent cycles, Area B = parasitoid regulates host population, and Area C = host population regulated by a combination of the parasitoid and other density dependent factors. This is based on a proportional refuge model, where the density of exploitation is the basic reproductive rate of the parasitoid and the uniformity of exploitation is the proportion of hosts exposed to attack (redrawn from Hochberg & Holt submitted).

parasitoid should go extinct, then in the time interval between extinction and recolonization, the host may be limited by other density dependent factors, some perhaps contributing to regulation, some not. If both species should go extinct, then the system can only re-establish itself if the host should re-invade from surrounding areas. Until the parasitoid subsequently joins the system, the host will (again) be under the reign of other density dependent forces.

Second, if the impact of the parasitoid is minor (*either* low density *or* high variability) then host densities will be limited by a combination of the parasitoid and other density dependent factors (Area C, Fig. 1). These include other natural enemies (parasitoids, predators, and microparasites) and resources (intraspecific competition and emigration). Though yet to be investigated, it is reasonable to expect that increasing numbers of density dependent factors should affect host dynamics as host densities take on larger and larger values. This would suggest that teasing-out the dynamic role of monophagous parasitoids should be most difficult when its role is potentially very subtle.

A unified concept is emerging, integrating elements of density dependent host limitation which is patchy in time and in space, and roles for natural enemies (including parasitoids) in promoting and detracting from regulation depend on when and where one looks (Hochberg 1996). Due to their tightly coupled nature, monophagous parasitoid-host associations should be the most likely to reflect the range of interaction types over their geographical range. This is because both biotic and abiotic environments contribute to the qualitative nature of the interaction, and these vary over large spatial scales.

3. RECENT STUDIES

Below, I briefly present five recent studies, supporting some of the contentions outlined above. The first is purely theoretical and uses simple mathematical formulae. The second study is also theoretical, but is based on a particular biological system of conservation. The third is an experimental study conducted in the laboratory on a stored products pest. The fourth is a comparative study of biological control attempts. Finally, the fifth study is a proposition for why pests may not readily evolve resistance to their insect parasitoids in biological control programs.

3.1 *Necessary and sufficient conditions for monophagous parasitoid-driven host regulation*

As I mentioned above, the great majority of theoretical studies on parasitoid-host population dynamics have delved into the role of exploitation uniformity, assuming that exploitation densities are high. Getz and Mills (1996) have reconsidered this logic by exploring models where exploitation density can be limited, due to either search or egg-limitation of the parasitoid. They show using very simple, but general modelling techniques that exploitation density has to be sufficiently greater than the productivity of the host population for the distribution of exploitation to be of substantial importance to population dynamics.

3.2 *Exploitation uniformity as a mediator of top-down versus bottom-up limitation in conservation biology*

One way to investigate in detail the role of insect parasitoids in the dynamics of their hosts is to employ the "case model" approach. The idea here is to develop mechanistic models, tailored to particular biological systems. Case models have two notable advantages over other modelling approaches, in being biologically realistic and testable on a site by site basis. Recently, the case model approach was used to investigate the roles of parasitoid search behaviour and host refuges from parasitoid attack in the population dynamics of the monophagous parasitoid,

Ichneumon eumerus (Wesmael), and its host, the lycaenid butterfly, *Maculinea rebeli* (Hir.) Details of the system biology and model development are presented elsewhere (Hochberg et al. 1996).

This butterfly lives in small, closed populations, rarely numbering over a few thousand individuals. Its parasitoid is found at only a small subset of butterfly-harboring sites. The computer analysis of a mechanistic model showed how parasitoid searching intensity and behaviour, and host refuges from parasitism affect the dynamics of the host-parasitoid association. When there is no explicit variation in the susceptibility of hosts to parasitism (i.e., high uniformity), the parasitoid was found to persist only for a very restricted range of search rates and searching behaviours (affecting exploitation diversity). Simulations which included refuges from parasitism, modelled as a cue-threshold phenomenon in the elicitation of intensive search for the host, were found to greatly expand the conditions for butterfly and parasitoid persistence. This is because refuges from parasitism take the system from one destabilised by the parasitoid (Area A, Fig. 1), to one in which highly stabilising intraspecific competition between *Maculinea* caterpillars dominate (Area C, Fig. 1). This study shows how deductions of the stabilising power of parasitoids from measures of exploitation uniformity alone can be spurious without information about the propensity of the parasitoid to dominate the limitation of its host, that is 'exploitation density'.

3.3 *Support for the role of exploitation uniformity in single and multiple natural enemy settings*

In a recent series of studies, Begon and colleagues have conducted laboratory experiments aimed at unravelling the ecological processes driving the population dynamics of two and three species systems involving the Indian meal moth (*Plodia interpunctella* (HÜbner), a parasitoid (*Venturia canescens*) and a granulosis virus (*Pi*GV). They found that when the host interacts with only one of its natural enemies, generation cycles in adult moth numbers are produced (Begon et al. 1996). These populations are not unstable, because their numbers are very regular from generation to generation.

Non-uniform exploitation would appear crucial to the ability of the parasitoid to regulate its host. The parasitoid has reduced access to deeply feeding hosts as compared to surface-feeders (Begon et al. 1995), increasing the heterogeneity of exploitation. The dynamic impact of the virus, however, is much less than that of the parasitoid, suggesting that its exploitation density is insufficient to make the distribution of infection (Sait et al. 1994) an important variable driving host dynamics.

Given that there are differences in exploitation tactics by the two natural enemies (i.e., host susceptibility to each is only slightly correlated, uncorrelated, or negatively correlated), the total exploitation distribution of the two natural enemies in the three-species system is more uniform than either of its two parts alone. It is this combined, higher uniformity which is the most reasonable explanation for the transition from stable generation cycles to unstable multi-generation cycles (leading to the extinction of the system in all the experimental boxes) observed in the system (from Area C or B to Area A in Fig. 1).

3.4 *The uniformity of pest exploitation explains economic success in biological control*

Hawkins et al. (1993) compiled a set of 74 biological control attempts employing parasitoids. They used the maximum observed level of parasitism by the introduced parasitoid as a measure of the uniformity of exploitation distribution. They found a highly statistically significant relationship between economic success in biological control programs and the proxy estimate of uniformity. The underlying message of their study is that a parasitoid's potential to limit its host population numbers is determined in part by the uniformity of host exploitation.

3.5 *The importance of parasitoid exploitation density and uniformity in the evolution of host resistance*

The evolution of pest resistance is a well established outcome of many chemical control attempts. Resistance to biological control attempts is much rarer, and even rarer still is the evolution of resistance to introductions of parasitoids. Holt and Hochberg (1997) have recently proposed that this phenomenon may not simply be a sampling artifact, but may have an ecological basis.

The basic idea is that strong population limitation need not lead to strong selection for resistance, because natural selection is always biased towards the environments most experienced by the population under selection. Consider the following argument. When exploitation density and uniformity are such that the parasitoid imposes itself strongly on most or all the members of the host population, we would expect unstable dynamics to ensue (Area A, Fig. 1). If, because of the instability, the host is able to escape control from the parasitoid frequently in space or in time; then globally, the parasitoid will be exerting less effective selection pressure than would be the case if outbreaks did not occur. This is because the highest selection pressure is affecting the relatively small numbers of hosts which are in localities that are crashing due to the local super-abundance of the parasitoid. Even when the parasitoid dominates the regulation of the host to more or less constant populations, the variability in the distribution of exploitation required to do this may mean that many hosts are not subject to significant selection pressure and host resistance is impeded (boarder of Areas B and C in Fig. 1). Host resistance in either of these top-down scenarios is, however, possible, but requires that costs to resistance be sufficiently small. In sum, then, low exploitation densities and/or high heterogeneities in exploitation (whether a product of dynamics, or the biologies of the species) tend to impede the evolution of resistance.

4 CONCLUSION

I have briefly presented how a small number of basic rules can be conceptually and quantitatively applied to the population ecology of monophagous insect parasitoids. In my view, a highly fertile area in the future will be the development of more elaborated paradigms of the population biology of monophagous parasitoids. These conceptual models should integrate driving forces acting at different levels of biological organisation, showing how they impinge the various other levels.

Despite the present attempt at synthesis, the role of parasitoids in tightly-coupled associations remains a virtual empirical black box. The best place to look for top-down effects is in biological control programs. If evolution does indeed work to lessen the top-down effects of monophagous parasitoids, then biological control as a model for community ecology is perhaps restricted to a narrow set of circumstances. The most evident is the series of events occurring when a natural enemy endemic to a regional pool, invades a locality from which it had never before been present (or had been absent for very long periods of time).

In contrast, the best place to look for bottom-up associations is in conservation programs. Here again, however, the fragile ecological or adaptative circumstances which surround the plight of an endangered species, may mean that the role of the parasitoid in the conservation target's dynamics is a reflection of a highly restricted range of community scenarios.

What we must now do is find out the following:

(1) Are parasitoids ever *necessary* for the regulation of their hosts?

(2) When parasitoid do contribute to regulation, to what extent do they dominate it and why?

(3) Does evolution work to reduce regulation by parasitoids or increase it?

(4) Can a single parasitoid assume qualitatively different dynamic roles depending on when and where one looks?

Although this agenda presents a considerable challenge, *elements* towards answers should be forthcoming using literature surveys (as a first pass at each question) and imaginative, carefully designed field and laboratory studies.

REFERENCES

Begon, M., Sait, S.M. & Thompson, D.J. 1995. Persistence of a parasitoid-host system: refuges and generation cycles. *Proceedings of the Royal Society of London B* 260:131-137.

Begon, M., Sait, S.M. & Thompson, D.J. 1996. Predator-prey cycles with period shifts between two- and three-species systems. *Nature* 381:311-315.

Getz, W.M. & Mills, N. 1996. Host-parasitoid coexistence and egg-limited encounter rates. *The American Naturalist* 148:333-347.

Hawkins, B.A., Thomas, M.B. & Hochberg, M.E. 1993. Refuge theory and classical biological control. *Science* 262:1429-1432.

Hochberg, M.E. 1996. An integrative paradigm for the dynamics of monophagous parasitoid-host interactions. *Oikos* 77:556-560.

Hochberg, M.E., Elmes, G.W., Thomas, J.A. & Clarke, R.T. 1996. The population dynamics of monophagous parasitoids: A case model of *Ichneumon euremus* (hymenoptera:ichneumonidae). *Philosophical Transactions of the Royal Society of London B* 351:1713-1724.

Hochberg, M.E. & Holt, R.D. submitted. The uniformity and density of host exploitation as guides to success in biological control. In B.A. Hawkins and H.V. Cornell (eds.). *Theoretical approaches to biological control.* Cambridge: Cambridge University Press.

Holt, R.D. & Hochberg, M.E. 1997. When is biological control evolutionarily stable (or is it)? *Ecology.*

Sait, S.M., Begon, M. and Thompson, D.J. 1994. Long-term populaiton dynamics of the Indian meal moth *Plodia interpunctella* and its granulosis virus. *Journal of Animal Ecology* 63: 861-870.

Taylor, A.D. 1993. Heterogeneity in host-parasitoid interactions: 'Aggregation of risk' and the 'CV^2>1 rule'. *Trends in Ecology and Evolution* 8: 400-405.

Population and Community Ecology for Insect Management and Conservation, Baumgärtner et al. (eds)
 ISBN 90 5410 930 0

Simulation of whitefly biological control, based on behaviour of individual parasitoids

H.J.W.van Roermund & J.C.van Lenteren
Department of Entomology, Wageningen Agricultural University, Netherlands

ABSTRACT: A model has been developed which simulates the temporal and spatial dynamics of greenhouse whitefly and the parasitoid *Encarsia formosa* in a tomato crop. The pest-parasitoid interaction is based on the local searching and parasitization behaviour of individual parasitoids in the whitefly-infested crop. The model comprises several submodels for (a) the parasitoid's foraging behaviour, (b) the whitefly and parasitoid population development, (c) the spatial distribution of whitefly and parasitoid in the crop, and for (d) leaf production. The model will allow us (1) to explain why the parasitoid can control whiteflies on some crops and not on others in large commercial greenhouses, (2) to improve introduction schemes of parasitoids for crops where previous control was ineffective, (3) to predict effects of changes in cropping practices (e.g. greenhouse climate, choice of cultivars) on the reliability of biological control, and finally (4) to develop criteria for the selection of natural enemies.

1. INTRODUCTION

Greenhouse and sweet potato whitefly (*Trialeurodes vaporariorum* and *Bemisia tabaci*) are very common, highly polyphagous pest insects worldwide. Biological control of greenhouse whitefly with the parasitoid *Encarsia formosa* has been applied with great commercial success during the past 20 years. One of the problems was that in some key greenhouse crops (e.g. tomato) results were very reliable, while on other important crops (e.g. cucumber) control often failed. Another, more recent problem, is that *E. formosa* does not seem to be a suitable parasitoid for the control of *Bemisia* and new natural enemies need to be evaluated. Modelling has always played an important role in the process of selecting and improving the efficacy of releases of natural enemies, but often biologically unrealistic simplifications were part of these models which strongly limited their predictive value.

We started a long-term project consisting of modelling and experimentation, with the aim to obtain quantitative understanding of this tritrophic system to explain failure or success of biological control. Because of the multitude of relationships between the trophic levels, the most important quantitative factors can only be traced after integration of all relevant processes. Systems analysis and simulation are powerful tools for this purpose. The study aims at integrating existing knowledge on the major processes known to affect the whitefly-parasitoid interaction in a crop by means of an explanatory simulation model. In this paper we summarize the work of the first phase, which was limited to the tritrophic system tomato-greenhouse whitefly-*Encarsia formosa*.

We developed a model which is unique in that it is individual-based (Judson 1994) and simulates the local searching and parasitization behaviour of individual parasitoids in a whitefly-infested crop. The model includes stochasticity and spatial structure based on location coordinates of plants and leaves. This model comprises several submodels for (a) the parasitoid's foraging behaviour, (b) the whitefly and parasitoid population development, (c) the spatial distribution of whitefly and parasitoid within and between plants in the crop, and for (d) leaf production. With the model we can simulate temporal and spatial dynamics of pest and natural enemy. The model will allow us (1) to explain why the parasitoid can control whiteflies on some crops and not on others in large commercial greenhouses, (2) to improve introduction schemes of parasitoids for crops where control was ineffective, and (3) to predict effects of changes in cropping practices (e.g. greenhouse climate, choice of cultivars) on the reliability of biological control, and finally (4) to develop criteria for the selection of natural enemies.

2. DIRECT OBSERVATION EXPERIMENTS ON FORAGING OF *E. FORMOSA*

When the present research project started, the behaviour of *E. formosa* had been observed in various experiments (for a review see Noldus & van Lenteren 1990). *E. formosa* is a solitary larval parasitoid: females lay one egg per host during an oviposition. Like in other synovigenic parasitoids new eggs mature when the egg load of the parasitoid drops below the storage capacity, which is 8-10 mature eggs for *E. formosa.* About ten days after oviposition the immature parasitoid pupates in the host pupa, which (in case of greenhouse whitefly) then turns black and parasitism can easily be seen from the outward appearance of the whitefly. Female parasitoids produce daughters parthenogenetically. Males are rarely observed.

The parasitoid searches for the sessile whitefly immatures by flying or hopping from leaf(let) to leaf(let), without distinguishing between infested and clean plants or leaves before landing. Once on the leaf she starts walking and drumming the leaf with her antennae. Hosts are encountered randomly and the walking pattern is not changed after an encounter with a host. After an encounter, four behaviours on that host can be distinguished: the parasitoid may reject the host after an inspection with the antennae (antennal rejection) or after insertion of the ovipositor (ovipositorial rejection), she may parasitize (oviposition) or she may use the host as a food source (host feeding) (van Lenteren et al. 1980).

However, these earlier experiments did not lead to a complete picture of foraging behaviour. Quantitative data on some aspects were lacking, such as the parasitoid's searching or walking activity between host encounters, and the effect of temperature on the foraging processes. In many of the earlier experiments parasitoids were confined to an experimental arena, and therefore little was known about the time allocation of the parasitoid on leaves, such as the time until leaving, the time spent on upper and lower leaf side, and how these are affected by encounters with or ovipositions in hosts.

Before developing simulation models these gaps in our knowledge were identified and studied experimentally. Experiments were carried out where individual parasitoids were observed continuously until they flew away, either on clean tomato leaflets, on leaflets with honeydew, or on leaflets with unparasitized and parasitized whitefly larvae (van Roermund & van Lenteren 1995a,b, van Roermund et al. 1994). The parasitoid's residence time on leaflets, its leaving tendency and effects of temperature and several experiences with hosts were quantified. Other basic aspects of foraging which where quantified were the parasitoid's walking speed and walking activity, the probability of each handling behaviour to occur after an encounter with a host and the host handling times. These data enabled quantification of the parasitoid's foraging process from landing on a leaf until its departure. This work resulted in the following conclusions:

(1) The parasitoid *E. formosa* searches at random without a preference for the edge or the middle of a leaf, or for the upper or lower leaf side, whereas its hosts, the whitefly immatures are present only on the lower side of a tomato leaflet.

(2) The median residence time of the parasitoid on uninfested tomato leaflets (or giving up time, GUT) is 18.6 min at 20, 25 and 30°C and equal to that on infested leaflets on which no hosts are encountered.

(3) Parasitoids are arrested on the leaf by encounters with, and especially by ovipositions in unparasitized hosts, by encounters with parasitized (unsuitable) hosts and by contact with honeydew. The time since latest host encounter until departure (GUT) is again 18.6 min, whether the hosts were already parasitized or not. It increases to 40 min after the first oviposition in an unparasitized host occurring on the leaflet.

(4) Parasitoids are arrested on the lower leaf side by encounters with hosts, and especially by ovipositions in unparasitized hosts. The median time since the arrival on a particular leaf side

or, if it occurred, since the latest host encounter on that leaf side until changing to the other side (time until changing, TUC) is 11.6 min, but drops to 5.7 min when the parasitoid has visited both sides of the leaflet. After the first oviposition in an unparasitized host, TUC on the lower leaf side (where hosts are present) becomes twice as long.
(5) Parasitoids usually leave from the upper leaf side, because no hosts will be encountered there.
(6) The leaving behaviour of the parasitoid on a leaf can be described by a stochastic threshold mechanism, which is characterized by a certain tendency (probability per time) to leave. The parasitoid leaves after the host encounter rate falls below a certain threshold (encounters per time, which is the reciprocal of GUT). This threshold is not fixed, but shows a great variation and is expressed as a probability.
(7) The parasitoid's walking speed increases linearly between 15 and 25-30°C.
(8) The parasitoid's walking activity (% of time walking on the leaf surface compared to the total time on the leaf without handling hosts) is very low at temperatures below 18°C and increases to about 75% at 20, 25 and 30°C. The walking activity is not affected by host encounters, but decreases with decreasing egg load after 4 ovipositions.
(9) The percentage of encounters resulting in an oviposition is about 75% for the most preferred stage (unparasitized L4 larva), but decreases with decreasing egg load.
(10) Host handling behaviour and handling time is not influenced by the host plant.
(11) The total handling time (including drumming etc.) for antennal rejection of an unparasitized hosts is about 20 s, for oviposition and for ovipositional rejection about 6 min, and for host feeding about 15 min. These handling times slightly differ when hosts are parasitized.
(12) Self-superparasitism is not observed. Conspecific-superparasitism occurs in 14% of the contacted hosts containing a parasitoid egg, but is not observed anymore when the parasitoid egg had hatched.
(13) No difference is observed in host handling behaviour between naive and experienced parasitoids.
(14) Many inactive parasitoids are observed when the barometric pressure had decreased over a time span of at least 12 h.

3. SIMULATION MODELS OF FORAGING BEHAVIOUR OF *E. FORMOSA*

The information described above was used as input in the simulation models of *E. formosa*'s foraging behaviour (van Roermund et al. 1996, 1997a,b). Foraging behaviour is analyzed using Monte Carlo simulation at three spatial scales: in a small experimental arena, on a tomato leaflet and on a tomato plant. Foraging behaviour is first studied at these small spatial scales, to better understand the quantitative effects of parasitoids on whitefly populations as observed in a crop.

The above simulation models are *mechanistic*, that is, they explain *how* parasitoids, in terms of searching efficiency, host handling and available eggs, realize the observed level of parasitism. Mechanistic explanations can help to understand failure or success of biological control in practice. The models do not provide a *functional* explanation of the observed behaviour.

The simulated number of hosts encountered, parasitized and killed by host feeding, and the residence times on leaflets were validated with experimental data and the simulation results agree well with these observations. According to the model, *E. formosa* can parasitize 16 hosts per day on average at 25°C on a tomato leaflet if they start searching with a full batch of mature eggs and if host density is not limiting. Thus about 7 new eggs mature during the day (16 h) at that temperature. From the second day onwards, the parasitoid can parasitize 11 hosts per day, due to egg limitation. If the parasitoid laid all eggs the preceding day, only 4 eggs mature during the night (of 8 h) at 25°C, so the parasitoids do not have a full batch of mature eggs the next morning.

The model shows that at a density of 1 L3 larva per tomato leaflet, 15.7% of the parasitoids discover the larva before they leave. Also at higher host densities, not all hosts are encountered and patches (leaflets) are not depleted after one visit. Variation in number of encounters and ovipositions between parasitoids is considerable, mainly caused by the random encounter of hosts, the variation in handling behaviour of an encountered host and by the variation in GUT and TUC.

In greenhouses, whiteflies show a clustered distribution over plants and leaves and average numbers are usually very low. The models show that at such conditions the number of parasitizations on tomato leaflets or plants is strongly affected by the leaf area, the parasitoid's walking speed and walking activity, the probability of oviposition after encountering a host, the initial egg load (egg load at the beginning of the experiment) and the ratio of search times on both leaf sides. At extremely high host densities, the egg storage capacity and the initial egg load of the parasitoid are most important, and on plants with a clustered host distribution also the parasitoid's GUT plays an important role.

At all spatial scales tested, the number of encounters, ovipositions and host feedings increase with host density with a decelerating rate until a maximum level is reached. This shape of the curves resemble a Holling Type II 'functional response', which is caused by the parasitoid's decreasing walking activity and smaller probability of oviposition after encountering a host, when egg load decreases. This is predominant at all levels, and even a change in GUT from 18.6 to 40 min after the first oviposition on the leaf does not result in an accelerating increase of the curve. The shape of the curves, describing the effect of host density on parasitism as a result of the basic processes, helps to understand the dynamics of the host-parasitoid interaction at the population level. In case of a Type II functional response, percentage parasitism declines with increasing host density and parasitism is inversily density dependent. A high host density thus reduces the per capita parasitization pressure caused by one parasitoid. According to theory, inversely density dependence tends to have a destabilizing effect on the dynamics of host and parasitoid (see, e.g., Murdoch & Oaten 1975; Oaten & Murdoch 1975). However, the functional response or the parasitization pressure caused by one parasitoid is only one factor in determining the dynamics at the population level. Another factor is the number of parasitoids on the leaf. For *E. formosa*, the effect on the population level depends on the balance between the parasitization pressure caused by one parasitoid and the arrestment and subsequent aggregation of parasitoids on leaves with high host density (see below).

4. LIFE-HISTORY PARAMETERS OF GREENHOUSE WHITEFLY AND *E. FORMOSA*

Life-history parameters of the greenhouse whitefly and *E. formosa* are reviewed in van Roermund & van Lenteren (1992a,b). Data from literature were selected on development rate of each immature stage, percentage mortality of each immature stage, sex ratio, longevity, pre-oviposition period, period of increase of daily oviposition, fecundity and oviposition frequency. Most of these experiments have focused on the effect of temperature with little attention to other environmental factors such as humidity or light. Using these data, the relationship between the life-history parameters and temperature was assessed by non-linear regression. For each life-history parameter five different curves were fitted, the best being selected on the basis of the coefficient of determination (r^2) and on visual comparison, which was necessary to check whether a curve was biologically realistic, particularly the tails. Coefficients to describe the mean of each life-history parameter as a function of temperature are summarized in these papers. Coefficients of variation (*cv*: sd/mean) among individuals are also given. These coefficients are used as input in the submodels of population development of whitefly and parasitoid (see below).

For greenhouse whitefly, the life-history parameters depend very much on the host plant. For *E. formosa*, data for several host plants were combined. The high r^2 values indicate that host plant effects can be disregarded for the parasitoid's life-history parameters, except for oviposition at low host densities, which is caused by differences in the parasitoid's walking speed and walking activity on leaves with a different morphology. The immature development rate and immature mortality of the parasitoid is strongly affected by the host instar which was parasitized.

The development rate is calculated as the reciprocal of the stage duration. For all immature stages of whitefly and parasitoid the relationship between development rate and temperature is described by the Logan curve: just above the lower threshold temperature, the development rate increases exponentially to an optimum, whereafter it declines sharply until the upper lethal temperature has been reached. The relationships of longevity, fecundity, and oviposition frequency with temperature are described by the Weibull curve: they increase exponentially from the lower lethal temperature to an optimum, whereafter they decrease exponentially. Only for *E. formosa*, the longevity decreases exponentially with temperature and an optimum was not found at greenhouse conditions. No relationship with temperature was found for the immature mortality, the sex ratio and the *cv* values of the life-history parameters of whitefly and parasitoid.

5. SIMULATION MODEL OF WHITEFLY-PARASITOID INTERACTION IN A CROP

The final model simulates the population dynamics of the pest insect-parasitoid interaction in a tomato crop (van Roermund 1995, van Roermund & van Lenteren 1995c, van Roermund et al. 1997c). The model is based on the parasitoid's searching and parasitization behaviour and on developmental biology of the two insect species. This model comprises several submodels, such as the submodel for whitefly population development, for parasitoid population development, for the parasitoid's foraging behaviour on tomato leaflets (van Roermund et al. 1997a), for spatial distribution of whitefly and parasitoid in the canopy based on dispersion of adult whiteflies and parasitoids from leaf to leaf, for leaf production and a submodel for checking simulation errors. Life-history parameters of van Roermund & van Lenteren (1992a,b) are used as input in the submodels for population development of whitefly and parasitoid on tomato. The model is unique in that it is an individual-based model which simulates local searching and parasitization behaviour of a large number of individual parasitoids in a whitefly-infested crop. The model includes stochasticity and spatial structure which is based on location coordinates of plants and leaves.

The model was validated with population counts from experiments on tomato with and without introduction of *E. formosa* in small greenhouse compartments and in a large commercial greenhouse. The simulated population increase of greenhouse whitefly in the *absence* of parasitoids agree well with the observations. This result can for an important part be explained by the accurate estimates of the life-history parameters, which are based on many experiments at a wide temperature range (see van Roermund & van Lenteren 1992a,b).

With these life-history parameters as input in the model, the intrinsic rate of increase (r_m) of both insect species was simulated. The r_m of *E. formosa* is much higher than that of the greenhouse whitefly above 14°C. The r_m of a parasitoid however, plays a limited role in biological control, because it is only valid when all parasitoids can lay their daily egg load. This can only happen at extremely high host densities when the parasitoids do not have to spent much time searching for hosts. In greenhouses whitefly densities are usually much lower and the realized whitefly density depends on the parasitoid's searching efficiency. Therefore, to evaluate and understand success or failure of biological control, r_m values are inappropiate and it is essential to built models which include searching and parasitization behaviour of the natural enemy at very low host densities.

Also in the *presence* of parasitoids, the simulation results agree well with greenhouse observations on tomato. Apparently, the hypothesized random host encounter of *E. formosa* in a tomato crop is reliable. In the model, the parasitoid does not distinguish between uninfested and infested leaflets before landing, the parasitoid searches randomly for hosts once on the leaflet, and shows a strong arrestment effect on the leaflet once a host is encountered. Simulations show that the adult parasitoid-whitefly ratio is very high and can even reach 250:1. As a result, whiteflies are suppressed rather than regulated by the parasitoids at extremely low host densities (<0.3 unparasitized pupae per plant), but never become extinct. These whitefly densities are much lower than the economic damage threshold for greenhouse whitefly. Percentage black pupae fluctuates between 40 and 70%. According to the model, the parasitoid adults reach high densities of 7.4 per plant, but due to the low whitefly density not more than 1% of the parasitoids is searching on infested leaflets.

The giving up times (GUT) of *E. formosa* vary to a large extend (van Roermund et al. 1994). The degree of whitefly control is very sensitive to those GUT's lower than 800 s of the parasitoids. The whiteflies are suppressed at much lower densities when the parasitoids stay *at least* five minutes on each leaflet (infested or uninfested) and after each host encounter. This minimum time increases the arrestment effect and the resulting percentage of parasitoids on infested leaflets, thereby reducing the chance that clustered hosts escape from parasitism. When variation in GUT is excluded in the model, the whitefly population becomes less stable and nearly goes extinct. Variation in GUT on leaflets induces host refuges from parasitoid attack. Also from more theoretical studies, host refuges are known to stabilize populations (Murdoch

& Oaten 1975; Chesson & Murdoch 1986).

Whitefly adults migrate to young leaves in the top of the plant. A slower leaf production results in a longer stay and more ovipositions of whitefly adults on a particular leaflet. Thus, the same number of hosts are distributed over fewer leaflets, resulting in a more aggregated host distribution. Whiteflies are then suppressed by *E. formosa* to much lower numbers, according to the model. Parasitism of one *E. formosa* female on a tomato leaflet is inversely density-dependent, which is caused by a decreasing walking activity and smaller probability of oviposition after encountering a host, when egg load decreases (van Roermund et al. 1996, 1997a,b). Host aggregation thus 'dilutes' the per capita parasitization pressure caused by one parasitoid on the leaflet. The effect on the population level however, depends on the balance between this 'dilution' effect and the strength of the arrestment and aggregation of *E. formosa*. Therefore, the stronger whitefly reduction when whiteflies are more aggregated is caused by a stronger parasitoid arrestment on infested leaflets. As a result, the relative number of parasitoids searching on these leaflets increases.

This shows that differences in whitefly distribution among crops are one factor in causing differences in success of biological control. Other factors are the size, number and surface (hairiness) of leaves in the canopy. Leaf size and total leaf area have a strong effect on whitefly control according to the model, caused by their direct effect on host density. Furthermore, leaf size and leaf surface strongly affect the efficiency of *E. formosa* by changing the parasitoid's arrestment effect (GUT) and the walking speed and activity, respectively.

Another important factor is the whitefly development duration on the crop. The model shows that plant resistance breeding aimed at an increase in egg-to-adult duration of the whiteflies is very efficient in causing a severe reduction of whitefly numbers, when biological control is applied. Observed development times of whitefly differ very little between tomato genotypes, and a much larger difference is found for whitefly longevity, oviposition rate and immature mortality. These parameters have a smaller effect on whitefly population development.

The important factors or crop properties affecting the success of biological control cannot be compared independently. For instance crops with large leaves usually have lower number of leaves which are produced at a lower rate than crops with small leaves or leaflets. It is particularly the combined effect of these important factors that can be tested with this model for different crops or plant varieties.

In biological control programs, parasitoids are usually tested in small-scale experiments at high host densities before introduction in the field. As a result, maximum daily oviposition of parasitoids is measured, whereas this study shows that egg storage capacity and egg maturation rate of *E. formosa* is not important for the level of whitefly control. In commercial greenhouses, whitefly densities have to be very low for biological control to be regarded as successful, therefore effective host searching is the most essential process. When selecting parasitoids for biological control, attention should be focused on the parasitoid's arrestment effect (minimum GUT), walking speed, walking activity, the probability of oviposition after encountering a host, the ratio of search times on both leaf sides and on longevity, when comparing different synovigenic and solitary parasitoid species with random search. These characteristic attributes of parasitoids are easily measured in laboratory studies. Again, they cannot be compared independently, however, because the attributes of natural enemies are often found in particular combinations. The combined effect of these important attributes of a parasitoid can be tested with this model.

6. CONCLUSIONS

The study aimed at integrating existing knowledge on the major processes known to affect the whitefly-parasitoid interaction in a crop. Because of the multitude of relationships between the three trophic levels (crop-pest-parasitoid), it was decided to follow a combined experimental-simulation approach. The goal was to obtain quantitative understanding of the tritrophic system to explain failure or success of biological control. With the model we now unravelled the ability of *E. formosa* to reduce whitefly populations on greenhouse tomato. The study resulted in increased understanding of the relative importance of basic processes that affect the population interaction of whitefly and natural enemy. The life-history of parasitoids, often summarized in a r_m value, are less important than the parasitoid's searching capacity. This shows that in addition to the traditional selection criteria, a criterion based on searching efficiency is essential. The study has further generated knowledge on foraging behaviour of *E. formosa.* The tremendous effect of variation in patch times of individual parasitoids on the whitefly population in the greenhouse shows that individual-based models which include stochasticity and local searching behaviour are a necessity when developing models of host-parasitoid interaction at extremely low host densities and aggregated host distributions.

One of the questions was to identify the main causal factors for differences among crops in success of biological control of whitefly. The parasitoid is more successful on tomato than on cucumber or gerbera. The present study showed that attention should be focused on differences in the parasitoid's arrestment effect (GUT), the parasitoid's walking speed and activity, the whitefly development duration and the number, size and production of leaves in the canopy. When adapting the parameters in the model for gerbera and cucumber we are able to (1) explain the lower ability of the parasitoid to reduce whitefly populations on these crops, (2) improve introduction schemes of parasitoids for these crops, and (3) predict effects of changes in cropping practices (e.g. greenhouse climate, choice of cultivars) on the reliability of biological control.

Furthermore, the model is being used to evaluate effects of temperature and daylength on biological control and yield of a tomato crop. As a result, greenhouse climate control may not only be based on crop characteristics (yield, quality, timing) and production costs, but also on prevention of production risks like the environment-friendly control of pests. With the model, recent problems with the use of *E. formosa* in greenhouses during autumn/winter will be tried to solve by evaluating effects of increased temperatures during short periods of the day. In another project, the model is being used to evaluate several parasitoids of the silverleaf whitefly (Drost et al. 1996). The model thus assists in identification of the characteristics which compose an efficient natural enemy. Good evaluation criteria allow for a choice between useless and potentially promising natural enemies. Systems analysis and simulation are important tools for integration of criteria, but go beyond the mere comparison of parasitoids, as they also take characteristics of the pest, the crop, and the environment into account.

REFERENCES

Chesson, P. & W.W. Murdoch. 1986. Aggregation of risk: relationships among host-parasitoid models. *Am. Nat.* 127: 696-715.

Drost, Y.C., A. Fadl Elmula, C.J.A.M. Posthuma-Doodeman & J.C. van Lenteren. 1996. Development of selection criteria for natural enemies in biological control: parasitoids of *Bemisia argentifolii. Proc. Exper. & Appl. Entomol., N.E.V. Amsterdam,* 7: 165-170.

Judson, O.P. 1994. The rise of the individual-based model in ecology. *Tree* 9: 9-14.

Murdoch, W.W. & A. Oaten. 1975. Predation and population stability. *Adv. Ecol. Res.* 9: 2-131.

Noldus, L.P.J.J. & J.C. van Lenteren. 1990. Host aggregation and parasitoid behaviour: biological control in a closed system. Mackauer, M., L.E. Ehler & J. Roland (eds.). *Critical Issues in Biological Control.* Intercept, Andover, pp 229-262.

Oaten, A. & W.W. Murdoch. 1975. Switching, functional response and stability in predator-prey systems. *Am. Nat.* 109: 299-318.

Roermund, H.J.W. van. 1995. *Understanding biological control of greenhouse whitefly with the parasitoid Encarsia formosa. From individual behaviour to population dynamics.* Ph.D. Thesis, Wageningen Agricultural University, The Netherlands, 243 pp.

Roermund, H.J.W. van & J.C. van Lenteren. 1992a. The parasite-host relationship between *Encarsia formosa* (Hymenoptera: Aphelinidae) and *Trialeurodes vaporariorum* (Homoptera: Aleyrodidae) XXXIV. Life-history parameters of the greenhouse whitefly, *Trialeurodes vaporariorum* as a function of host plant and temperature. *Wageningen Agric. Univ. Papers* 92.3: 1-102.

Roermund, H.J.W. van & J.C. van Lenteren. 1992b. The parasite-host relationship between *Encarsia formosa* (Hymenoptera: Aphelinidae) and *Trialeurodes vaporariorum* (Homoptera: Aleyrodidae) XXXV. Life history parameters of the greenhouse whitefly parasitoid *Encarsia formosa* as a function of host stage and temperature. *Wageningen Agric. Univ. Papers* 92.3: 103-147.

Roermund, H.J.W. van & J.C. van Lenteren. 1995a. Foraging behaviour of the whitefly parasitoid *Encarsia formosa* on tomato leaflets. *Entomol. Expl. Appl.* 76: 313-324.

Roermund, H.J.W. van & J.C. van Lenteren. 1995b. Residence times of the whitefly parasitoid *Encarsia formosa* Gahan (Hym., Aphelinidae) on tomato leaflets. *J. Appl. Ent.* 119: 465-471.

Roermund, H.J.W. van & J.C. van Lenteren. 1995c. Simulation of biological control of greenhouse whitefly with the parasitoid *Encarsia formosa* on tomato. *Proc. Exper. & Appl. Entomol., N.E.V. Amsterdam,* 6: 153-159.

Roermund, H.J.W. van, L. Hemerik & J.C. van Lenteren. 1994. Influence of intra-patch experiences and temperature on the time allocation of the whitefly parasitoid *Encarsia formosa* (Hymenoptera, Aphelinidae). *J. Insect Behav.* 7: 483-501.

Roermund, H.J.W. van, J.C. van Lenteren & R. Rabbinge. 1996. Analysis of the foraging behavior of the whitefly parasitoid *Encarsia formosa* (Hymenoptera: Aphelinidae) in an experimental arena: a simulation study. *J. Insect Behav.* 9: 771-797.

Roermund, H.J.W. van, J.C. van Lenteren & R. Rabbinge. 1997a. Analysis of the foraging behavior of the whitefly parasitoid *Encarsia formosa* on a leaf: a simulation study. *Biological Control* 8: 22-36.

Roermund, H.J.W. van, J.C. van Lenteren & R. Rabbinge. 1997b. Analysis of the foraging behaviour of the whitefly parasitoid *Encarsia formosa* on a plant: a simulation study. *Biocontrol Science and Technology* 7: 131-151.

Roermund, H.J.W. van, J.C. van Lenteren & R. Rabbinge. 1997c. Biological control of greenhouse whitefly with the parasitoid *Encarsia formosa* on tomato: an individual-based simulation approach. *Biological Control* 9: 25-47.

Population and Community Ecology for Insect Management and Conservation, Baumgärtner et al. (eds)
© 1998 Balkema, Rotterdam, ISBN 90 5410 930 0

Complex population dynamics in a host-pathogen-parasitoid interaction

S.M.Sait, M.Begon & D.J.Thompson
Population Biology Research Group, School of Biological Sciences, University of Liverpool, UK

ABSTRACT: Theoretical studies that have encompassed multispecies interactions have enhanced our understanding of population and community ecology, but empirical studies of these complex systems are rare. Here, the population dynamics of the Indian meal moth, *Plodia interpunctella*, and two of its natural enemies, a baculovirus and the solitary endoparasitoid *Venturia canescens*, are described. Replicated populations consisting of host, host-pathogen, host-parasitoid and host-pathogen-parasitoid combinations were established and the patterns in abundance of each species monitored. The host alone and each of the two-species systems predictably exhibited long term persistence and cyclic fluctuations of approximately one host generation in length. In striking contrast, the three-species populations were characterized by instability and a marked propensity to go extinct. Furthermore, the increase from two to three species resulted in a remarkable shift in period of the host and parasitoid dynamics, which exhibited multigeneration cycles.

1 INTRODUCTION

Traditionally, population ecologists have attempted to unravel the complexity that exists between species in the natural world by investigating the dynamics of single species or pairs of species, in the expectation that general patterns and principles can nonetheless be identified. More recently, there has been considerable interest in moving beyond these simple systems to explore the dynamics of multispecies interactions. This approach, primarily theoretical, has demonstrated that the incremental step from two to three-species interactions is accompanied by a rich variety of dynamical patterns, with broad implications for our understanding of population and community ecology. A number of studies have examined the dynamics of a single prey species that is attacked by two or more predators, parasites or pathogens (e.g. May & Hassell 1981; Hassell & May 1986; Hochberg & Holt 1990; Hochberg *et al.* 1990; Briggs 1993; Hassell *et al.* 1994), or when these natural enemies attack two or more prey species (e.g. Roughgarden & Feldman 1975; Comins & Hassell 1976; Holt 1977; Holt & Pickering 1985; Begon *et al.* 1992; Holt & Lawton 1994).

Empirical studies in insect population ecology beyond two species are rare, but experiments that have examined multispecies interactions in laboratory microcosms, or model systems, have always been illuminating (e.g. Park 1948; Huffaker 1958). These systems serve as a valuable intermediate between the full complexity of the field and abstract mathematical models that seek to describe this complexity (Kareiva 1989; Lawton 1995).

One such model system investigated here is an interaction between a lepidopteran host and two of its natural enemies, one a viral pathogen, the other a parasitoid wasp. The host is the

Indian meal moth *Plodia interpunctella* (Lepidoptera; Pyralidae) (Hübner), a pest of stored products world-wide. It has five larval instars and a generation time of around 42 days (Sait *et al.* 1994a). The pathogen is a granulovirus (GV), which belongs to the baculovirus family. Baculoviruses in general only infect invertebrates, primarily Lepidoptera, and often have narrow host ranges. As in many host-baculovirus interactions, mortality of *P. interpunctella* increases with dose of GV, whilst susceptibility decreases with host age, such that the final instar is effectively invulnerable (Sait *et al.* 1994b). Infection with the pathogen only occurs in the larval stage, following ingestion of virus particles with contaminated food or through cannibalism of virus-killed cadavers, and the infection is always lethal once disease symptoms are overt. The parasitoid is the solitary endoparasitoid *Venturia canescens* (Hymenoptera; Ichneumonidae) (Gravenhorst). The parasitoid synchronizes its development with that of the host, taking between 21 and 25 days depending on the instar parasitized (Harvey *et al.* 1994). In contrast to the pathogen, first instar larvae appear to be invulnerable to attack by the parasitoid, due to their small size. It will parasitize second to fifth instar larvae (Harvey *et al.* 1994) but prefers larger hosts (Sait *et al.* 1997) since fitness costs to the parasitoid are considerably less if it attacks these later stages (Sait *et al.* 1995).

In this paper the population dynamics of the host-pathogen-parasitoid interaction are described, and contrasted with the single and component two-species systems in order to place the more complex interaction in its true context.

2 POPULATION STUDIES

Replicated host alone, host-pathogen and host-parasitoid cultures were set up using established

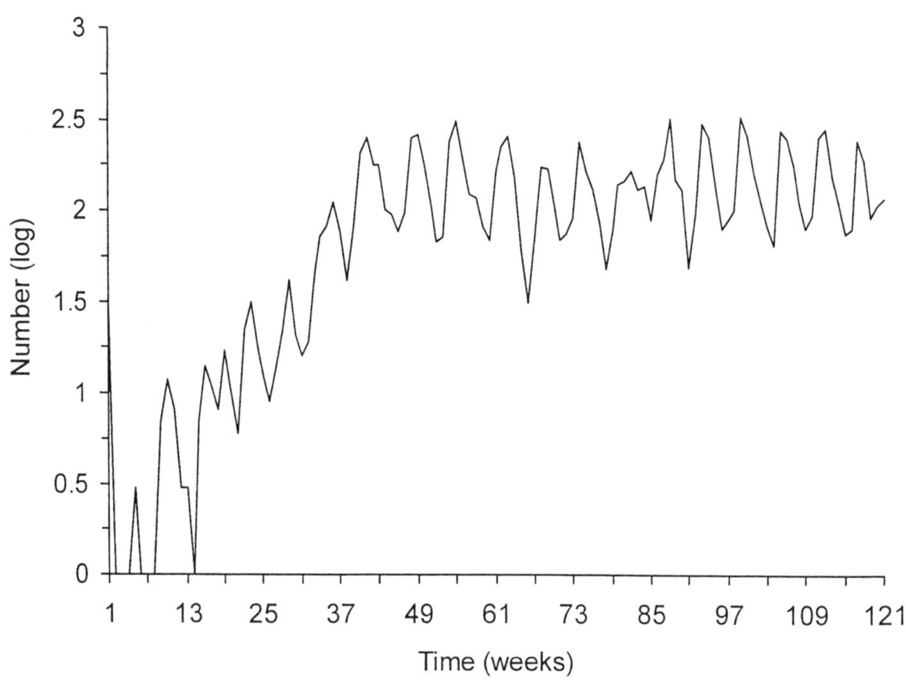

Figure 1. The population dynamics of *P. interpunctella* alone, measured as the number of dead adults per weekly census. The mean density (± s.e.) of this representative culture, after an initial settling down period of 37 weeks, was 147.5 ± 7.6. The cycle period was 6 and 6.1 weeks (from the peak ACF and spectral analysis respectively) and the value of the peak ACF (a measure of the strength of the periodicity) was 0.61 (P<0.05).

protocols [for full details of moth diet, insect rearing conditions and maintenance procedures see Sait *et al.* (1994a) and Begon *et al.* (1995)]. The three-species system was initiated by adding the parasitoid to an established host-pathogen population (Begon *et al.* 1996). Where appropriate, the number of dead adult moths and parasitoids were counted each week. In addition, the number of infected larvae retrieved from one sixth of the host's food medium were counted in order to give a measure of the prevalence of the GV infection in the host population. Autocorrelation functions (ACFs) were calculated to determine the period and strength of the population cycles and spectral analysis performed where possible to determine the cycle period more precisely. In all cases, analyses were carried out on logged data, but for the infected larvae analysis was of the differences between successive logged data points in order to de-trend a significant increase in the level of infection.

Each population combination was replicated but since the consistency between them was high, one representative culture is shown here for brevity. Complete population trajectories and analyses can be found in Begon *et al.* (1995, 1996 and 1997).

2.1 *Host alone dynamics*

After an initial settling down period of population growth, the dynamics of the host were highly consistent between the three replicates. The abundance of adult moths fluctuated around a mean of 120, with a typical range of between 60 and 230 (Figure 1). Details of the strength of the cycle period for the representative population (the value of the peak ACF) and the length of the period (the lag at which the peak ACF occurred and, where possible, the peak in the spectral analysis) are given throughout in the relevant figure legends. The host-alone populations

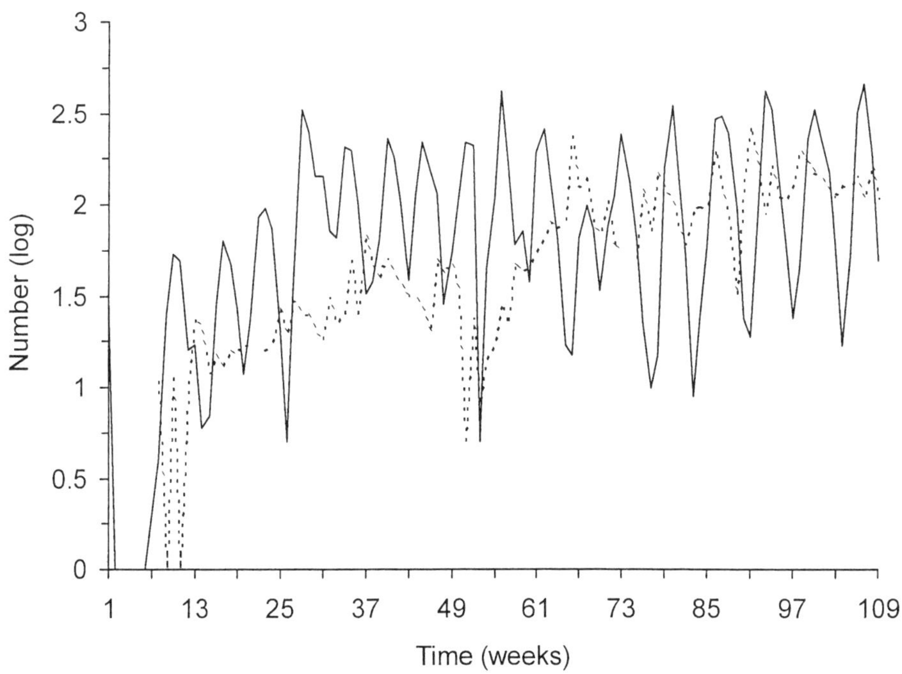

Figure 2. The dynamics of *P. interpunctella* (thin line) and GV-infected larvae (dotted line), measured as the number of dead adult moths and infected larvae per weekly census. The mean host density (± s.e.) of this population, after a settling down period of 26 weeks, was 134.7 ± 11.6. The cycle period was 6 and 6.4 weeks (ACF and spectral analysis respectively), and the peak ACF was 0.52 (P<0.05). Other than a steady increase in the level of infection, there was no obvious pattern in the dynamics of the pathogen.

exhibited clear cycles with periods of between six and seven weeks, approximately the period of one host generation.

In all species combinations, the ACF trajectories (not presented here) were strongly damped with increasing period, suggesting an endogenous, population-related cause of the cycles ("phase-forgetting"), rather than an exogenous, environment-related cause ("phase-remembering") (Turchin & Taylor 1992). Generation cycles have been predicted in single species populations when larval competition acts immediately and is strongly asymmetric (Gurney & Nisbet 1985), as it is here (Sait *et al.* 1994a). One other point of note was the long term persistence of these cultures, which were either used for other experimental purposes or ultimately abandoned whilst still extant.

2.2 *Host-pathogen dynamics*

In the three replicate populations the host again exhibited generation cycles, but whilst the average abundance did not differ from the host-alone cultures (around 150; t=1.58, P=0.19), the amplitude of the cycles was noticeably greater (typical range between 40 and 320) and coefficients of variation of abundance significantly higher (t=2.90, P<0.05; Figure 2). Briggs and Godfray (1995) predict generation cycles in host-pathogen interactions when the host suffers explicit stage-structured mortality, as here (Sait *et al.* 1994b). The dynamics of the GV showed no discernible pattern, perhaps as a consequence of it being long-lived compared to the host. Once again, the three replicate populations were persistent.

2.3 *Host-parasitoid dynamics*

Both the host and parasitoid displayed regular generation cycles as above. The abundance of the host, however, was significantly reduced (mean of around 10), and the amplitude of the cycles greatly enhanced (range between 0 and 100) (Figure 3). One host generation cycles have been predicted to arise in both the host and parasitoid (Godfray & Hassell 1989; Gordon *et al.* 1991) under the conditions pertaining in the system decribed here: i) the reproductive period of the host and parasitoid are relatively short [4 and 2 days respectively (Begon *et al.* 1995)], ii) the pre-adult periods of the host and parasitoid should be in the ratio of approximately 2:1 or 2:3 [42 days for the host (Sait *et al.* 1994a), 23 days for the parasitoid (Harvey *et al.* 1994)], iii) there should be strong density dependence in the rate of parasitism (Hughes *et al.* 1994). As above, the three replicate populations were persistent.

2.4 *Host-pathogen-parasitoid dynamics*

The three-species interaction contrasted markedly with the simpler single and two-species systems (Figure 4). The first notable point is the lack of persistence of all of the replicates (8 in total). In each case, the host went extinct first, followed rapidly by both of the natural enemies. Secondly, and perhaps most strikingly, both the host and parasitoid exhibited multigeneration cycles, with periods of 3-4 host generation cycles (20 weeks). The replicate culture given in Figure 4 passed through four such multigeneration cycles before extinction, whilst other replicates more typically passed through 1 or 2 cycles in abundance (with periods ranging between 17 and 26 weeks) before the host and its natural enemies disappeared (full details in Begon *et al.* 1996 and 1997).

3 SUMMARY AND CONCLUSIONS

The single and two-species systems were characterized by long term persistence and stable dynamics, in the sense that they fluctuated with regular host generation length cycles. The intrinsic cyclic dynamics of the host were affected to a moderate extent by its interaction with the pathogen, whilst the parasitoid exerted a considerable influence on both host abundance and cycle amplitude.

Despite detailed knowledge of the component systems at the individual and population level, a surprising and contrasting picture emerged from the three-species interaction. In all cases there was no long term persistence, extinction of the host being followed by each of the parasites. Interestingly, neither of the natural enemies were unilaterally excluded, so the system was never observed reverting to either of the simpler two-species combinations. Furthermore, whilst still extant, the host and parasitoid exhibited multigeneration cycles approximately three to four host generation cycles in length. These cycles were also of such amplitude that extinction was likely following any one of them, and the three species systems typically only persisted for one or two such cycles of abundance.

One of the proposed mechanisms for the multigeneration cycles was the lack of age-structured mortality caused by the natural enemies when acting in combination, since all larval stages were exposed to attack from either or both of the parasites (the pathogen exploiting the early larval instars, the parasitoid preferring the later stages). The multigeneration cycles observed here have indeed been predicted in models lacking age structure, contrasting with the clear age-dependent vulnerability of the host to parasitism and viral infection which lead to the observed generation cycles, also predicted by theory.

Hochberg *et al.* (1990) explored the population and community dynamics of a host-pathogen-parasitoid interaction and predicted that a number of dynamic patterns were likely, such as unilateral exclusion of either species, or three-species co-existence with constant, cyclic (including multigeneration cycles) or chaotic populations. The outcome depended on the

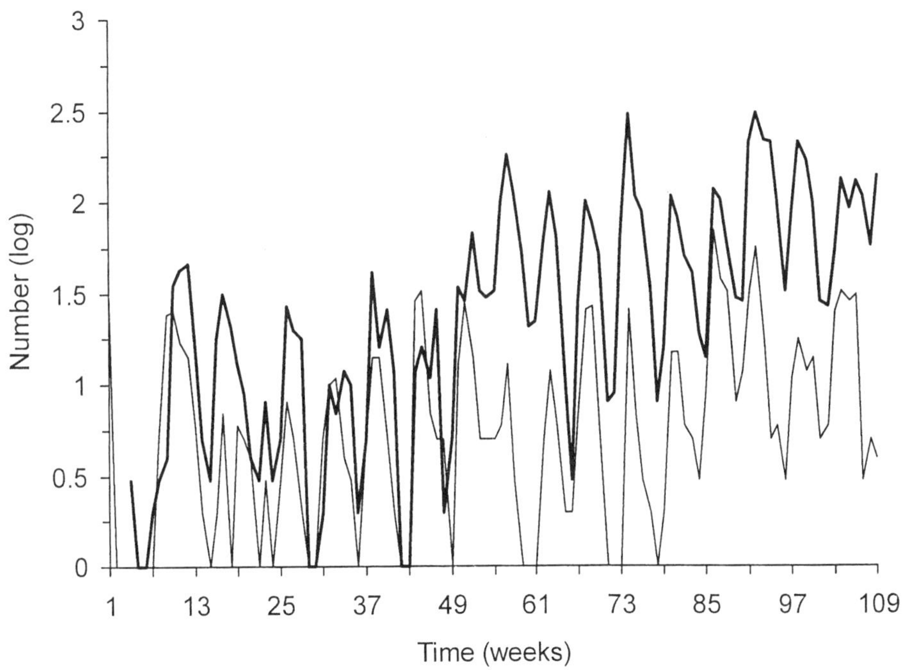

Figure 3. The dynamics of *P. interpunctella* (thin line) and *V. canescens* (bold line), measured as the number of dead adult moths and parasitoids per weekly census. Following a settling down period of 7 weeks, the mean host density (± s.e.) of this population was 9.6 ± 1.2; the mean parasitoid density was 55.4 ± 6.5. The cycle periods were 6 and 6.3 weeks for the host; 6 and 6.5 weeks for the parasitoid (ACF and spectral analysis). The peak ACFs were 0.53 and 0.56 for the host and parasitoid respectively ($P<0.05$).

relative competitive abilities within co-infected hosts and their relative extrinsic potentials, such as searching and transmission efficiency. In the GV-*V. canescens* competitive interaction, as in others (see review in Brooks 1993), the pathogen is the superior within-host competitor, whilst the parasitoid is better able to exploit susceptible hosts in the environment (Begon *et al.* 1997). Yet, the theoretical prediction of persistent two- or three-species coexistence was not demonstrated experimentally here.

Similarly, in analagous investigations, several multiparasitoid-prey models have been developed (May & Hassell 1981; Kakehashi *et al.* 1984; Hassell & May 1986; Briggs 1993; Hochberg 1996) which suggest that, as above, coexistence between natural enemies is enhanced if there is an element of resource partitioning. This may take the form of preference for different host developmental stages, or of asymmetric competitive abilites in co-parasitized hosts. Armstrong and McGehee (1980) show that coexistence is also possible if the resource exhibits limit cycles, as it does here (in this case, the host).

The same principles apply to the host-pathogen-parasitoid interaction, where the pathogen attacks the early larval stages of the host whilst the parasitoid prefers the later stages. However, the greater complexity did not promote stability and coexistence, and the three-species interaction exhibited a marked propensity to go extinct. This observation seems to contradict the apparently prevailing natural situation where species coexist in a complex web of interactions. Clearly, interactions with other species in more complex systems, as in nature, perhaps leads to the persistence and coexistence that was absent here. The spatial element, not examined explicitly in this study, could also be a crucial factor. For example, a physical refuge, afforded in this case by the depth of food, determines the vulnerability of the prey to parasitism (e.g. Begon *et al.* 1995), but it will also influence the spatial distribution of the pathogen and, hence, the probability that susceptible hosts become infected. Perhaps more important is a

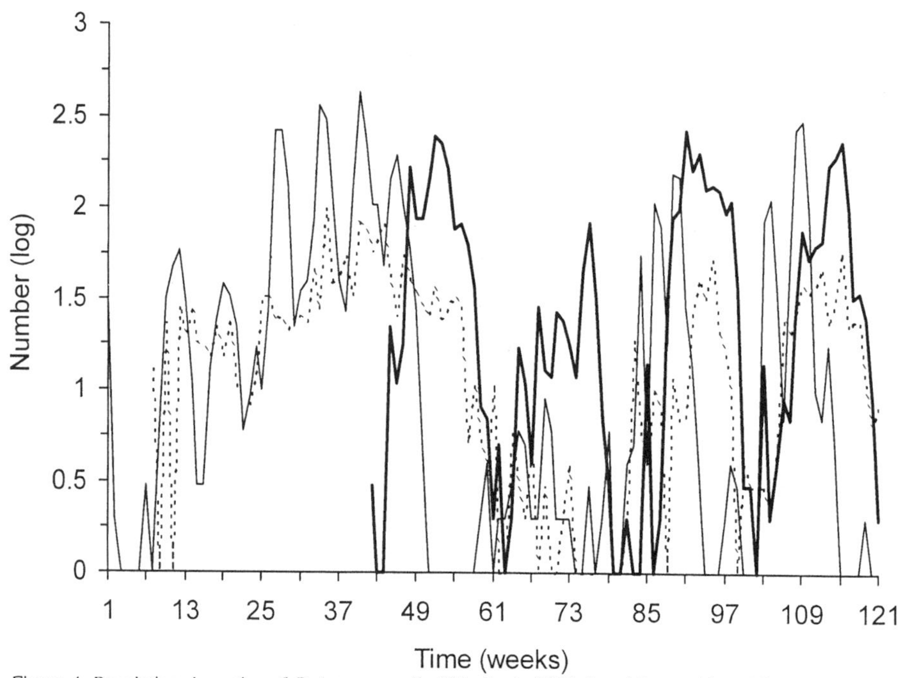

Figure 4. Population dynamics of *P. interpunctella* (thin line), GV-infected larvae (dotted line) and *V. canescens* (bold line), recorded as a weekly census of dead adults and infected larvae, following the introduction of the parasitoid (after 42 weeks) to an established host-pathogen culture. The cycle periods (host, parasitoid and pathogen respectively) were 20, 20 and 64 weeks. Similarly, the peak ACFs were 0.24, 0.53 and 0.15 ($P<0.05$). The data series were too short to perform spectral analysis.

larger spatial scale that encompasses metapopulations (Hanski & Gilpin 1997, and references therein). In such a scenario, dispersal between subpopulations, a key process in metapopulation dynamics, leads to the re-colonization of habitat patches left vacant following the extinction of a particular species. Such a process, if frequent enough, could also prevent extinction in the first place. Thus, complex multispecies systems, which may be inherently unstable, as here, continue to persist as a result of large spatial processes (or are seen to persist at the spatial scales observed by ecologists).

This study highlighted the role of the pathogen as a crucial factor in the resulting predator-prey dynamics, and served to illustrate that seemingly strongly coupled predator-prey interactions in nature may in fact be an integral part of a more complex interaction with a pathogen, the observed dynamics being relevant *only* in the context of the multispecies interaction. The ecology of baculoviruses in the field remains poorly understood (Godfray 1995; Cory *et al.* 1997), possibly because they are difficult to identify and observe. However, since their effects can nonetheless be dramatic, as this study suggests, seemingly inapparent pathogen infections may have a fundamental role to play in population and community ecology.

REFERENCES

Armstrong, R.A. & McGehee, R. 1980. Competitive exclusion. *Am. Nat.* 115:151-170.

Begon, M., Bowers, R.D., Kadianakis, N. & Hodgkinson, D.E. 1992. Disease and community structure: the importance of host self-regulation in a host-host-pathogen model. *Am. Nat.* 139:1131-1150.

Begon, M., Sait, S.M. & Thompson, D.J. 1995. Persistence of a parasitoid-host system: refuges and generation cycles? *Proc. R. Soc. Lond. B* 260:131-137.

Begon, M., Sait, S.M. & Thompson, D.J. 1996. Predator-prey cycles with period shifts between two- and three-species systems. *Nature* 381:311-315.

Begon, M., Sait, S.M. & Thompson, D.J. 1997. Two's company, three's a crowd: host-pathogen-parasitoid dynamics. In A.C. Gange & V.K. Brown (eds.), *Multitrophic Interactions in Terrestrial Systems*: 307-332. Oxford: Blackwell Science.

Briggs, C.J. 1993. Competition among parasitoid species on a stage-structured host and its effect on host suppression. *Am. Nat.* 141:372-397.

Briggs, C.J. & Godfray, H.C.J. 1995. The dynamics of insect-pathogen interactions in stage-structured populations. *Am. Nat.* 145:855-887.

Brooks, W.M. 1993. Host-parasitoid-pathogen interactions. In N.E. Beckage, S.N. Thompson & B.A. Federici (eds.), *Parasites and Pathogens of Insects,* Volume 2, Pathogens: 231-272. New York: Academic Press.

Comins, H. & Hassell, M.P. 1976. Predation in multi-prey communities. *J. Theor. Biol.* 62:93-114.

Cory, J.S., Hails, R.S. & Sait, S.M. 1997. Baculovirus ecology. In L.K. Miller (ed.), *The Baculoviruses*: 301-339. New York: Plenum Press.

Godfray, H.C.J. 1995. Field experiments with genetically manipulated insect viruses: ecological issues. *Trends Ecol. Evol.* 10:465-469.

Godfray, H.C.J. & Hassell, M.P. 1989. Discrete and continuous insect populations in tropical environments. *J. Anim. Ecol.* 58:153-174.

Gordon, D.M., Nisbet, R.M., De Roos, A., Gurney, W.S.C. & Stewart, R.R. 1991. Discrete generations in host-parasitoid models with contrasting life cycles. *J. Anim. Ecol.* 60:295-308.

Gurney, W.S.C. & Nisbet, R.M. 1985. Fluctuation periodicity, generation separation, and the expression of larval competition. *Theor. Popul. Biol.* 28:150-180.

Hanski, I. & Gilpin, M.E. 1997. *Metapopulation Biology: Ecology, Genetics and Evolution*. New York: Academic Press.

Harvey, J.A., Harvey, I.F. & Thompson, D.J. 1994. Flexible larval growth allows the use of a range of host sizes by a parasitoid wasp. *Ecology* 75:1420-1428.

Hassell, M.P. & May, R.M. 1986. Generalist and specialist natural enemies in insect predator-prey interactions. *J. Anim. Ecol.* 55:923-940.

Hassell, M.P., Comins, H.M. & May, R.M. 1994. Species coexistence and self-organizing spatial dynamics. *Nature* 370:290-292.

Hochberg, M.E. 1996. Consequences for host population levels of increasing natural enemy species richness in classical biological control. *Am. Nat.* 147:307-318.

Hochberg, M.E. & Holt, R.D. 1990. The coexistence of competing parasites. I. The role of cross-species infection. *Am. Nat.* 136:517-541.

Hochberg, M.E., Hassell, M.P. & May, R.M. 1990. The dynamics of host-pathogen-parasitoid interactions. *Am. Nat.* 135:74-94.

Holt, R.D. 1977. Predation, apparent competition, and the structure of prey communities. *Theor. Popul. Biol.* 12:197-229.

Holt, R.D. & Pickering, J. 1985. Infectious disease and species coexistence: a model in Lotka-Volterra form. *Am. Nat.* 126:196-211.

Holt, R.D. & Lawton, J.H. 1994. The ecological consequences of shared natural enemies. *Ann. Rev. Ecol. Syst.* 25:495-520.

Huffaker, C.B. 1958 Experimental studies on predation: dispersion factors and predator-prey oscillations. *Hilgardia* 27:343-383.

Hughes, J.P., Harvey, I.F. & Hubbard, S.F. 1994. Host searching behaviour of *Venturia canescens* (Grav.) (Hymenoptera: Ichneumonidae): interference - the effect of mature egg load and prior behaviour. *J. Insect Behav.* 7:433-454.

Kakehashi, M., Suzuki, Y. & Iwasa, Y. 1984. Niche overlap of parasitoids in host-parasitoid systems: its consequences to single versus multiple introduction controversy in biological control. *J. Appl. Ecol.* 21:115-131.

Kareiva, P. 1989. Renewing the dialogue between theory and experiments in population ecology. In J. Roughgarden, R.M. May & S.A. Levin (eds.), *Perspectives in Ecological Theory*: 68-88. New York: Princeton University press.

Lawton, J.H. 1995. Ecological experiments with model systems. *Science* 269:328-331.

May, R.M. & Hassell, M.P. 1981. The dynamics of multiparasitoid-host interactions. *Am. Nat.* 117:234-261.

Park, T. 1948. Experimental studies of interspecific competition. I. Competition between populations of the flour beetles *Tribolium confusum* Duval and *Tribolium castaneum* Herbst. *Ecol. Monogr.* 18:267-307.

Roughgarden, J. & Feldman, M. 1975. Species packing and predation pressure. *Ecology* 56:489-492.

Sait, S.M., Begon, M. & Thompson, D.J. 1994a. Long-term population dynamics of the Indian meal moth, *Plodia interpunctella*, and its granulosis virus. *J. Anim. Ecol.* 63:861-870.

Sait, S.M., Begon, M. & Thompson, D.J. 1994b. The influence of larval age on the response of *Plodia interpunctella* to a granulosis virus. *J. Invertebr. Pathol.* 63:107-110.

Sait, S.M., Andreev, R.A., Begon, M., Thompson, D.J., Harvey, J.A. & Swain, R.D. 1995. *Venturia canescens* parasitizing *Plodia interpunctella*: host vulnerability - a matter of degree. *Ecol. Entomol.* 20:199-201.

Sait, S.M., Begon, M., Thompson, D.J., Harvey, J.A. & Hails, R.S. 1997. Factors affecting host selection in an insect host-parasitoid interaction. *Ecol. Entomol.* 22:225-230.

Turchin, P. & Taylor, A.D. 1992. Complex dynamics in ecological time series. *Ecology* 73:289-305.

3 Community ecology

Population and Community Ecology for Insect Management and Conservation, Baumgärtner et al. (eds)
 ISBN 90 5410 930 0

The relevance of life-history theory for carabid species of Western Europe*

P.J.den Boer
Wijster, Netherlands

Th.S.van Dijk
Biological Station of the Agricultural University Wageningen, Wijster, Netherlands

ABSTRACT: By assuming that natural selection controls life-history traits evolutionary biologists try to construct evolutionary processes. Models should show how life-history traits that promote the fitness of individuals must lead to the life histories observed at present as well as those to expect in future.

The authors of this paper compare the generally occurring life histories among West European carabid species, and conclude that the traits governing these life histories differ significantly from those usually advanced by evolutionary biologists. Next to 'time of reproduction' especially 'dispersal power' and 'turnover frequency' of population units appear to be significant. They try to explain these departures from the generally accepted schemes, and emphasize the need to do more comparative investigations of life histories among groups of related species. As an example of the result to expect they give a provisional scheme of the dominant life history patterns of the carabid species of Drenthe in the North of the Netherlands.

1 INTRODUCTION

Evolutionary biologists try to understand evolutionary processes by imagining how natural selection most probably would lead to life histories that will maximize fitness of the individuals of the populations concerned. Which life-history traits can be assumed to most contributing to individual fitness? It is quite logical to suppose that traits which under the given conditions would maximize reproduction will give the individual the best chance to propagate its genes, and thus to best contribute to the fitness of the population. Under highly unstable conditions a maximum reproduction can be reached by producing as many eggs (or young) as possible, in order to compensate for unpredictable losses in future; this means: maximizing productivity, i.e. r-selection. Under stable conditions it seems better to give the young the best possible start, even when this would mean the production of a few young only, in order to provide them with the highest fitness possible under competition; this means maximizing efficiency, i.e. K-selection. These ideas of MacArthur & Wilson (1967) were extended, diversified and adapted to special circumstances in what is called 'life-history theory' (Stearns 1992; Roff 1992).

**Communication No. 588 of the Biological Station Wijster*

2 LIFE-HISTORY THEORY

In the course of time most evolutionary biologists agreed that the most important life-history traits promoting fitness must be (1) brood size, (2) relative size of the young, (3) age distribution of reproduction, i.e. a single reproductive period per female : semelparity or more reproductive periods per female: iteroparity, (4) the interaction of reproductive effort with other energy-asking processes and thus with adult mortality, i.e. trade-offs between fecundity and survival of adults, (5) age of first reproduction. Depending on the general circumstances under which certain animals are living one or more of these five life-history traits would be promoted by natural selection. Efficient combinations of life-history traits are sometimes called 'strategies'. As most animals do not actively choose the most efficient combinations of life-history traits, in my opinion, the term 'strategy' (derived from military operations) gives a wrong suggestion. Most animals are passively manipulated by natural selection into a certain way of living, in which learning by experience allows only restricted modifications of the genetically fixed pattern. Therefore, I prefer to call combinations of life-history traits life-history patterns, i.e. combinations of traits that by 'trial and error' are brought together by natural selection in the genomes of individuals.

However, the development of life-history theory was mainly theoretical and stimulated the construction of models for special situations. This trend was heavily criticized by Stearns (1976), but apparently he let persuade himself by his colleagues that his criticisms were wrong, for in his book of 1992 he associates himself with the general trend in literature. In 1976 Stearns' criticism mainly concerned the fact that theory developed without sufficiently testing hypotheses with adequate field data. As a field ecologist this criticism of Stearns (1976) appealed to me.

3 LIFE-HISTORY TRAITS AMONG CARABID BEETLES

For a long time carabidologists have been fascinated by differences in wing development in relation to life histories and habitat selection. Larsson (1939) was the first to discriminate between spring and autumn breeders. Darlington (1943) following Darwin (1859) evaluated wing development in relation to dispersal and habitat selection. Lindroth (1945, 1949) brought life histories, habitat selection and wing development together in an imaginative picture of the development of the carabid fauna of Fennoscandia after the last Ice Age. Den Boer (1977) tried to give a comparable picture for the cultivated countryside Drenthe in the Netherlands, and Thiele (1977) composed an excellent review of our knowledge of the adaptations of carabid beetles in relation to habitat selection. Stimulated by the paper of Stearns (1976) and because I had available life-history data of more than 150 carabid species, at a carabid conference in 1978 in Rees-Grietherbusch I gave a provisional impression of the most important life-history traits among the carabid species of Drenthe by comparing the way of living of forest species with that of species of unstable habitats (Den Boer 1979). These data were supplemented by den Boer et al. (1980) for wing development and by Den Boer & Den Boer-Daanje (1990) for reproduction periods. Recently all this was worked out more completely by Den Boer & Van Dijk (1996), and I want to tell you what are our most important conclusions.

First of all we concluded that 'brood size' could not be a very important life-history trait among carabid species, because in poikilothermic animals reproduction is highly affected by temperature and amount of high quality food, also among carabids (Van Dijk 1983, 1986). Even under constant conditions individual egg production varies with a factor ten (Van Dijk 1979, 1994; Nelemans 1987). Hence, a high reproductive potence can hardly be expected ever to result in a correspondingly high production of young, the more so as larval mortality usually is very high and unpredictable. Second, most carabid species, at least in western Europe, are iteroparous, showing more reproductive periods in succeeding years; the few semelparous species, so far discovered, do not live under special conditions. Further, there were no 'trade-offs' between egg production and survival (Van Dijk 1979, 1994), nor between egg production and development of wings (Aukema 1991, 1994). On the contrary, the highest reproducing females survived best, and females that developed functional wings and wing-muscles produced more eggs than wingless females. The size of eggs differs between species, but there could not be found any relationship with dominating conditions: species producing a few big eggs both inhabit stable habitats like forest or old heath areas, and unstable habitats; the same applies to species producing many small eggs, and to the relationship between size of the beetles and size of reproduction.

The only life-history trait favoured by the evolutionary biologists of the life-history theory that appears to be important among carabid species is 'age of first reproduction'. Autumn breeders, which reproduce immediately after emergence, prefer stable habitats and usually produce many small eggs, i.e. potentially they have a high level of egg production. This is necessary to compensate for the usually very high mortality of young larvae under winter conditions (e.g. Van Dijk & Den Boer 1992). Spring breeders, which reproduce after having hibernated, usually produce less eggs and the eggs often are somewhat bigger, but under field conditions there is no actual difference in size of reproduction between spring breeders of stable and of unstable habitats.

An important life-history trait among carabid species appears to be 'dispersal power', a trait not appreciated by evolutionary biologists. Most species of stable habitats have low powers of dispersal, either being wingless or wing-dimorphic with a low fraction of fully winged individuals, or, if fully winged, not showing a high level of flight activities. Species of unstable habitats usually are constantly fully winged and show frequent flight activities. Dispersal power links 'turnover of local populations' with habitat preference. In unstable habitats carabid species show both high powers of dispersal and a high turnover of local populations, i.e. local populations only survive for a few years, but are frequently founded or refounded. In stable habitats carabid species show both low powers of dispersal and a lower turnover of local populations. In accordance with life-history theory in the most stable habitats, old forests and especially on mountains, we find a number of carabid species that produce low numbers of rather big eggs (*Carabus*, *Abax* and *Molops* species), and which often show some simple kind of parental care (Brandmayr & Zetto-Brandmayr 1979). But they form only part of the carabid fauna of old forest; there are also carabid species that lay many small eggs living in stable habitats, old forests included. Moreover, in other stable habitats carabid beetles with quite different reproductive potentials can be found. Hence, we can conclude that among

Table 1. Provisional scheme of a classification of life-history patterns of carabid species of Western Europe.

(1). Spring breeding (summer larvae)
well dispersing species with a high turnover of populations
- a. macropterous species of unstable or temporary habitats
- b. macropterous or wing-dimorphic species of unstable, stable or transitional and/or changing habitats

(2) Spring breeding (summer larvae)
rather badly dispersing species with a generally not very high turnover of populations and occupying stable habitats
- a. macropterous species, possibly with low powers of dispersal
- b. wing-dimorphic species with low powers of dispersal
- c. brachypterous species with poor powers of dispersal

(3) Species with a complex reproduction cycle and variable developmental periods and occupying stable habitats

(4) Summer or autumn breeding (winter larvae)
well dispersing species with a high turnover of populations
- a. macropterous species of unstable or temporary habitats
- b. wing-dimorphic species of unstable and/or transitional or changing habitats

(5) Summer or autumn breeding (winter larvae)
rather badly dispersing species with a not very high or low turnover of populations
- a. macropterous species of stable and/or transitional habitats
- b. wing-dimorphic species of stable habitats
- c. brachypterous species of stable habitats

(6) Late autumn breeding with summer diapause
low powers of dispersal and a low turnover of populations

(7) Winter or early spring breeding
inhabitants of open, sandy sites with a high turnover of populations
- a. macropterous species with good powers of dispersal
- b. wing-dimorphic species with rather good powers of dispersal

(8) Winter or early spring breeding
inhabitants of open, sandy sites with a not very high turnover of populations
- a. wing-dimorphic or brachypterous species with low powers of dispersal

carabid species of Western Europe especially dispersal power, turnover frequency and habitat selection are closely associated, i.e. they form a life-history pattern that is weakly associated with 'age (time) of first reproduction'.

Therefore, for the carabid species of western Europe I consider 'time of first reproduction' as the primary life-history trait, among which we can distinguish again differences in dispersal power, turnover frequency and habitat selection. A more detailed treatment and a complete scheme of life-history patterns for 174 carabid species can be found in Den Boer & Van Dijk (1996) and is summarized in Fig. 1 and the tables 1 and 2 of this paper.

4 CAUSES OF DEVIATING LIFE-HISTORY PATTERNS AMONG CARABID BEETLES

We may wonder why our experiences with the dominant life-history traits of carabid beetles deviate to such a high degree from the predictions of life-history theory. In our opinion this is connected with a too simple idea of the effects of natural selection. Following Charles Darwin too strictly natural selection is often thought to be 'survival of the fittest'. But do only the 'fittest' survive and are the non-survivors always 'unfit'? Den Boer (in Den Boer & Reddingius 1996: Ch. 3) explains that natural selection is the 'non-survival of the non-fit', because only the individuals being 'non-fit' in some respect and those being 'unlucky' will die (being selected away) under a certain set of conditions. All other individuals will survive and reproduce, i.e. are positively selected. During its life time each individual passes through a number of such selective events, and during each event it may be 'fit' or 'unfit', or belong to the 'lucky' chance survivors or to the 'unlucky' chance victims. Among the reproducing individuals there may be some that under other sets of conditions would have been 'unfit' and others that were simply 'lucky' enough to escape so far from the current adverse conditions.

Therefore, under the complicated and variable conditions in the field we may not expect that females producing the highest numbers of eggs will always have the greatest progeny. Moreover, before these eggs have produced fertile females again many environmental processes and chance events may have changed this advantage to such an extent that it may have disappeared completely. And, as mentioned above, egg production among carabid beetles (and most other arthropods) is highly variable, by which females potentially able to produce a high number of eggs may not be able to realize this potential under field conditions. Therefore, it can be expected that quite different life-history traits will determine the fate of such species, as was demonstrated for carabid beetles of western Europe by Den Boer & Van Dijk (1996). Just as in the discussions around 'regulation of numbers' (Den Boer & Reddingius 1996) we cannot get an adequate picture of natural processes without taking the variability of the more important natural processes into account, i.e. models and ideas based upon deterministic thinking cannot be expected to give us a realistic insight into natural processes, in spite of the often appealing logic of such deterministic thoughts. Therefore, in my opinion life-history theory as it is presented by Stearns (1992) and Roff (1992) does not apply to animal populations (especially not to invertebrates) under natural conditions.

5 REQUEST TO ENTOMOLOGISTS

I told all this not only because our findings about the life-history patterns among carabid species remarkably deviate from the conclusions of life-history theory, but more to stimulate entomologists to think about the most important life-history patterns dominating other groups of arthropods. It would be highly interesting to compare the empirically found life-history patterns in many different groups of arthropods, and in this way to contribute to an improvement of life-history theory, making this theory less theoretical and more empirical. For, it is beyond dispute that

Table 2. Provisional classification of the carabid species of Drenthe (the Netherlands) according to the life-history patterns in the scheme of Table 1 The numbers behind the species are used in Fig. 1 and refer to Den Boer (1977: Table 2 and Appendix A, Part I).

A. the most abundant species (mentioned in Fig. 1).

---

(1). a. *Agonum sexpunctatum* (18: wet), *Amara aenea* (21: agr.), *A. communis* (26: agr.), *A. famelica* (30: wet), *A. familiaris* (31: agr.), *A. plebeja* (35: agr.), *Anisodactylus binotatus* (42: agr.), *Asaphidion flavipes* (44: agr.; Coll.), *Harpalus affinis* (99: agr.), *Loricera pilicornis* (112: wet; Coll.), *Pterostichus minor* (135: wet; 95% macr.), *P. nigrita/rhaeticus* (137: wet; see Koch & Thiele, 1980), *P. quadrifoveolatus (angustatus)*(130: burn), *P. vernalis* (140: wet).

b. *Agonum fuliginosum* (9: wet, forest borders; 28% macr.), *Amara lunicollis* (34: wet, grass; macr.), *Bembidion lampros* (54: agr., wet; 17% macr.), *Clivina fossor* (86: agr., wet; 90% macr.), *Dyschirius globosus* (95: agr., wet; 0.1% macr.), *Notiophilus palustris* (122: forest borders; 7% macr.), *Pterostichus diligens* (133: wet, grass; 4% macr.), *P. strenuus* (139: meadows, forest; 23% macr.).

(2). a. *Agonum assimile* (6: forest borders), *Notiophilus rufipes* (124: forest; Coll.), *Pterostichus oblongopunctatus* (138: forest), *P. versicolor (coerulescens)* (132: heath, poor meadows).

b. *Syntomus (Metabletus) foveatus* (114: sandy heath; 1.2% macr.).

c. *Agonum ericeti* (8: peat moor), *Carabus nemoralis* (78: forest), *C. arvensis* (75: heath), *Pterostichus lepidus* (134: sandy heath)*.

(3). *Abax parallelepipedus (ater)* (1: forest; brach.), *Calathus rotundicollis (piceus)* (73: light forest; 93% macr.), *Notiophilus aquaticus* (119: heathy areas; 1.2% macr.; Coll.), *N. biguttatus* (120: light forest; 74% macr.; Coll.), *Carabus cancellatus* (76: heathy areas; brach.).

(4). a. *Harpalus rufipalpis (rufitarsis)* (106: poor grassland) *H. rufipes (pubescens)* (104: agr.).

b. *Calathus fuscipes* (69: trans.; 0.9% macr.), *Pterostichus melanarius (vulgaris)* (141: agr.; 2% macr.), *Trechus obtusus* (146: forest borders; 3% macr.).

(5). a. *Amara brunnea* (25: light forest), *A. equestris* (29: heath), *Broscus cephalotes* (66: sand), *Harpalus latus* (103: forest, heath), *H. quadripunctatus* (105: forest borders), *H. solitaris (fuliginosus)* (102: heath), *Nebria salina* (118: trans.), *Pterostichus niger* (136: all habitats).

b. *Agonum obscurum* (17: wet grass; 0.1% macr.), *Calathus erratus* (68: sand; 0.2% macr.), *C. melanocephalus* (70: sandy heath; 0.2% macr.), *Cymindis vaporariorum* (89: heath; 6% macr.), *Notiophilus germinyi* (121: sandy heath; 3% macr.), *Olisthopus rotundatus* (125: heath; 21% macr.).

c. *Carabus problematicus* (80: dry forest), *Cymindis macularis* (88: sand), *Pterostichus lepidus* (134: sandy heath)*, *Trechus secalis* (148: forest).

(6). *Leistus rufomarginatus* (110: forest; macr.), *L. terminatus (rufescens)* (109: grass, trans.; macr.), *Nebria brevicollis* (117: light forest, trans.).

(7). a. *Bradycellus ruficollis (similis)* (65: heath), *Trichocellus cognatus* (149: heath), *T. placidus* (150: light forest).

b. *Bradycellus caucasicus (collaris)* (62: heath), *Bradycellus harpalinus* (64: grassy heath).

(8). a. *Amara infima* (33: sand; 1.3% macr.), *Bembidion nigricorne* (56: sand; brach.).

**Pterostichus lepidus* seems to be both a spring and an autumn breeder (Van Dijk, pers. comm., and Paarmann, 1990).

B. less abundant species

(1) a. *Acupalpus brunnipes, A. consputus, A. dubius, A. exiguus, A. flavicollis, A. meridianus, A. parvulus (dorsalis) Agonum albipes (ruficorne), A. dorsalis, A. gracile, A. marginatum, A. muelleri, A. piceum, A. thoreyi, A. versutum, A. viduum, Amara anthobia, A. ingenua, A. ovata, A. similata, A. spreta, Bembidion assimile, B. bruxellense (rupestre), B. doris, B. femoratum, B. obliquum, B. properans, B. quadrimaculatum, B. varium, Blethisa multipunctata, Dyschirius aeneum, D. luedersi, D. politus, D. thoracicus (arenosus), Elaphrus cupreus, E. riparius, Omophron limbatus, Oodes helopioides, Stenolophus mixtus, S. teutonus.*

b. *Amara convexior, Badister dilatatus, Bembidion guttula, B. tetracolum (ustulatum), Calosoma inquisitor, Chlaenius nigricornis, Cicindela campestris, C. hybrida, Demetrias atricapillus, Dromius agilis, D. angustus, D. melanocephalus, D. quadrimaculatus, D. spilotus (quadrinotatus), Harpalus anxius, Notiophilus substriatus, Panageus cruxmajor.*

(2) a. *Agonum krynickii, A. livens, A. moestum, Anisodactylus nemorivagus, Badister bullatus (bipustulatus), B. sodalis, B. unipustulatus, Bembidion humerale, Lebia chlorocephala, Odocantha melanura.*

b. *Carabus clathratus, C. granulatus, Syntomus (Metabletus) truncatellus, Pterostichus anthracinus.*

c. *Bembidion mannerheimi (unicolor), Carabus nitens, Cicindela germanica, Cychrus caraboides rostratus.*

(3). *Carabus coriaceus.*

(4) a. *Amara apricaria, A. aulica, A. bifrons, A. consularis, A. convexiuscula, A. fulva, A. majuscula, Asaphidion pallipes, Calathus ochropterus (mollis), Harpalus distinguendus, H. rubripes, H. smaragdinus, H. tardus, Trechus quadristriatus, T. discus.*

b. *Calathus cinctus (erythroderus).*

(5) a. *Amara pseudocommunis, A. kulti, A. praetermissa, A. quenseli, Miscodera arctica, Nebria livida.*

b. *Calathus ambiguus, Synuchus nivalis.*

c. *Calathus micropterus, Masoreus wetterhalli, Patrobus atrorufus (excavatus), Stomis pumicatus.*

(6). *Leistus spinibarbis.*

(7) a. *Bradycellus verbasci.*

b. *Bradycellus csikii.*

(8) a. *Bradycellus sharpi.*

Between brackets behind the species an indication of the preferred habitat
macr. = macropterous; brach. = brachypterous; agr. = occupying agricultural fields and other ruderal or disturbed sites; wet= occupying banks of pools and rivers and other wet sites; burn= reproduces at sites where woods or remnants of wood has been burnt; grass= prefers dense grass vegetations; trans. = occupying transitional sites between forest and heath or grassy vegetations; sand= occupying blown sand areas with only little vegetation; Coll. = specialized in the hunting of Collemboles.

For estimates of dispersal power and turnover frequencies of the most abundant carabid species of Drenthe see Den Boer (1990a: Table 1; 1990b: Table 4).

species	no.	habitat	dp	Mar Apr May June July August Sept Oct Nov Dec Jan Feb 1 2 3 4 5 6 7 8 9 10 11 12 13 14 15 16 17 18 19 20 21 22 23 24 25 26
Asaphidion flavipes	44	H+A	m	◇◇◇◇◇◇◇◇◇◇
Pterostichus nigrita/rhaeticus	137	W	m	◇◇◇◇◇◇◇◇◇
Pterostichus diligens	133	W	d	◇◇◇◇◇◇◇◇◇◇◇◇◇◇◇
Pterostichus minor	135	W	d	◇◇◇◇◇◇◇◇◇◇◇◇
Pterostichus strenuus	139	F+W	d	◇◇◇◇◇◇◇◇◇◇◇
Pterostichus oblongopunctatus	138	F	(m)	◇◇◇◇◇◇◇◇◇◇◇◇◇
Pterostichus quadrifoveolatus	130	B	m	◇◇◇◇◇◇◇◇◇◇◇◇◇
Syntomus foveatus	114	H+S	(d)	◇◇◇◇◇◇◇◇◇◇
Agonum assimile	6	F	(m)	◇◇◇◇◇◇◇◇
Agonum ericeti	8	H+P	b	◇◇◇◇◇◇
Amara communis	26	A+R	m	◇◇◇◇◇◇◇◇◇
Amara familiaris	31	A+R	m	◇◇◇◇◇◇◇◇
Carabus nemoralis	78	F	b	◇◇◇◇◇◇◇
Loricera pilicornis	112	W	m	◇◇◇◇◇◇◇◇◇◇◇◇◇
Amara famelica	30	W	m	◇◇◇◇◇◇◇◇◇◇
Carabus arvensis	75	H	b	◇◇◇◇◇◇◇
Amara aenea	21	S+A	m	◇◇◇◇◇◇◇◇◇◇◇
Dyschirius globosus	95	H+A	(d)	◇◇◇◇◇◇◇◇◇◇◇◇◇
Pterostichus vernalis	140	W	m	◇◇◇◇◇◇◇◇◇◇◇◇◇
Pterostichus versicolor	132	H	(m)	◇◇◇◇◇◇◇◇◇◇◇◇◇
Notiophilus aquaticus	119	H	(d)	◇◇◇◇◇◇◇◇◇◇◇◇◇◇◇◇◇◇◇◇◇◇◇◇◇◇
Notiophilus palustris	122	F	d	◇◇◇◇◇◇◇
Notiophilus rufipes	124	F	(m)	◇◇◇◇◇◇◇◇
Agonum sexpunctatum	18	W	m	◇◇◇◇◇◇
Clivina fossor	86	H+A	d	◇◇◇◇◇◇◇
Harpalus affinis X	99	A	m	◇◇◇◇◇◇◇◇◇◇◇◇
Harpalus latus •	103	H+F	(m)	◇◇◇◇◇◇◇◇◇◇◇◇◇◇◇◇
Notiophilus biguttatus X	120	F	d	◇◇◇◇◇◇◇◇◇◇◇◇ ◇◇◇◇◇◇◇◇◇◇◇◇◇◇
Carabus cancellatus X	76	H	b	◇◇◇◇◇◇◇ ◇◇◇◇◇◇◇◇◇◇◇
Calathus rotundicollis •	73	F	d	◇◇◇◇◇◇ ◇◇◇◇◇◇◇
Cymindis macularis •	88	S	b	◇◇◇◇◇◇◇◇◇◇◇◇◇◇◇◇◇◇◇◇◇
Bembidion lampros X	54	W+A	d	◇◇◇◇◇◇◇◇◇◇◇
Harpalus quadripunctatus •	105	F	m	◇◇◇◇◇◇◇◇◇◇
Abax parallelepipedus •	1	f	b	◇◇◇◇◇◇◇◇◇◇◇◇◇
Agonum fuliginosum X	9	F	d	◇◇◇◇◇◇◇◇◇◇◇
Anisodactylus binotatus X	42	A	m	◇◇◇◇◇◇◇
Harpalus solitaris •	102	H	m	◇◇◇◇◇◇◇
Agonum obscurum •	17	H	(d)	◇◇◇◇◇◇◇◇◇◇◇◇◇◇
Pterostichus lepidus X	134	H+S	b	◇◇◇◇◇◇◇◇◇◇◇◇◇◇
Harpalus rufipes •	104	A	m	◇◇◇◇◇◇◇◇◇◇◇◇◇◇◇◇
Amara plebeja X	35	A	m	◇◇◇◇◇◇◇◇◇◇
Amara lunicollis X	34	H+W	m	◇◇◇◇◇◇◇◇◇◇◇◇
Harpalus rufipalpis •	106	R	m	◇◇◇◇◇◇◇◇◇
Pterostichus melanarius •	141	A	(d)	◇◇◇◇◇◇◇◇◇◇◇◇◇◇
Trechus secalis •	148	F	b	◇◇◇◇◇◇◇◇◇◇◇◇◇
Cymindis vaporariorum •	89	H	d	◇◇◇◇◇◇◇◇◇
Leistus terminatus •	109	F+H	(m)	◇◇◇◇◇◇◇◇◇◇◇◇◇◇◇◇◇
Calathus erratus •	68	H+S	(d)	◇◇◇◇◇◇◇◇◇◇◇◇◇◇
Broscus cephalotes •	66	S	m	◇◇◇◇◇◇◇◇◇◇
Amara brunnea •	25	F	m	◇◇◇◇◇◇◇◇◇
Calathus melanocephalus •	70	H	(d)	◇◇◇◇◇◇◇◇◇◇◇◇◇◇◇◇◇
Amara equestris •	29	H	(m)	◇◇◇◇◇◇◇◇◇◇
Pterostichus niger •	136	F+R	(m)	◇◇◇◇◇◇◇◇◇◇
Trechus obtusus •	146	F+R	d	◇◇◇◇◇◇◇◇◇◇◇◇
Notiophilus germinyi •	121	S	d	◇◇◇◇◇◇◇◇◇◇◇
Olisthopus rotundatus •	125	H	(d)	◇◇◇◇◇◇◇◇◇◇◇◇◇

Species	no.	habitat	dp			
Calathus fuscipes *	69	S+R	d		◇◇◇◇◇◇◇	
Carabus problematicus *	30	F	b		◇◇◇◇◇◇◇◇	
Nebria brevicollis *	117	F	m		◇◇◇◇◇◇◇◇◇◇◇◇◇◇	
Nebria salina *	118	H	m		◇◇◇◇◇◇◇◇◇	
Leistus rufomarginatus *	110	F	(m)		◇◇◇◇◇◇◇◇◇◇	
Trichocellus placidus X	150	F	m	◇◇◇◇◇◇◇◇	◇◇◇◇◇◇◇◇◇◇	◇◇◇◇◇
Trichocellus cognatus X	149	H	m	◇◇◇◇◇	◇◇◇◇◇◇◇◇◇◇	◇◇◇◇◇◇◇
Bradycellus ruficollis X	65	H	m	◇◇◇◇◇◇◇	◇◇◇◇◇◇◇◇◇◇◇◇◇	◇◇◇
Bradycellus harpalinus *	64	H	d	◇◇◇◇◇	◇◇◇◇◇◇◇◇◇◇◇◇◇◇	◇
Amara infima X	33	S	(d)	◇◇◇◇◇◇◇	◇◇◇◇◇◇◇◇◇	◇◇◇◇◇
Bembidion nigricorne X	56	S	b	◇◇◇◇◇◇	◇◇◇◇◇◇◇	
Bradycellus caucasicus X	62	H	d	◇◇◇◇◇◇◇	◇◇◇◇◇◇◇◇◇◇	◇◇◇◇

Fig.1. Reproductive periods of the 68 most abundant carabid species of Drenthe (the Netherlands) ranked from early reproducers (March-May) until late reproducers (October and in winter, sometimes continued in early spring). For each species the expected dispersal power (dp) is indicated: m= fully winged (macropterous), (m)= wings relatively small, often unsuitable for flight, d= wing- dimorphic with a high fraction of fully winged adults (> 2%), (d)= wing-dimorphic with a low fraction of fully winged adults (< 2%), b= unwinged (brachypterous). Also the preferred habitat of each species is indicated: W= wet, open habitats, e.g. banks of pools and rivers, B= sites where forest or peat has been burned, R= ruderal and other waste or disturbed sites, A= agricultural fields, S= blown sand fixed by vegetation and dry heathland, H= moist heathland, P= peat moor, F= forest. Species indicated by an asteriks must be considered autumn breeders with winter larvae, those indicated by X possibly are also autumn breeders, but the occurrence of winter larvae is not convincingly shown. Also, the species numbers (no.) are given, which were used by Den Boer (1977: table 2 and Appendix A, part 1).

life-history theory might be important for our understanding of evolutionary processes, but then life-history theory should be realistic and based on well-documented facts in the first place.

REFERENCES

Aukema, B. 1991. Fecundity in relation to wing-morph of three closely related species of the *Melanocephalus* group of the genus *Calathus* (Coleoptera, Carabidae). *Oecologia* 87: 118-126.

Aukema, B. 1994. The evolutionary significance of wing dimorphism in carabid beetles (Coleoptera, Carabidae). In *Dispersal polymorphism of insects, its adaptation and evolution*: 173-181. *Proceedings of the Memorial and international Symposium in Okyama University, June 30-July 1, 1994.*

Brandmayr, P. & Zetto-Brandmayr, T. 1979. The evolution of parental care phenomena in Pterostichine ground beetles with special reference to the genera *Abax* and *Molops* (Coleoptera, Carabidae). In P.J. den Boer, H.U. Thiele and F. Weber (eds), *On the evolution of behaviour in carabid beetles*: 35-49. *Miscellaneous Papers Landbouwhogeschool Wageningen*, 18.

Darlington, P.J. 1943. Carabidae of mountains and islands. *Ecological Monographs* 13: 37-61.

Darwin, Ch. 1859. *The Origin of species by means of Natural Selection or the Preservation of favoured Races in the Struggle for Life*. John Murray, London.

Den Boer, P.J. 1977. *Dispersal and survival. Carabids in a cultivated countryside. Miscellaneous Papers Landbouwhogeschool Wageningen* 14.

Den Boer, P.J. 1979. Some remarks in retrospect. In P.J. den Boer, H.U. Thiele and F. Weber (eds), *On the evolution of behaviour in carabid beetles*: 213-222. *Miscellaneous Papers Landbouwhogeschool Wageningen*, 18.

Den Boer, P.J. 1990a. The survival value of dispersal in terrestrial arthropods. *Biological Conservation* 54: 175-192.

Den Boer, P.J. 1990b. Density limits and survival of local populations in 64 carabid species with different powers of dispersal. *Journal of Evolutionary Biology* 3: 19-40.

Den Boer, Pieter J., Van Huizen, Taco H.P., Den Boer-Daanje, Wilhelmina, Aukema, Behrend & Den Bieman, Cees F.M. 1980. Wing polymorphism and dimorphism in ground beetles as stages in an evolutionary process (Coleoptera, Carabidae). *Entomologia Generalis* 6: 107-234.

Den Boer, P.J. & Den Boer-Daanje, W. 1990. On life history tactics in carabid beetles: are there only spring and autumn breeders? In Nigel E. Stork (ed.), *The role of ground beetles in ecological and environmental studies*: 247-258. Intercept, Andover, Hampshire.

Den Boer, P.J. & Van Dijk, Th.S. 1996. Life-history patterns among carabid species. *Tijdschrift voor Entomologie* 139: 1-16.

Den Boer, P.J. & Reddingius, J. 1996. *Regulation and stabilization. Paradigms in population ecology*. Chapman & Hall, London.

Koch, D. & Thiele, H.U. 1980. Zur Ökologisch-physiologischen Differenzierung und Speziation bei Laufkäfer-Art *Pterostichus nigrita* (Coleoptera, Carabidae). *Entomologia Generalis* 6: 135-150.

Larsson, S.G. 1939. Entwicklungszeiten und Entwicklungstypen der dänischen Carabiden. *Entomologiske Meddelelser* 20: 227-547.

Lindroth, C.H. 1945. *Die Fennoskandischen Carabidae, I. Spezieller Teil*, Elanders Boktryckeri Aktiebolag, Göteborg (translated by Smithsonian Institution Libraries, Washington in English in 1992).

Lindroth, C.H. 1949. *Die Fennoskandischen Carabidae, III, Allgemeiner Teil*, Bröderna Lagerström Boktryckare, Stockholm (translated by Smithsonian Institution Libraries, Washington in English in 1992).

MacArthur, R.H. & Wilson, E.O. 1967. *The theory of island biogeography*. Monographs in Population Biology. Princeton, New York.

Nelemans, M.N.E. 1987. On the life-history of the carabid beetle *Nebria brevicollis* (F.). Egg production and larval growth under experimental conditions. *Netherlands Journal of Zoology* 37: 26- 42.

Paarmann, W. 1990. *Poecilus lepidus* Leske (Carabidae, Coleoptera) a species with the ability to be a spring and autumn breeder. In N.E. Stork (ed.), *The role of ground beetles in ecological and environmental studies*: 259-267. Intercept, Andover, Hampshire.

Roff, D.A. 1992. *The evolution of life histories: theory and analysis*. Chapman & Hall, New York.

Stearns, S.C. 1976. Life-history tactics: a review of the ideas. *Quarterly Review of Biology* 51: 13-47.

Stearns, S.C. 1992. *The evolution of life histories*. Oxford University Press, Oxford.

Thiele, H.U. 1977. *Carabid beetles in their environments. A study on habitat selection by adaptations in physiology and behaviour*. Springer Verlag, Berlin, Heidelberg, New York.

Van Dijk, Th.S. 1979. On the relationship between reproduction, age and survival in two carabid beetles: *Calathus melanocephalus* L. and *Pterostichus coerulescens* L. (Coleoptera, Carabidae). *Oecologia* 40: 63-80.

Van Dijk, Th.S. 1983. The influence of food and temperature on the amount of reproduction in carabid beetles. In P. Brandmayr, P.J. den Boer, and F. Weber (eds). *Ecology of carabids: the synthesis of field study and laboratory experiment*: 105-123. PUDOC, Wageningen.

Van Dijk, Th.S. 1986. How to estimate the level of food availability in field populations of carabid beetles? In P.J. den Boer, M.L. Luff, D. Mossakowski, and F. Weber (eds), *Carabid beetles, their adaptations and dynamics*: 371-384. Gustav Fischer Verlag, Stuttgart.

Van Dijk, Th.S. 1994. On the relationship between food, reproduction and survival of two carabid beetles: *Calathus melanocephalus* and *Pterostichus versicolor*. *Ecological Entomology* 19: 263-270.

Van Dijk, Th.S. & Den Boer, P.J. 1992. The life histories and population dynamics of two carabid species on a Dutch heathland. 1. Fecundity and the mortality of immature stages. *Oecologia* 90: 340-352.

Population and Community Ecology for Insect Management and Conservation, Baumgärtner et al. (eds)

Randomization test for the similarity index of communities

Y. Natuhara & C. Imai
Osaka City Institute of Public Health and Environmental Sciences, Japan

ABSTRACT: We tested similarity indices C_λ and C'_λ of communities by randomization tests. Two kinds of null-hypothesis are assumed in randomization tests of community indices. One of them is that individuals are recorded independently, at random and the other is that samples are taken at random. Randomness of the dispersion of individuals was tested with the dispersion index, m*/m. Individuals were not recorded at random in some field data examined, in which case we did not use the former as a null-hypothesis. In core samples of oribatid mites from five plots with a gradient of vegetation, communities in adjacent plots were not significantly different, but communities of remote plots were significantly different. In repeated transect records of butterflies, both the effects of recorders and time of day were simultaneously tested. The comparison of the similarity index with those of random rearrangement of records is useful for the comparison of communities.

1 INTRODUCTION

Community indices are difficult to estimate with confidence intervals and statistical tests because datasets in community ecology contain some discrepancies from statistical assumptions. For example, species are often distributed contiguously, species richness in a sample has a rarefaction effect, etc.

Randomization tests require fewer assumptions than standard statistical tests. This is important for community data where scientists face problems such as unequal variance, etc. Furthermore, randomization tests solve problems for which no satisfactory solution has previously been proposed (Manly 1991).

Similarity indices have these problems. These are often estimated without confidence limits and significance levels, though the jackknife method has been suggested for use with these indices (Smith et al. 1979). We tested the similarity indices, C_λ (Morisita 1959) and C'_λ (Morisita 1971) with randomization tests.

2 WHAT IS RANDOMIZED?

The sampling procedure in community ecology is shown at Figure 1. We randomly take samples in a local community and these samples include several individuals belonging to some species. Thus two kinds of null-hypothesis are assumed. One of these is that individuals are recorded at random or independently, and the other is that samples are taken at random. The first null hypothesis is that all individuals were taken independently from the same population and allocated to samples; this is a kind of randomization community. Solow (1993) used randomized communities to test for community change. He pooled individuals belonging to two communities and allocated these randomly to two groups and compared indices of the observed communities to those of the randomization communities. The second null hypothesis is that samples were taken at random. Thus, similarity among samples within a local community is as same as that of randomly allocated samples.

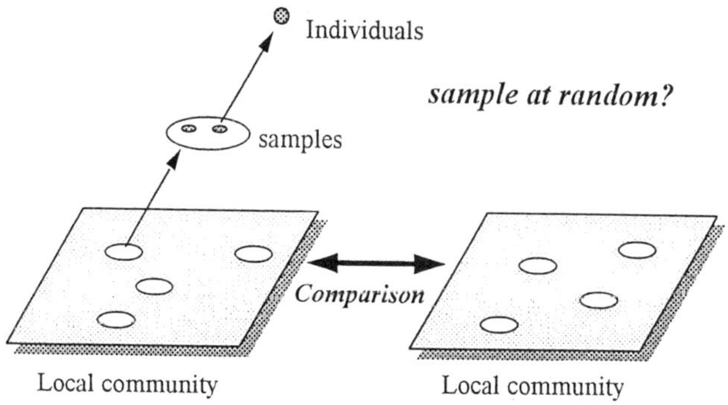

Figure 1. Sampling procedure and units of randomization

2.1 *Do individuals recorded/collected at random?*

To answer this question, we examined dispersion of individuals in core samples of Oribatid mites and in transect records of birds and butterflies. Harada and Aoki (1986) collected eight random samples (100 cm^2 each) from each of eight different vegetation types on a sand beach. We use samples from five plots of these plots, in which there is a gradient from (A) potted seedling of evergreen broad-leaved trees to (E) pine forest. In each sample, the abundance of every species was recorded.

In samples from the plot (E), the dispersion index, m*/m (Lloyd 1967), indicates a contagious distribution of individuals in every species (Figure 2), though the value of index varies from species to species. The median and range of the index and dispersion pattern are shown in Table 1.

In transect data, the strength of temporal variability was measured by the dispersion index (Table 1). Examined data were repeated records from the same population. Natuhara et al. (1996) recorded butterflies along a transect route (2.2 km) in an urban park. Four recorders counted butterflies along the same transect route for three times on May 9, 1995. Sakuratani et al. (unpublished) examined the precision of transect records of butterfly in a greenhouse, under controlled conditions. Mazda (1995) recorded abundance of bird species along a transect route in a urban park Rokugien, Tokyo Japan, for every few days. We extracted data in the breeding season and winter from this bird record. Then dispersion of each species among the days was estimated.

The sampling distributions of similarity indices depends on the spatial distributions of the species, and on the sampling methods. We only found a random distribution of species in the samples in the greenhouse. Thus the null-hypothesis of the random collection of individuals cannot be used for test for community difference.

3 TEST FOR SIMILARITY INDICES

3.1 *Test for similarity between two communities*

First, we consider a test for the similarity index, C_λ, between two Oribatid communities, with eight samples from each. C_λ is defined,

$$C_\lambda = \frac{\Sigma\left((n_{1i}/N_1)(n_{2i}/N_2)\right)}{(\lambda_1 + \lambda_2)/2} \tag{1}$$

where

$$\lambda_1 = \frac{\sum n_{1i}(n_{1i}-1)}{N_1(N_1-1)}, \quad \lambda_2 = \frac{\sum n_{2i}(n_{2i}-1)}{N_2(N_2-1)} \tag{2}$$

Table 1. Distribution pattern of individuals in replications of transect data

	Taxonomic group	Sorce of variation	Number of samples	m*/m Median (Min-Max)	Distribution (no. of species) Random	Contagious	Uniform
A	Oribatid mite		8 x 5	1.74 (0.92-6.86)	19	51	0
B	Butterfly	recorders, time of day	12	1.06 (0.28-3.63)	9	4	0
C	Butterfly	recorders, replication	15	0.99 (0.29-1.24)	13	0	1
D	Butterfly	recorders, replication	15	0.98 (0.14-9.08)	12	1	1
E	Bird ('84 summer)	days	18	1.08 (0.65-1.28)	4	3	1
F	Bird ('85 summer)	days	20	1.32 (0.83-2.02)	2	6	0
G	Bird ('84 winter)	days	14	1.41 (0.00-3.23)	3	11	2
H	Bird ('85 winter)	days	12	1.37 (0.00-3.15)	6	8	1

A: Harada and Aoki (1986), B: Natuhara et al. (1996), C and D: Sakuratani et al. (unpublished), E-H: Matsuda (1995)

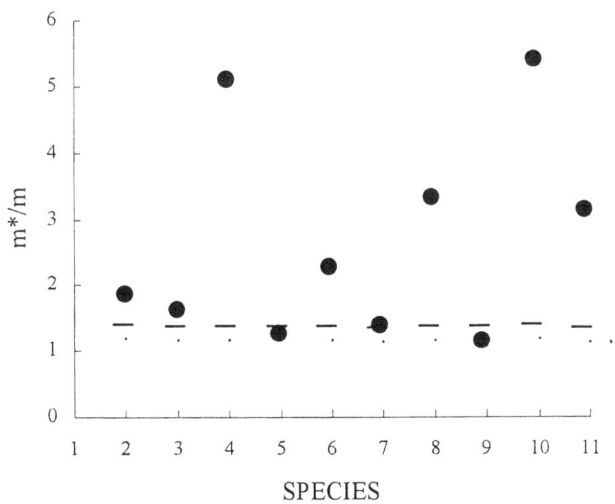

Figure 2. Dispersion index m*/m of Oribatid species (●) and randomized individals (—).

n_{1i} is the abundance of the ith species in the 1st community, and N_1 is the total abundance in the 1st community. The null-hypotheses is that the 16 samples were randomly allocated to the two communities from the same population. Eight samples were randomly taken from the 16 and the similarity index between two communities was calculated. In this way the randomization distribution of C_λ could be estimated. The similarity between the original pair was compared to this randomization distribution.

Plots A and B were neighbors and these were similar in vegetation; young trees of *Euonymus japonicus* Thunb., *Pittosporum tobira* Ait., *Elaeagnus umberrata* Thunb., etc. These plots were different only in these location, that is, A was more close to the sea than B was. The similarity index, C_λ, between A and B was 0.928 (Table 2). On the other hand, the vegetation of E was mature woodland of *Pinus thunbergii* Parlat. C_λ between A and E was 0.174.

The number of combinations of 8 items from 16 is ${}_{16}C_8$ = 12870. We estimated randomization distribution for the 12870 permutations. The observed C_λ was 674th smallest in the 12870 permutations of pooled A-B and its significance level was 0.0524. On the other hand, the significance level of A-E combination was 0.0001. This is a multiple comparison and 10 tests were done, then the correct $\alpha' = \alpha/10 = 0.005$. The null-hypothesis could not be rejected for A-B, but could be rejected for A-E. A gradient pattern was detected in these five communities (Table 3).

Table 2. C_λ for the simmilarity between two oribatid communities and the results of resampling

	A-B	A-E
Original	0.928	0.174
No. resampling	12870	12870
Average	0.975	0.873
SD	0.024	0.104
*P**	0.0524	0.0001

* Proportion of resampling values lower than the original one

Table 3. Matrix of similarity and probability of significance

	A	B	C	D	E
A	-	*0.056*	0.000	0.000	0.000
B	*0.928*	-	0.001	0.000	0.001
C	0.256	0.340	-	*0.146*	*0.014*
D	0.151	0.223	*0.852*	-	*0.018*
E	0.174	0.204	0.620	0.602	-

Lower left: C_λ, upper right: *P* values

Table 4. Mean similarity, C'_λ between 5 communities

	C'_λ
Original	0.422
No. resampling	5000
Average	0.866
SD	0.075
*P**	0.0002

* Proportion of resampling values lower than the original one

3.2 *Test for similarity among k communities*

For *k*-samples comparisons, C'_λ should be estimated. C'_λ is a mean similarity among *k* communities (Morisita 1971), defined by

$$C'_\lambda = \left(\sum_{j=1}^{L-1} \sum_{k>j}^{L} \sum_i n_{ij} n_{ik} \Big/ \sum_{j=1}^{L-1} \sum_{k>j}^{L} N_j N_k \right) / \bar{\lambda}_g \tag{3}$$

where

$$\bar{\lambda}_g = \frac{\sum_{j=1}^{L} \sum_i n_{ij}(n_{ij}-1)}{\sum_{j=1}^{L} N_j (N_j - 1)} \tag{4}$$

L is the number of communities, n_{ij} is the abundance of the *i*th species in the *j*th community, and N_j is the total abundance in the *j*th community.

We estimated C'_λ for the original data, then pooled the 8 x 5 = 40 samples, and resampled 8 samples x 5 communities 5000 times. C'_λ for the original five communities was 0.422 and the significance level in the randomization distribution was 0.0002 (**Table 4**).

Table 5. Mean similarity, C'_λ between transect data.

	C'_λ	
	Recorder	Time of day
Original	1.022	0.973
No. resampling	5000	5000
Average	0.996	0.996
SD	0.020	0.019
P*	0.941	0.108

* Proportion of resampling values lower than the original one

3.3 *Transect records of butterfly*

Butterfly counts may be influenced by recorder differences and the time of day. To test the effects of recorders, similarities (C'_λ) among samples recorded by the same recorder were estimated. The null-hypotheses was that 12 records were randomly allocated to the recorder-time combinations from the same population, then similarity index was not lower between recorders or between time of day than for random samples. When 12 samples are randomly allocated without blocking, the number of permutation is 12! = 479,001,600. We therfor carried out the procedure with Monte Carlo instead of using all permutations. In each resampling, C'_λ between times of day and C'_λ between recorders were estimated simultaneously.

For the recorder, the rank of observed C'_λ in 5000 (4999 randomization and 1 observed data) was 4705 and 94.1% of randomization values were lower (i.e. more dissimilar) than the observed one (Table 5). This is enough to accept the null-hypotheses in which three records were randomly allocated from the 12 records. For time of day, the rank was 540 (P = 0.108).

On this butterfly assemblage, the number of individuals recorded was significantly different among the four recorders, species richness was significantly different among times of day, and the diversity index was not significantly different among the 12-transects data (Natuhara et al. 1996). An increase of species richness in later counts suggested that possibility of daily activity of butterflies affecting the discovery rates. However, results with the similarity index suggest that species composition does not vary significantly both between recorders and between times of day.

4 REFERENCES

Harada, H. and J. Aoki 1986. Oribatid mite communities in the soil of sandbreak plantation alonb the Shonan beach. *Bulletin of the Institute of Environmental Science and Technology, Yokohama National University*. 13:207–215

Lloyd, M. 1967. Mean crowding. *J. Anim.Ecol*. 36: 1–30.

Manly, B. F. J. 1991 *Randomization and Monte Carlo Methods in Biology*. Chapman and Hall: London.

Mazda, M. 1995. *The Birds of Rikugien Park*. M. Mazda: Tokyo.

Morisita, M. 1959. Measureing of interspecific association and similarity between communities. *Mem. Fac. Sci. Kyushu Univ. Ser. E (Biol.)* 3: 65–80.

Morisita, M. 1971. Composition of the Id-index. *Researches on Population Ecology* 13: 1–27.

Natuhara, Y., C. Imai, M Ishii, Y. Sakuratani, and S. Tanaka 1996. Reliability of transect-count method for monitoring butterfly communities 1. Repeated counts in an urban park. *Japanese Journal of Environmental Entomology and Zoology*. 8: 13–22

Smith, W. D., D. Kravitz, J. F. Grassle 1979. Confidence intervals for similarity measures using the two sample jackknife. In L. Orloci, C. R. Rao, and W. M. Stiteler (eds), *Multivariate Methods in Ecological Work*: 253–262. International Cooperative Publishing House: Fairland.

Solow, A. R. 1993 A simple test for change in community structure. *Journal of Animal Ecology*. 62: 191–193.

Population and Community Ecology for Insect Management and Conservation, Baumgärtner et al. (eds)

Mechanisms and properties of ratio-dependent multitrophic models with particular reference to logistic food webs

Alan Berryman & Xin Chen
Department of Entomology, Washington State University, Pullman, Wash., USA

ABSTRACT: Ratio-dependent multitrophic models differ from classic Lotka-Volterra models and their allies in being built upon the economic principle of "diminishing returns" rather than the chemical law of "mass action". One variant of the ratio-dependent school is the generalized logistic food web model. This model meets with all the criteria for a credible predator-prey model and describes a rich array of complex dynamic behavior. It embeds competition and cooperation in the trophic-interaction structure and is consistent in describing species occupying different trophic levels. Although the logistic model makes no explicit reference to functional responses, type II and III functional response emerge implicitly from the structures of trophic interactions. We discuss some of the insights that emerge from this model with respect to the biological control of pest organisms.

1 INTRODUCTION

Multitrophic population dynamics models, or what are more usually called food web models, evolve from the generalization of models involving two interacting species (i.e., predator-prey models) to systems of many interacting species. The first predator-prey model, the so-called Lotka-Volterra or LV model, was formulated as continuous-time (differential) equations (Lotka 1925, Volterra 1928). Shortly afterwards, however, Nicholson and Bailey (1935) formulated a discrete-time (difference) equation model for interactions between insect parasitoids and their prey. Royama (1971) has shown that the NB and LV models are conceptually identical. These models have been widely applied in the study of food webs and biological control, and have been extensively modified in attempts to remove some of their unreasonable biological properties (Hassell 1978, Berryman 1992).

Recently a different approach to predator-prey and food web modeling has been developed under the title of "ratio-dependent" (RD) predation (Arditi and Ginzburg 1989, Berryman 1992, Gutierrez 1992). Although the RD approach is not necessarily new (see Leslie 1948), it has not previously occupied a central place in ecological theory. Yet RD models solve many of the problems and paradoxes associated with LV models (Berryman 1992, Berryman et al. 1995a). In this paper we will discuss the mechanistic basis and emergent properties of RD multitrophic models, with particular reference to models emerging from generalization of the logistic equation, and also some of the implications of RD theory for the biological control of pest organisms.

2 MECHANISMS AND MODELS

The basic mechanism underlying LV theory is the chemical "law of mass action" which describes the random encounter between particles in a homogeneously mixed gas or liquid. Under this assumption, the number of prey killed by an individual predator, N_a (the functional response), is dependent on the density of prey, N, but is independent of predator density, P; i.e.,

$$N_a = bN \tag{1}$$

Obviously this is biologically impossible because the predator has an infinite appetite; i.e., $N_a \to \infty$ as prey density gets very large. To correct this, the functional response is usually formulated as a saturating function of prey density, the simplest of which is the "disk equation" (Holling 1959, 1966):

$$N_a = \frac{bN}{1+N} \tag{2}$$

Several other mathematical formulations have been used to describe the functional response of predators (see, e.g., Royama 1971 for a review), but we will not concern ourselves with them here, except to say that they all attain a constant attack rate at high prey densities due to predator satiation.

In contrast to LV theory, RD models rest, mechanistically, on the economic "law of diminishing returns" which describes the increasing difficulty a consumer has in meeting its energy demands as its population density rises or the supply of resources falls. Under this assumption, the *fitness cost*, in terms of net reproduction, depends directly on the demand/supply ratio; increasing demand (more predators) raising the cost, and increasing supply (more prey) lowering the cost. Hence, the fitness cost depends directly on the demand/supply, or predator/prey, ratio (or, conversely, the fitness gained depends directly on the prey/predator ratio). Arditi and Ginzburg (1989) suggest modifying the LV model by substituting the prey/predator ratio for prey density;

$$N_a = \frac{bN/P}{1+N/P} = \frac{bN}{P+N} \tag{3}$$

Other authors have also suggested similar functional response models to account for interference between predators (e.g., DeAngelis et al. 1975).

The effects of "diminishing returns" can be modeled mechanistically by considering the geometry of consumers searching randomly for resources (prey) in a finite environment. With some mathematical analysis (Royama 1992, Berryman et al. 1995a), it is possible to obtain a simple equation for the net reproductive rate (numerical response) of a population of consumers

$$R_p = a_p - b_p P_{t-1} / N_{t-1} \tag{4}$$

where $R_p = \ln P_t - \ln P_{t-1}$ is the instantaneous rate of change of the consumer population estimated over the discrete time period $t-1$ to t, and N_{t-1} is the fixed density of prey. In this equation the net rate of growth (and by inference the attack rate N_a) of an individual predator ($P_{t-1} = 1$) rises to a constant rate (a_p) as the density of the prey rises. In other words it has the same general form as the functional response curve. This model is identical to Leslie's (1948) adaptation of the classical "logistic" equation to describe predator dynamics, and is commonly used in the literature to model predator populations (May 1973, Tanner 1975). Notice that the maximum fitness of the average predator is reduced in proportion to the predator/prey ratio.

If $\ln P_t - \ln P_{t-1}$ is substituted for R_p in equation (4) we get

$$\begin{aligned} P_t &= P_{t-1} e^{a_p(1-P_{t-1}/K_p)} \\ K_p &= a_p N / b_p \end{aligned} \tag{5}$$

which is the well-known discrete "logistic equation" (Cook 1965), or

$$P_t = P_{t-1}\lambda_p e^{-\alpha_p P_{t-1}}$$

$$\lambda_p = e^{a_p} \tag{6}$$

$$\alpha_p = b_p / N$$

which is the "Ricker model" commonly used by fisheries ecologists (Ricker 1954).

It is fairly straight forward to extend equation (4) to trophic chains and webs (Berryman 1992, Berryman et al. 1995b)

$$R_i = a_i - b_i N_i - \frac{N_i}{\sum_j c_{ji} N_j F_j^{r_i}} - \frac{\sum_k d_{ik} N_k F_k^{c_i}}{N_i} \tag{7}$$

$$F_j^{r_i} = \frac{q_{ji} N_i F_i^{c_j}}{\sum_m q_{jm} N_m F_m^{c_j}}, \quad F_k^{c_i} = \frac{p_{ik} N_i F_i^{r_k}}{\sum_n p_{nk} N_n F_n^{r_k}}$$

where the terms on the right hand side of the equation are interpreted as follows:

1. The first term, a_i, is the maximum per-capita rate of change of species *i* in a given physical environment when food and space are superabundant and predators are absent (maximum fitness),
2. The second term, $b_i N_i$, is the reduction from the maximum rate of change due to competition for fixed resources (mainly space), with b_i the marginal effect of adding another consumer to the population of species *i* (fitness cost due to limited space).
3. The third term, $N_i / \sum_j c_{ji} N_j F_j^{r_i}$, is the reduction due to competition for food, with c_{ji} the value of food species *j* to the fitness of species *i*, and $F_j^{r_i}$ the fraction of population N_j that can be used by species *i* because it has not been consumed by other competing species.
4. The fourth term, $\sum_k d_{ik} N_k F_k^{c_i} / N_i$, is the reduction due to predation, with d_{ik} the cost of an individual of predator species *k* to the fitness of species *i*, and $F_k^{c_i}$ is the fraction of predator population N_k that actually attacks *i* because it is not attacking other species.
5. The fraction $F_j^{r_i}$ defines the indirect effect of other consumers on the fitness of species *i* and therefore models the effects of interspecific competition for food, with q_{ji} the relative competitive ability of consumer *i* for prey *j* (Fig. 1a).
6. The fraction $F_k^{c_i}$ defines the indirect effects of other prey on the fitness of prey *i*, and models the effects of interspecific cooperation (in the sense that other prey distract predators from prey *i*). The parameter p_{ij} defines the relative apparency of prey *i* to consumer *j* (Fig. 1b).

Because equation (7) derives directly from the venerable logistic equation (Verhulst 1838, Leslie 1948), we refer to it as the *logistic food web model* (Berryman et al. 1995b). Before proceeding to a discussion of the properties of this model, we should mention one other kind of RD formulation. Gutierrez and his coworkers (e.g., Gutierrez 1992) have developed models from the basic physiological processes of energy acquisition and allocation between trophic levels, and their energy transfer functions have RD structure. Gutierrez's models, as well as the RD models of other authors (e.g., DeAngelis et al. 1975, Metzgar and Boyd 1988, Arditi and Ginzburg 1989), seem to have the same general properties as equation (7). In this paper, however, we will be mainly concerned with logistic webs (7) because this is the model we are most familiar with. What we have to say, however, probably applies in general to the other

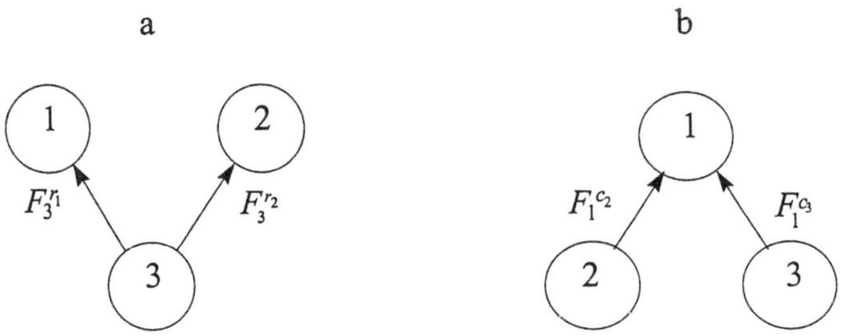

$$F_3^{r_1} = \frac{q_{31}N_1}{q_{31}N_1 + q_{32}N_2} \qquad F_1^{c_2} = \frac{p_{21}N_2}{p_{21}N_2 + p_{31}N_3}$$

$$F_3^{r_2} = \frac{q_{32}N_2}{q_{31}N_1 + q_{32}N_2} \qquad F_1^{c_3} = \frac{p_{31}N_3}{p_{21}N_2 + p_{31}N_3}$$

Fig. 1. a) When 2 predators (species 1 and 2) compete for a common prey (species 3), the fraction $F_3^{r_1}$ represents the proportion of the prey population available to predator 1 and, hence, the indirect effect of interspecific competition with predator 2. b) When 2 prey (species 2 and 3) are consumed by a common predator (species 1), the fraction $F_1^{c_2}$ represents the proportion of the prey 2 that is actually consumed by predator 1 and, hence, the indirect effect of mutualism between prey 2 and 3 (as prey 3 reduces the impact of the predator on prey 2, the indirect effect can be considered as inadvertent cooperation).

RD formulations, because they all seem to have similar dynamic properties (Berryman et a. 1995a).

3 CREDIBILITY

We first examine the properties of equation (5) that, we believe, make it a credible food web model. First, if a mathematical model is to be "believed", it should adhere to at least a minimal set of biological standards. Following Royama (1992), Berryman et al. (1995a) listed 4 biological conditions that all trophic models should adhere to:

1. When the prey population is constant, the per-capita rate of increase of *prey* should decline monotonically with increasing predator density (because predators, by definition, must have a negative impact on their prey) (Fig. 2a).
2. When the predator population is constant, the per-capita rate of increase of *prey* should rise at first with prey density (because survival rises as predators become satiated), and then decline as prey become very dense (because intra-specific competition for resources amongst the prey will increase) (Fig. 2b).
3. When the prey population is constant, the per-capita rate of increase of *predators* should decline monotonically with increasing predator density (because of increased intra-specific competition for prey) (Fig. 2c).
4. When the predator population is constant, the per-capita rate of increase of *predators* should rise with prey density (because of decreased intra-specific competition for prey) to a maximum positive value (because predators have limited appetites and fecundity) (Fig. 2d).

Berryman et al. (1995a) found that all RD models obey these simple rules, while most other models of similar complexity usually violate one or more of them (particularly condition 2). In fact Royama (1992) found condition 2 so difficult to satisfy that he remarked that "no single model may be satisfactory". That RD models satisfy this condition without added detail, is one of their strengths.

Because RD equations contain the concept of "diminishing returns", they should be considered as *overall* rather than *instantaneous* predator/prey models. Royama (1971) argued

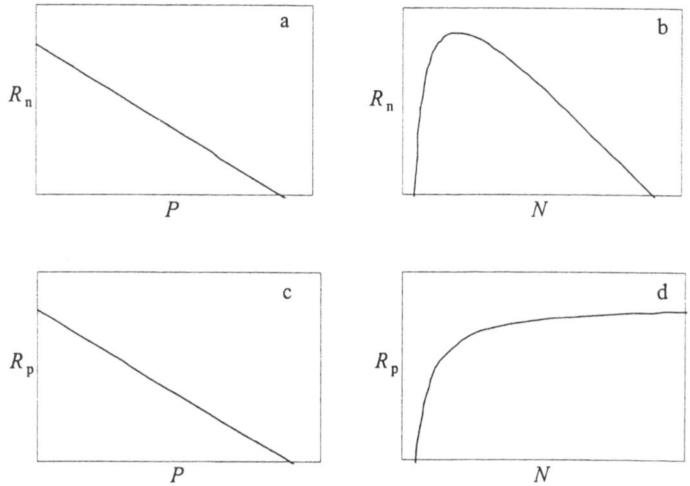

Fig. 2. Conditional per-capita rate of change functions (R-functions) for; a) prey population with density constant but predator density varying, b) prey population with predator density constant but prey density varying, c) predator population with prey density constant and predator density varying, and d) predator population with predator density constant and prey density varying (after Royama 1992, Berryman et al.1995a).

that instantaneous equations, such the classical LV model, only apply to situations where prey and predator populations are fixed during the time interval of change. This may not be a problem if births and deaths occur simultaneously and if their rates do not change dramatically with time, as may be true in tropical aquatic systems or laboratory chemostats. Overall equations, on the other hand, apply to situations where prey are removed during the time period of change (this is what gives rise to "diminishing returns"), and where births and/or deaths occur seasonally, as is true for many temperate insects. Hence, overall equations are more suitable for modeling trophic interactions where the objects of interest sleep, hibernate or reproduce synchronously in response to diurnal and seasonal rhythms (Berryman et al. 1995a). In Royama's own words: "The essential difference between the overall and instantaneous equations is that the former involves the effect of diminishing returns while the latter holds only at an instant and so does not involve this effect. If one intends to build a model to study a predator-prey interacting system, what is needed, from a theoretical point of view, is the overall hunting equation, since this is the equation which provides the estimates of the number of prey killed and of the final density of the prey at the end of the hunting period. The former estimate gives a basis for calculating the number of predator's progeny and the latter the number of prey's progeny." (Royama 1971, page 16). Because of these considerations, we believe that R-D equations are more appropriate for studying the dynamics of animals that breed seasonally, particularly univoltine insects.

4 UNIFORMITY

A general model for trophic interactions should be applicable to any organism in the web, for all organisms can be both predators on lower trophic levels (even plants "prey" on sunlight and nutrients) and, at the same time, prey to higher trophic levels (except, perhaps, the top predators). They can also be competitors (for food) and cooperators (to escape predators) with other organisms in the same trophic level. On the other hand, many trophic models like the L-V and its relatives require structurally different equations for predator and prey populations, and for the processes of predation and interspecific competition and cooperation. In contrast, equation (7) applies to any organism, regardless of its position in the food web or its feeding preferences, and also describes interspecific competition and cooperation (Fig. 1), without the need for additional detail (Berryman et al. 1995b). We note, in passing, that the

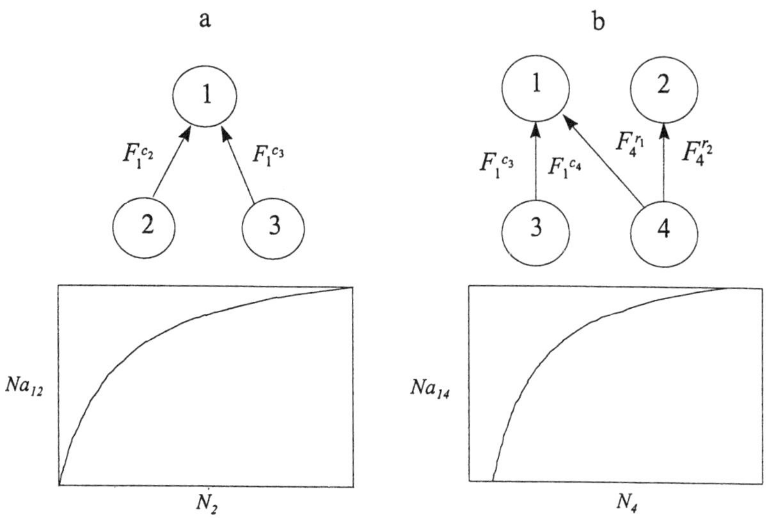

Fig. 3. Functional responses emerging from logistic food webs: a) Type II functional response of predator 1 to prey 2 when prey 3 is constant. b) Type III functional response of predator 1 to prey 4 when prey 4 is shared with predator 2.

inter-specific competition model arising from logistic webs is identical to the LV equations for competition between two species of consumers but that these equations are incompatible with LV predator-prey models.

It is interesting and perhaps significant that, unlike traditional predator-prey models, equation (7) makes no explicit reference to the functional response of the predator. At first this may seem alarming, but in retrospect it seems to be a reasonable property. The functional response describes the behavior of a single predator searching in a local area (often an arena) over a short period of time, and is therefore more appropriate in instantaneous equations. It is interesting, however, that *implicit* functional responses are actually present within equation (7), and that the form of this response depends on the structure of the web. For example, Holling type II functional responses emerge from systems in which the predator is dynamically coupled to one prey species but has a constant supply of alternative prey (Fig. 3a), while type III responses emerge when the predator shares, dynamically, one of its prey with another predator (Fig. 3b).

5 UTILITY

To have practical utility, food web models should not only be useful for studying the properties of hypothetical ecosystems, but should also be useful for predicting and managing the dynamics of real ecosystems. Although equations (7) is highly nonlinear in terms of its variables, it is linear in the ratios and, therefore, can be fit to real data by standard regression techniques. For a simple two-species predator-prey system equation (7) reduces to

$$R_n = a_n - b_n N - d_n \frac{P}{N+F} \qquad \text{(8a, prey)}$$

$$R_p = a_p - b_p P - c_p \frac{P}{N+F} \qquad \text{(8b, predator)}$$

where F is the sum of all other prey, assumed constant in the two-species model.
The parameters of these equations are easily estimated from time-series data like that in figure 4 by multiple regression (Berryman 1994). Data such as this are commonly obtained from

routine annual surveys or harvest records. Models derived in this way from real data have considerable practical significance, for the fitted equations can be used to study the static and dynamic properties of the system, to predict future population trends, and to identify effective resource or pest management strategies (see below).

6 STATICS

All models with predator-prey (or prey-predator) fractions in their arguments seem to have similar isocline structures (statics). For example, in a food chain with two trophic levels, as in equation (8), the zero-growth isoclines (nullclines) have the following properties (Fig. 5):

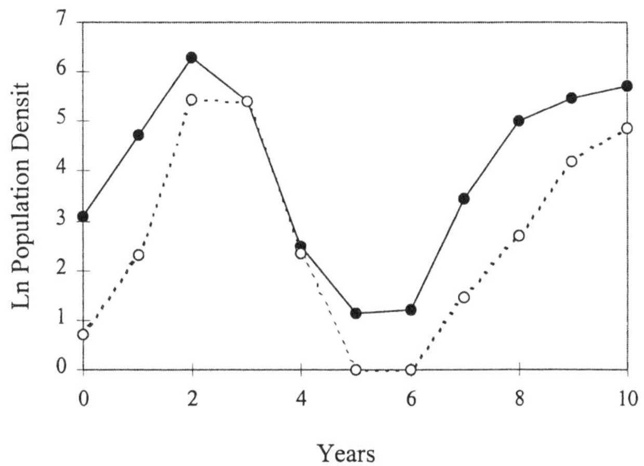

Fig. 4. Time series data for blackheaded budworm larvae (solid line) and larval parasitoids (broken line) per 10 m^2 of balsam fir foliage (Morris 1959). Fitting equation (8) to this data yielded the following model parameters: $a_n = 2.152$, $b_n = 0.003$, $d_n = 4.131$, $r^2 = 0.924$; $a_p = 2.442$, $b_p = 0.001$, $c_p = 5.535$, $r^2 = 0.955$ (after Berryman 1991, 1994).

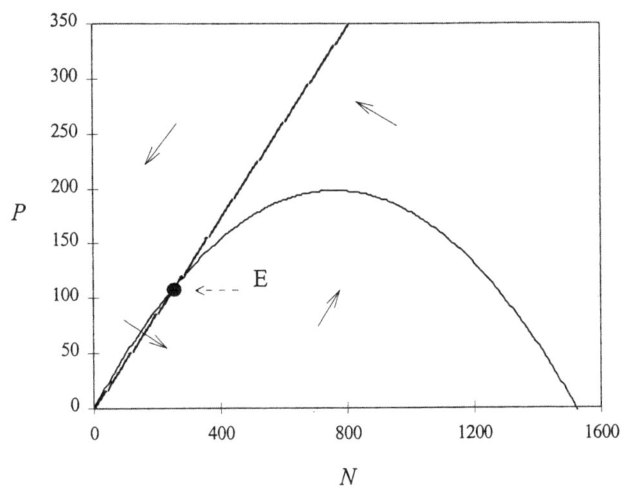

Fig. 5. Zero-growth isoclines for the logistic predator-prey model (Eqn. 8) fit by multiple linear regression to the blackheaded budworm-parasitoid time series (see Fig. 4 for details). *E* is the community equilibrium.

1. The isocline of the prey (e.g., the budworm) is parabolic, intersecting the prey axis at its "carrying capacity" (K), and the predator axis at a point reflecting the availability of alternative prey (F) (in the case of the budworm, $F = 0$) (see Arditi and Ginzburg 1989, Gutierrez et al. 1994, for another possible form, however).
2. The isocline for the predator (e.g., the parasitoids) is always right-slanted, rising with finite positive slope with respect to prey density (May 1973, Tanner 1975, Arditi and Ginzburg 1989, Arditi and Berryman 1991, Berryman 1992) and intersecting the predator axis at a point determined by alternative prey.

Isocline structures for three-dimensional systems are also similar for most ratio-dependent models (see Gutierrez et al. 1995). The isocline for the first trophic level is usually parabolic, as in the two trophic level system (Fig. 5), the second trophic level has a C-shaped isocline, and the top trophic level a right-slanted isocline (Fig. 6). The three trophic level system can have both a stable (E) and an unstable (U) community equilibrium (see Metzgar and Boyd 1988, Gutierrez et al. 1994).

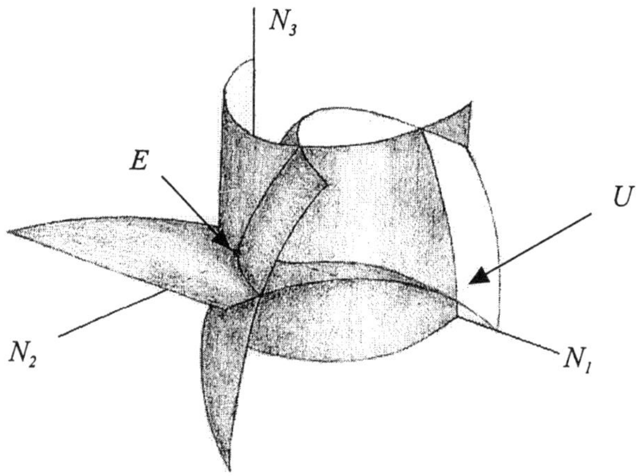

Fig. 6. Zero-growth isoclines for logistic model with 3 trophic levels. N_1 is the bottom, N_2 the middle and N_3 the top species. E is a stable equilibrium and U an unstable one.

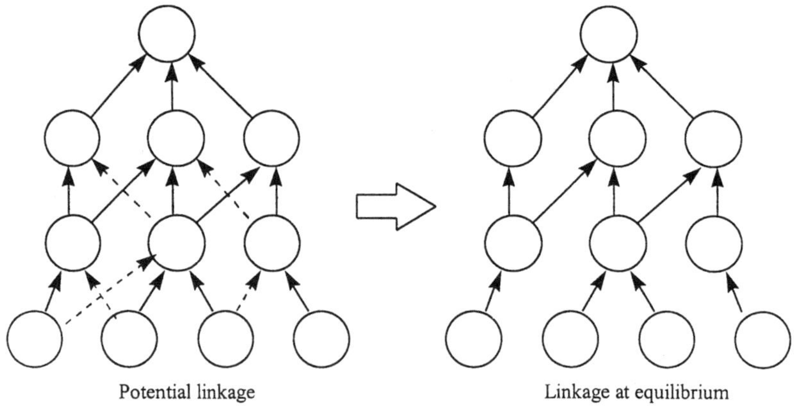

Fig. 7. Illustration of the phenomenon of food wed simplification in RD models; left = food web architecture far from equilibrium with 17 connections, and right = food web architecture at equilibrium with only 12 functional connections.

Of course, it is impossible to visualize isocline structures for food webs with dimension greater than three, but we can find the equilibrium points by numerical analysis. Because of the presence of implicit Type III functional responses, due to predator switching in complex food webs (Fig. 3), it is possible to have more than one stable community equilibrium separated by unstable boundaries, or *basins of attraction*. Another interesting property that emerges as a result of predator switching is *food web simplification* (Michalski and Arditi 1995). For example, if we start with a community of 11 species with 17 interactions (Fig. 7, left), some of the interactions become effectively zero as the system approaches equilibrium, so that we are left with 11 species with 12 *effective* interactions (Fig. 7, right). This happens because some predators that feed on several prey species (generalists) cannot compete successfully with others for particular prey species as competition becomes intense near the community equilibrium.

7 DYNAMICS

The dynamics of two-species food webs depend on the location of the community equilibrium (E) on the prey isocline, with rapid oscillations (first-order or period-2 oscillations) being more evident when E is close to K (on the descending arm of the prey's isocline [Fig. 8a]), and slow oscillations (second-order or period-3+ oscillations, often termed "cycles") being more evident as E moves to the left (on the ascending arm of the prey's isocline (Fig. 8b). The period and amplitude of the predator-prey oscillation (or cycle) is directly related to the efficiency and reproductive rate of the predator and inversely to the reproductive rate and vulnerability of the prey (Berryman 1996, Chen unpublished). It is also possible for unstable predator-prey systems to attain a *stochastic equilibrium* (Arditi and Berryman 1991) (Fig. 8c). In this case there is no community equilibrium and so deterministic coexistence between the two populations is impossible. In a stochastic system, however, coexistence can occur because the trajectory to extinction travels very close to the two isoclines and small increases in the prey population (say due to immigration) can initiate another cycle.

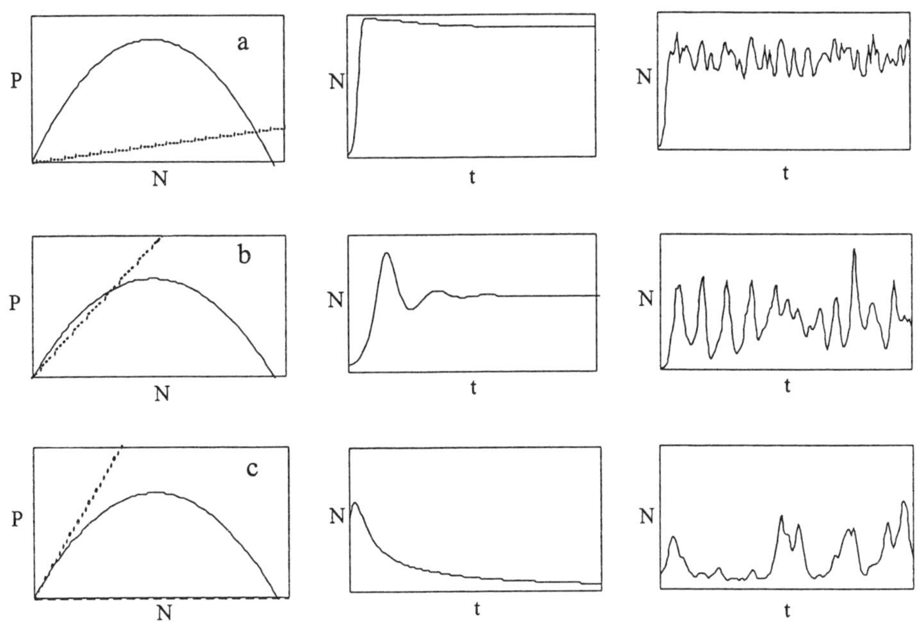

Fig. 8. Dynamics of the logistic predator/prey model (equation [8] with parameters: $a_n = 1$, $b_n = -0.003$, $d_n = -4$; $b_p = 0$, $c_p = -6$) as the per capita rate of increase of the predator increases from a) $a_p = 0.1$, b) $a_p = 1$, and c) $a_p = 1.5$. Left column shows the predator-prey isoclines, middle a deterministic simulation, and left a stochastic simulation (after Berryman 1996).

The dynamics of the three trophic level interactions depend on the location of the community equilibria (E and U). For a very simple tritrophic food chain, there may exist at least one interior fixed point $\{N_1*, N_2*, N_3*\}$ ($N_1^*, N_2^*, N_3^* \neq 0$) and three boundary fixed points: 1) $\{0, 0, 0\}$ (the origin), saddle node; 2) $\{K_1, 0, 0\}$ (only N_1 persists), saddle node; 3) $\{N_1*, N_2*, 0\}$ (N_1 and N_2 persist), stable two-species interaction. The stability of the interior equilibrium depends on its location in reference to the three boundary fixed points (Fig. 9). The structure of the orbit connections between the interior and boundary fixed points in the R^3 flow determines the dynamic characteristics of the system. When the interior equilibrium is sufficiently close to a saddle node boundary fixed point, a link similar to homoclinic orbit (a closed path which leaves the saddle node in one direction and returns to it in another

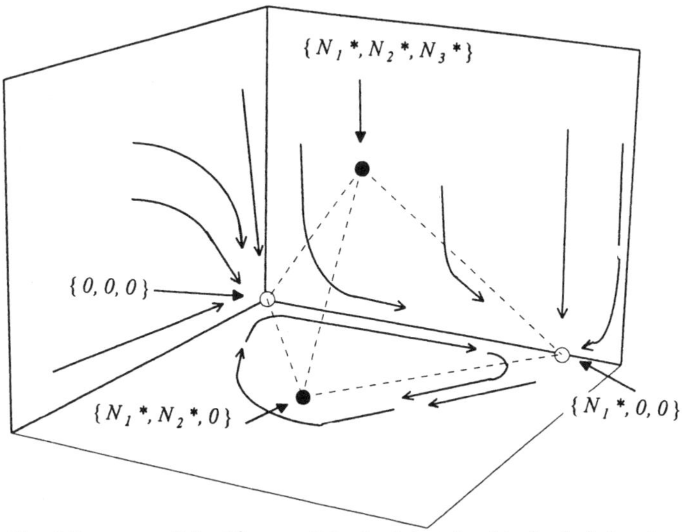

Fig. 9 Structure of the R^3 map of the logistic tritrophic food chain, showing relationships of the interior and boundary fixed points. The dynamic features of the system are greatly affected by the relative distance of the interior fixed point from each of the boundary points.

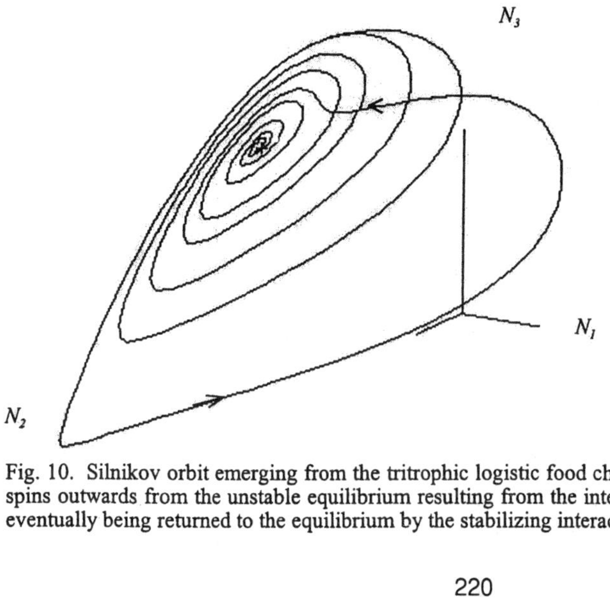

Fig. 10. Silnikov orbit emerging from the tritrophic logistic food chain, showing how the trajectory spins outwards from the unstable equilibrium resulting from the interaction between N_1-and N_2, but eventually being returned to the equilibrium by the stabilizing interaction with N_3.

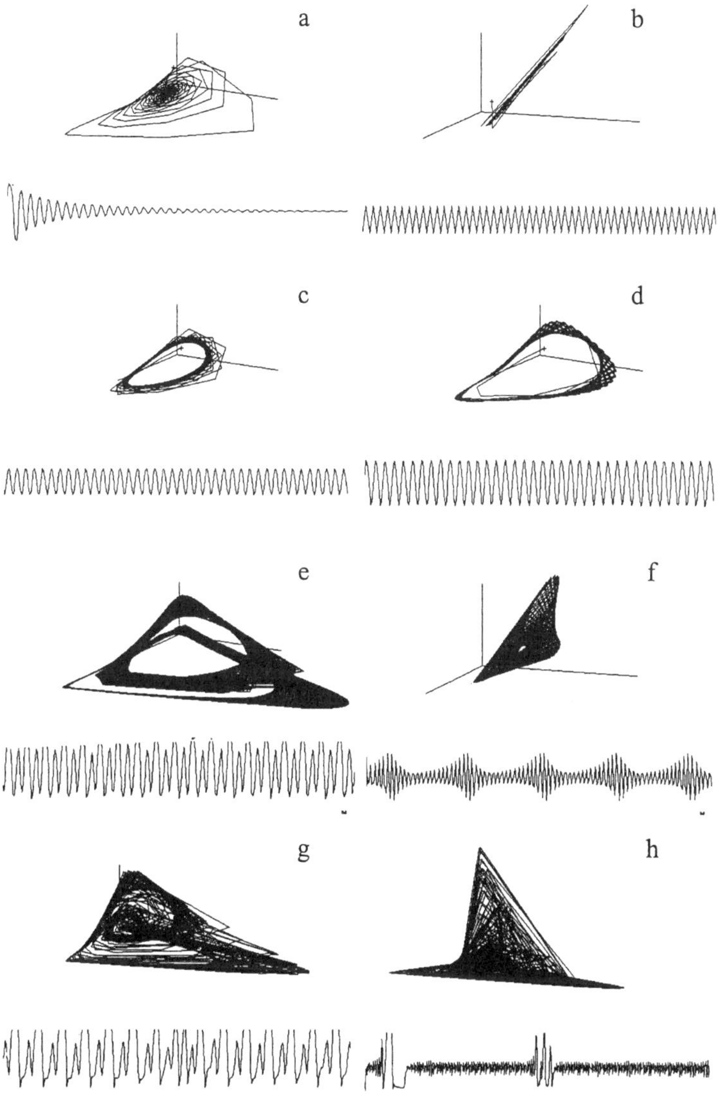

Fig. 11 Dynamical behavior of a tritrophic logistic food chain, showing qualitative changes that occur as the maximum per-capita rate of increase of the middle and top species N_2 and N_3 increase. a) stable equilibrium point; b) rapid (first order) oscillation; c) limit cycle; d) torus; e) twisted torus; f) complex harmonics; g) Silnikov oscillation; h) chaos.

direction) in continuous time systems may be formed around the interior equilibrium. Complex dynamics, such as limit cycles and chaos, usually emerge when the force drawing the orbits towards the interior equilibrium is counteracted or balanced by forces imposed by the boundary fixed points and, as a result, the interior equilibrium becomes locally unstable while the trajectories of the system remain bounded. The general geometric scenario depicting the dynamical complexity of the tritrophic food chain is that the two-dimensional locally unstable manifold is contained in a plane which will eventually line up to the x-y plane, and the one-dimensional stable manifold is contained in a curve which eventually lines up parallel to the z-axis, i.e. the bottom and middle species make an unstable pair such that the trajectories spin outwards from the equilibrium, while the top predator plays a stabilizing role so that the trajectories return to the vicinity of the equilibrium point (Fig. 10). The dynamics of the tritrophic food chain changes with the value of its parameters through a series of qualitative bifurcations (Fig. 11).

8 IMPLICATIONS FOR BIOLOGICAL CONTROL

The development of food web models with reasonable biological attributes enables us to ask questions about real ecosystems: For example, if we are interested in the biological control of pest organisms we may wish to ask questions such as:

1. *What is the best way to employ chemical and biological control in an integrated pest management (IPM) program?* Suppose we have an effective parasitoid that normally regulates a pest insect at a low density; i.e., the community equilibrium is on the ascending arm of the pest isocline (Fig. 5). In a variable environment, the pest population will cycle around the community equilibrium and may occasionally exceed the economic damage level (EDL). The question is; what kind of chemical and/or biological control should be used to minimize the probability of exceeding the EDL in the long-term? The first thing to notice is that the further the system is away from the community equilibrium, the wider will be the cyclical orbit, and the greater the likelihood of the pest population exceeding EDL. Hence, the pest manager should try to keep the system as close to the equilibrium point as possible, say within the circular target in figure 12. To do this requires tactics that change with the current state of the system. For example, if the populations are at positions A, then we require a treatment that reduces the pest population and increases the parasitoid population; at the very least, the treatment should not kill parasitoids. Thus a selective insecticide is imperative, and an inoculative release of parasitoids would also be helpful. On the other hand, if we were at positions B, then we would need to reduce the pest population *and* the parasitoid population to reach the target. Hence a less selective pesticide could be justified. Finally, if the system was at position C,

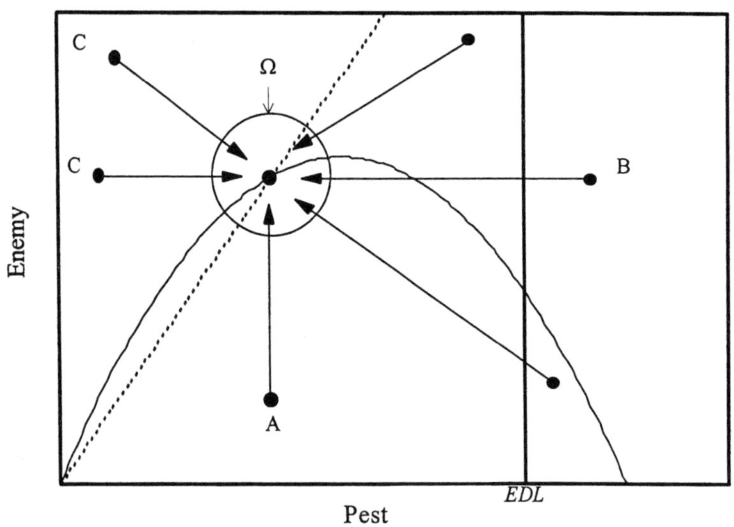

Fig. 12. Zero growth isoclines for a pest-enemy interaction: *EDL* = economic damage level; Ω=target area that minimizes the risk of the pest population exceeding the economic damage level; A's = positions where a selective pesticide that only kills the pest and/or inoculative releases of enemies should be used; B's = positions where a nonselective pesticide that kills both species could be used; C's = positions where a specific pesticide that kills enemies and/or inoculative releases of the pests could be employed. Note that pesticides should be used to effect optimal kill not maximum kill because overkill can reduce the populations below the target area, resulting in future outbreaks- what has been called "the boomerang effect" ,"pest resurgence" or "bounce-back".

a reduction of the parasitoids and an inoculative release of the *pest* would minimize the probability of a future outbreaks! (see Maksimovic et al. 1970 for an example of the application of such a counterintuitive tactic). Of course, if we can find the isoclines from real data, as we did for the blackheaded budworm and its parasitoids (Fig. 5), then it would be possible to determine the optimal IPM tactics, in terms of pesticide selectivity and toxicity, for any given pest and parasitoid densities. Notice that very toxic pesticides should be avoided because they can reduce the pest population below the target area and initiate a violent outbreak cycle; what is called in pest management jargon, bounce-back or pest resurgence.

2. *Is it better to release a one or more than one specialist parasitoid in a traditional biocontrol program?* Suppose we release two parasitoids (species 2 and 3) to control one pest (species 1). According to Eqn. (7), the R-functions for the three species are

$$R_1 = a_1 - b_1 N_1 - \frac{d_{12}N_2 + d_{13}N_3}{N_1}$$

$$R_2 = a_2 - b_2 N_2 - \frac{q_{21}N_2 + q_{31}N_3}{c_{12}q_{21}N_1} \tag{9}$$

$$R_3 = a_3 - b_3 N_3 - \frac{q_{21}N_2 + q_{31}N_3}{c_{13}q_{31}N_1}$$

There are three possible outcomes of releasing two parasitoids species to control a single prey species: 1) both parasitoids coexist in dynamic equilibrium (Fig. 13a); 2) one of the

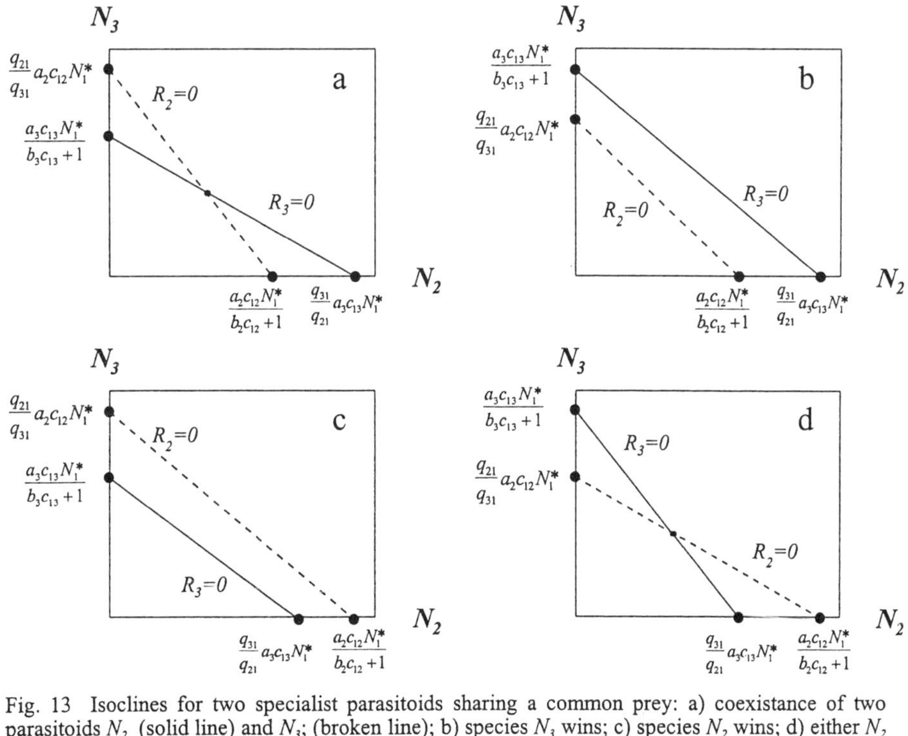

Fig. 13 Isoclines for two specialist parasitoids sharing a common prey: a) coexistance of two parasitoids N_2 (solid line) and N_3; (broken line); b) species N_3 wins; c) species N_2 wins; d) either N_2 or N_3 wins depending on the initial conditions. N_1*= equilibrium density of prey; see Eqns. (7) and (9) for the other symbols.

parasitoids is driven to extinction by the other (Fig. 13b,c); 3) one or the other of the parasitoids goes extinct depending on the initial conditions (Fig. 13d). The necessary and sufficient condition for coexistance between the two species is that intra-specific competition must be stronger than inter-specific competition; i.e.,

$$\frac{q_{31}}{q_{21}} a_3 c_{13} N_1^* > \frac{a_2 c_{12} N_1^*}{b_2 c_{12} + 1} \quad \text{and} \quad \frac{q_{21}}{q_{31}} a_2 c_{12} N_1^* > \frac{a_3 c_{13} N_1^*}{b_3 c_{13} + 1} \tag{10}$$

where N_1* is the equilibrium density of prey, and the other symbols have the same meaning as in Eqn. (7). Rearranging the above, we have

$$\frac{q_{21} a_2 c_{12}}{q_{31} a_3 c_{13}} - b_2 c_{12} < 1 \quad \text{and} \quad \frac{q_{31} a_3 c_{13}}{q_{21} a_2 c_{12}} - b_3 c_{13} < 1 \tag{11}$$

from which we see that coexistence between two parasitoids is independent of the prey, depending only on parasitoid parameters. The only case where the release of two specialist parasitoids is better than one is when

$$d_{12} N_2^* + d_{13} N_3^* > d'_{12} N_2'^* \quad \text{and} \quad d_{12} N_2^* + d_{13} N_3^* > d'_{13} N_3'^* \tag{12}$$

where N_2* and N_3* are the equilibrium densities of the two parasitoid species when coexisting, and $N_2'^*$ and $N_3'^*$ are their equilibrium densities in the absence of the other. From the above discussion, we see that there is no easy answer to the question posed at the beginning of this section. For cases 2 and 3, where one of the parasitoids is driven to extinction by the other, irrespective of initial densities, the system can be simplified, without loss of generality, by setting $b_2=b_3=0$, from which we find that parasitoid 2 wins if $q_{21} a_2 c_{12} > q_{31} a_3 c_{13}$, and vice versa for parasitoid 3. If we let the control effectiveness of the wining parasitoid be represented by the equilibrium prey population, then

$$N_1^* = \frac{a_1}{b_1} - \frac{a_3 c_{13} d_{13}}{b_1} \quad \text{or} \quad N_1^* = \frac{a_1}{b_1} - \frac{a_2 c_{12} d_{12}}{b_1} \tag{13}$$

which states that the equilibrium prey density is the difference between the equilibrium density in the absence of parasitism (a_i/b_i = carrying capacity) and a value related to the parameters defining the parasitoid-prey interaction divided by the intra-specific competition coefficient. Hence, the effectiveness of a parasitoid depends on the relative value of the parameter set $a_i c_{ij} d_{ij}$; i.e., species 2 is more effective if $a_2 c_{12} d_{12} > a_3 c_{13} d_{13}$ and vice versa. Because the conditions for the competitive exclusion of species 2 ($q_{31} a_3 c_{13} > q_{21} a_2 c_{12}$) and for species 2 being more efficient ($a_2 c_{12} d_{12} > a_3 c_{13} d_{13}$),can hold at the same time, then it is possible for a superior biological control agent to be excluded by an inferior one in a multiple release situation.

9 CONCLUSIONS

Ratio-dependent logistic models, which arise from the principle of "diminishing returns", seem to create biologically reasonable descriptions of the dynamics of natural food webs. Not only do they have credible biological properties, but they can easily be fit to real data and are then useful for forecasting changes in ecological systems and designing management strategies. It is also interesting, and perhaps significant, that logistic trophic equations emerge from much more complicated physiological and behavioral models (Rapport and Turner 1975, Gutierrez 1992), from mechanistic models of foraging in finite space (Royama 1992), and from models designed for practical purposes (Metzgar and Boyd 1988).

REFERENCES

Arditi, R. and A.A. Berryman. 1991. The biological control paradox. *Trends in Ecology and Evolution* 6: 32.

Arditi, R. and L.R. Ginzburg. 1989. Coupling in predator-prey dynamics: ratio-dependence. *Journal of Theoretical Biology* 139: 311-326.

Berryman, A. A. 1991. Population theory: an essential ingredient in population prediction, management, and policy making. *American Entomologist* 37: 138-142.

Berryman, A. A. 1992. The origins and evolution of predator-prey theory. *Ecology* 73:1530-1535.

Berryman, A. A. 1994. *Population Analysis System, Two-Species Analysis (Version 4.0)*. Pullman WA: Ecological Systems Analysis.

Berryman, A. A., A.P.Gutierrez and R. Arditi. 1995a. Credible, parsimonious, and useful predator-prey models: a reply to Abrams, Gleeson, and Sarnelle. *Ecology* 76: 1980-1985.

Berryman, A. A., J. Michalski, A.P.Gutierrez and R. Arditi. 1995b. Logistic theory of food web dynamics. *Ecology* 76: 336-343.

Berryman, A. A. 1996. Periodicity of population cycles depends on the relative dominance of bottom-up or top-down control. Submitted.

Cook, L. M. 1965. Oscillation in the simple logistic growth model. *Nature* 207: 316.

DeAngelis, D. L., R. A. Goldstein, and R. V. O'Neill. 1975. A model for trophic interaction. *Ecology* 56: 881-892.

Ginzburg, L. R. and H. R. Akcakaya. 1992. Consequences of ratio-dependent predation for steady state properties of ecosystems. *Ecology* 73: 1536-1543.

Gutierrez, A. P. 1992. The physiological basis of ratio-dependent predator-prey theory: A metabolic pool model of Nicholson's blowflies as an example. *Ecology* 73: 1552-1563.

Gutierrez, A.P., N. J. Mills, S. J. Schreiber, and C. K. Ellis. 1994. A physiologically based tritrophic perspective on bottom-up top-down regulation of populations. *Ecology* 75: 2227-2242.

Hassell, M. P. 1978. *The dynamics of arthropod predator-prey systems*. New Jersey: Princeton University Press

Holling, C. S. 1959. The components of predation as revealed by a study of small mammal predation of the European pine sawfly. *The Canadian Entomologist* 91:293-320.

Holling, C. S. 1966. The functional response of invertebrate predators to prey density. *Memoirs of the Entomological Society of Canada* 48: 1-86.

Leslie, P. H. 1948. Some further notes on the use of matrices in population mathematics. *Biometrica* 35: 213-245.

Lotka, A. J. 1925. *Elements of physical biology*. Baltimore: Williams & Wilkins.

Maksimovic, M., P. L. Bjegovic, and L. Vasiljevic. 1970. Maintaining the density of the gypsy moth enemies as a method of biological control. *Zastita Bilja* 107: 3-15.

May, R. M. 1973. *Stability and complexity in model ecosystems*. Princeton, New Jersey: Princeton University Press.

Metzgar, L. H. and E. Boyd. 1988. Stability properties in a model of forage-ungulate-predator interactions. *Natural Resource Modeling* 3: 3-43.

Michalski, J. and R. Arditi. 1995. Food web structure at equilibrium and far from it: is it the same? *Proceedings of the Royal Society of London B*,

Nicholson, A. J. and V. A. Bailey. 1935. The balance of animal populations. *Proceedings of the Zoological Society of London* 3: 551-598.

Rapport, D. J. and J. E Turner. 1975. Feeding rates and population growth. *Ecology* 56: 942-949.

Ricker, W. E. 1954. Stock and recruitment. *J. Fish. Res. Bd. Canad.* 11: 559-623.

Royama, T. 1971. A comparative study of models for predation and parasitism. *Researches on Population Ecology, Supplement No.* 1: 1-91.

Royama, T. 1992. *Analytical population dynamics*. London: Chapman and Hall.

Tanner, J. T. 1975. The stability and intrinsic growth rates of prey and predator populations. *Ecology* 56: 855-867.

Verhulst, P. F. 1838. Notice sur la loi que la population suite dans son accroissement. *Correspondence Mathematique et Physique* 10: 113-121.

Volterra, V. 1928. Variations and fluctuations of the number of individuals in animal species living together. Translated in R. N. Chapman. 1931. *Animal Ecology*. NY: Arno Press.

Population and Community Ecology for Insect Management and Conservation, Baumgärtner et al. (eds)
 ISBN 90 5410 930 0

Carabid population and community features as an 'adaptation' to the landscape system: Importance of the ecotope as a landscape unit

Pietro Brandmayr, Stefano Scalercio, Tullia Zetto & Roberto Pizzolotto
Dipartimento di Ecologia, Università della Calabria, Arcavacata di Rende, Italy

ABSTRACT: Carabid community features have been studied from 53 forest and open habitat stands along a 70 km topographical profile from the western to the eastern coast of Calabria, Southern Italy. At regional (geographic) scale species diversity changes are easier to understand if climax habitats (forests) are dealt with separately from other open habitats, and especially from azonal ones. At landscape-ecological scale a new "three layer" model to organize species biodiversity data is proposed, based on the introduction of a new level, the η-diversity, between α and γ, that corresponds to the ecotopes. These are seen as geomorphic units guesting more animal communities strictly related each other also by syndynamic links. The results of a study on three ecotopes of the Sila Greca "pilot landscape" are reported, and compared to other animal and plant taxa. The distribution of some population- and life history traits of carabid beetles, as time of larval development and dispersal power, show that the ecotopes could be considered important "adaptation" units within the landscape.

1 INTRODUCTION

To attain a survey of the animal biodiversity at country ("regione", "Land") level and to produce reliable maps of the species richness and of its related contents an approach based on landscape ecology is needed. The core of the problem is given here by the necessity to discover the links between the population density (or abundance, activity-density, etc.) values found in faunal (taxon) assemblages and the environment (ecosystem) parametres and their changes in space and time, and, especially, at what scale they are easier to understand. The problem of scale is perhaps today the most strenghtened by modern landscape ecologists (Forman & Godron, 1986; many contributions in Hudson, 1991). During long-lasting studies in Italy and Southern Europe, and especially in Calabria (1987-1996) the first author and his coworkers accumulated some hundreds of year samples of carabid communities (species/abundance patterns), and their analysis revealed that Carabid beetle assemblages seem adapted not only to the habitat features, but also to the landscape unit where the biotope is placed. Moreover, in most cases the relevant landscape units were recognizable by geomorphlogical parameters, a fact that fits well with the basic adaptations of the "ground beetles".

This theme was investigated in Calabria with the founding of italian C.N.R. and of the DG XII scientific programme "Medimont", coordinated by I.C.A.L.P.E., (Int. Centre for Alpine Environments). The object of the study was the **mediterranean landscape**, a landscape where mountains largely predominate.

An ecologically significant landscape profile was chosen 1989, along a transect of about 70 km length, that links the Thyrrhenian coast of Calabria to the Ionian Sea, crosses the large valley of the Crati river and the granitic mountain tableland called Sila with its nearby 2000 metres high tips (M.te Botte Donato, m 1928 a.s.l.). Some aspects of the profile have been described in Brandmayr et al., 1991, climate factors seem to be of great importance for the understanding of vegetation and faunation changes from coast to coast. Northwestern winds and oceanic humid air accumulate heavy yearly rain amounts on the thyrrhenian slope of the Coastal Range ("Catena Costiera"), with precipitations varying here from mm 1100 to peaks of 1800 and more, a secondary peak is found on the Sila plateau (mm 1500), on the ionian (Sila Greca) side the rain amount is strongly decreasing, with a minimum of 700-600 millimetres on the coast.

The transect covers two very different biomes: the sclerophylls, mostly remnants of holly and cork oak forests, and the biome of deciduous temperate forests, better conserved and represented especially by pubescent oak forests, beech woods and, on the Sila, by *Pinus laricio* paraclimax formations. On the Western Coastal range the holly-beech forest (*Aquifolio-Fagetum*) is found at unusually low altitudes, and begins often at only 800-900 metres a.s.l., probably in connection with a strongly developed fog belt, that enhances the survival of beech and turkey-oak trees. Many anthropogenous habitats have been included in the investigation, also because in arid-mediterranean parts of the transect practically no climax vegetation survived. The biodiversity survey covered not only carabid beetles, but also: song birds, small mammals and butterflies.

In the eastern part of the transect profile an area of about 1000 kmq, that steeply drops to the Ionian Sea, the "Sila Greca" Medimont Pilot zone, has been more carefully surveyed, also in order to reconstruct the landscape-biodiversity relationships in a model area, better known in its ecosystem components: climate, vegetation, soil, geomorphology, man impact, degradation, soil use intensity and trends (Caloiero et al., 1990; CNR, 1995; Fascetti et al., 1988; Cavazzani, 1990). In the coastal landscape system of Sila Greca also the phanerogamous plant communities have been described, surveying as far as possible the same stands visited for animals.

2 THE PROBLEM OF VARIATION IN SPACE OF CARABID COMMUNITIES

The aim of this paper is to demonstrate that it is possible to follow variation of carabid communities from at least three points of view: i) studying large geographic gradients, mostly determined by climate and/or bedrock changes; ii) analyzing landscape systems, where the problem is to use a landscape unit of adequate size and correctly defined; in this frame, some adaptational features of carabid populations will be examined; iii) at biotope level, the approach preferred by ecologists and "carabidologists" who use "year samples" obtained by pit fall-traps to describe insect community structure in selected habitat types, an approach that allows not only to measure biodiversity, but also population abundance (activity density) at the same time.

The study of large geographic gradients can be performed in various modes, e.g. by comparing the same habitat or a restricted group of habitats in different points of the gradient, but also by comparing a sequence of landscape systems, if the research effort is sustainable. The diversity value measured at landscape level is mostly defined as γ-diversity by the authors (Whittaker, 1973). Greater problems arise, especially for zoologists, if we have to analyse parts or elements of a landscape system and then try to reconstruct or a mapping of the communities censused. During the survey of the Sila Greca landscape system, we constated that geologic and geomorphologic features of the area, for its greatest pat occupied by the Trionto river basin, are quite simple. From the highest elevations to the coast granites, metamorphic rocks, limestones, sandstones and Pliocene clays replace each other in rapid sequence, and if we try to superpose to the lithologic variation the vegetation map, a possible range of some hundreds of "carabid habitats" could be individuated. But, if we look

at the geomorphic units drawn on the "Geomorphologic map of the Trionto River Basin" (CNR, 1995), a noticeable reduction of the sampling sites seemed possible. For this reason, and starting from our previous experience in Northern Italy, we organised the samplings in **ecotopes**, the experimental design being that of the following model (figure 1).
The ecotope is, basing on the definition of Troll and more recent landscape ecologists, "**the smallest landscape unit that's still a landscape**", and geomorphic units are in fact small landscapes that build a complex landscape. Ecotope definitions are given by several authors, see e.g. the clear figure of Leser (1976, page 208, redrawn from Haase, 1967). Another use of the word is found in some ecologists, meaning the full range of adaptations of a species to external factors (Lincoln et al., 1993), but this second meaning coincides practically with Hutchinson's niche concept. In this work ecotope is defined as following: "**a basic landscape unit delimited mostly by its particular geomorphology, and guesting a certain number of more or less closely interrelated ecosystem (vegetation, faunal community) types, that built a well defined community complex**". This concepts is for some aspects very similar to the geocomplex or "geobiocoenosis" of Easterneuropean botanists and landscape ecologists (see the very useful work of Petch & Kolejka, 1993), but here the strict geomorphic boundaries are emphasized, as in Brandmayr (1982) for the zoocoenotic complexes of the Eastern Po Plain. Vegetation complexes and their relation to ecotopes are dealt with also by italian botanists (Pignatti, 1980; 1994). Finally, ecotopes may be **zonal** or **azonal**, i.e. their geomorphologic features can be more or less favourable to the complete development of ecological successions under a given climate. In other words, in zonal ecotopes seres run up to the final stage of the ecosystem determined by climate (the "climax", in Europe mostly a forest), in azonal ones substrate and other physical conditions as subsoil water or erosion restrain the development of ecosystem/vegetation and zoocoenosis to early succession stages, as we see for example on sand dunes, in river beds, on rock cliffs and so on.

Of course, to describe the biodiversity of an entire landscape given by more ecotopes, at least three levels of diversity measurements are needed: the a level (biotope level) is well known in the literature, and also the g level (an entire landscape), for the ecotope level a new and hitherto unused Greek letter is proposed: the **η-diversity**.

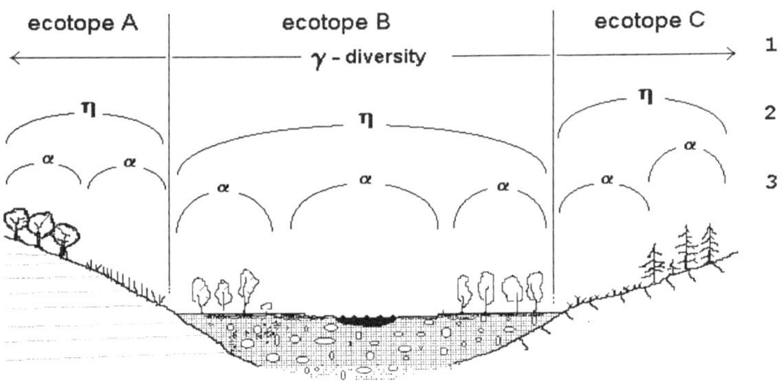

"The ecotope is a basic landscape unit, prevailingly defined by geomorphologic features, and inhabited by elementary animal communities, more or less similar each other and related by syndynamic ties, forming an unique zoocoenotic complex" (from Brandmayr, 1990).

"the smallest holistic land unit, characterized by homogeneity of at least one land attribute of the geosphere (athmosphere, vegetation, soil, rock) and with non-excessive variations in other attributes"... (Zonneveld, 1972, in Naveh & Liebermann, 1990).

"the smallest possible land unit, that is still a holistic unit"... (Forman & Godron, 1986).

Figure 1. Three neighbouring ecotopes and the levels α, η and γ at which diversity can be measured.

These three levels have been combined into an unique new model, the "**three layer model of landscape biodiversity**". The model is very simple in its construction, and requires only the basic knowledge of the α-diversity values of as many as possible year samples caught in all the habitat types recognizable in an homogeneous ecotope. The η-level is calculated by putting together all the alpha-values by the following set-theoretical equation:

$$|\eta| = |\alpha_1| \cup |\alpha_2| \cup |\alpha_3| \cup \ldots |\alpha_n|$$

This can be solved through an inclusion-exclusion equation. In other words, if in an ecotope three main types of habitat are found, habitat 1 inhabited by x species, habitat 2 by y, habitat 3 by z, the eta-diversity is given by x + y +z, less the species shared by habitat 1 and 2, and that what habitat 3 bears in common with the formers. If the habitat sampled are representative of the variation found in the ecotope, the η-(species) diversity obtained on this way is comparable with values obtained for other ecotopes sampled with the same criterion.

3 METHODS

The data presented refer to year samples of carabid beetles in phytosociologically careful defined biotopes, obtained by 5-8 pit-fall traps emptied monthly. The traps were plastic vessels, upper diameter 9 cm, filled with wine vinegar added with a 5% of formalin, the same used hitherto for the most carabid samplings in Italy. The results (species/individuals composition) of each stand have been calculated as aAD (mean annual activity density, expressed as indivduals/trap in the standard period of 10 days, calculated over the entire trapping season). The data set concerning the entire transect is enclosed in a species/stand matrix of 53 columns (stands) and 170 species lines ("Excel"-Microsoft file) based on the capture and counting of about 90,000 beetles, the data set of the Sila Greca landscape concerns only 67 species and the relevant data are listed in table 1.

Finally, also "multitaxon" comparisons were possible, because the entire research program has been devoted to a certain number of "key taxa": birds, micromammals, butterflies, for some sector of the transect also amphibians, reptiles and phanerogamous plants (see Cagnin et al, 1991; Pizzolotto et al., 1991; Brandmayr et al., 1996.

4 RESULTS

4.1 *The transect (geographical dimension).*

Along the transect about 170 Carabid species have been censused in 7 years sampling, the expected number lies around 300, when all the habitats will have been sampled in at least one "typical stand". The communities so far recognized are about 20 (see many of them in Brandmayr & Pizzolotto, 1995). A diversity of 270 carabid species seems to be not very far from the supposed "ε-diversity" (regional diversity) of Calabria or at least of the Cosenza Province, and equals more or less one fourth of the Italian fauna (Vigna-Taglianti, 1993) or the species number of some countries of Northern Europe: United Kingdom, Denmark, etc.

In figure 2 the altitude of the 53 stands sampled is plotted as a background of the Carabid species number found in one year. At first glance on gets the impression that species richness in a biotope (α-diversity) is higher at lower altitudes, but in fact there is no clear relationship to this factor, although lower habitats can reach S-diversity values higher than high-altitude ones.

A first clear insight in the problem is reached if the α-diversity values are subdivided into habitat categories (figure 3). Here the transect clearly reveals its ecosystem heterogeneity, and each habitat-type shows certain peculiarities. Especially in forests, that are the remnants of the climax ecosystems in Calabria, the specie number is highly variable, going from a minimum in degraded sclerophylls (less than 5) to a maximum of 22 in *Quercus cerris* (turkey oak) stands at medium elevations (see stands Ce, Tce1 in the figures).

On average, the forests of the Western Coastal chain are species-richer, and this seems to be due to a climate factor, the greater air humidity and the fogs generated by the dominant NW-winds of the thyrrhenian side. A very similar situation has been described for the Sicilian coastal ranges (Pizzolotto & Brandmayr, 1990), and in both cases a crowding of "historical indicators" in the community is observed, these are mostly endemic carabid species of the Westmediterranean basin or of the Appennine range: *Laemostenus acutangulus*, *Leistus sardous*, *Pseudomasoreus canigouensis*, *Platyderus canaliculatus*, etc. Another group of

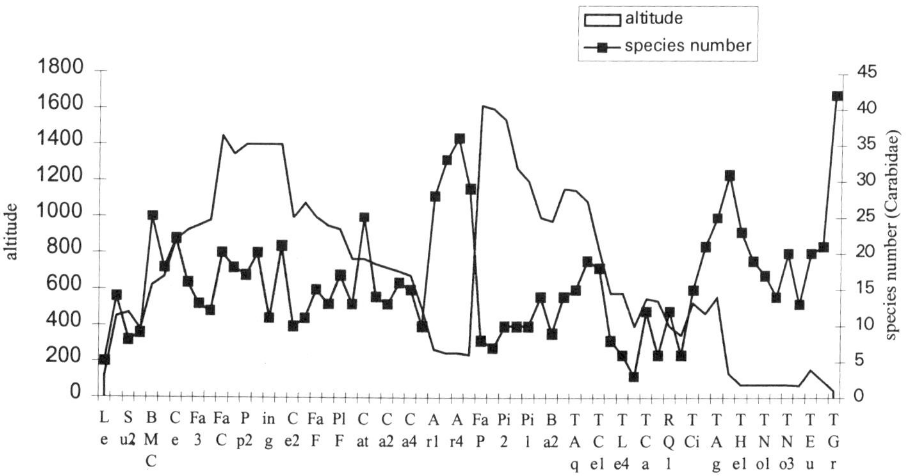

Figure 2. Species number of Carabidae in the Paola-Calopezzati transect of Calabria, plotted on the altitude profile. The distance between the Western limit (Thyrrhenian Sea) and the Eastern (Ionian Sea) is of about 70 kilometres.

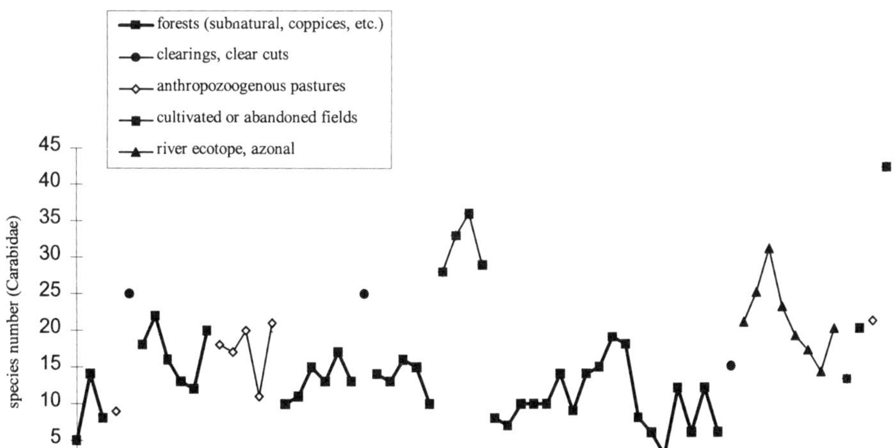

Figure 3. Species diversity of Carabidae in the Paola-Calopezzati transect of Calabria, sampling sites subdivided in main habitat types. The category "river ecotope" encloses all habitat types found in the Trionto fiumara.

species found here is clearly tied to oceanic climates in many Westerneuropean countries, e.g.: *Trichotichnus nitens*, *Leistus spinibarbis*. While many carabid species are found on the western side of the transect, no forest species has been found that shows a clear preference for the eastern slope, but many cases are known of "equal distribution" on both sides, as it could be stated for the oak forest dweller *Notiophilus rufipes* (see figures 4 and 5).

A statistical analysis of species diversity in forests of the western and eastern side (20 stands versus 19 ones) gave a mean species diversity of 13.75 and 9.95 respectively (SD: 3.97; 4.65). Wilcoxon-Mann-Whitney test and Student t-test both indicate that the two groups do not come

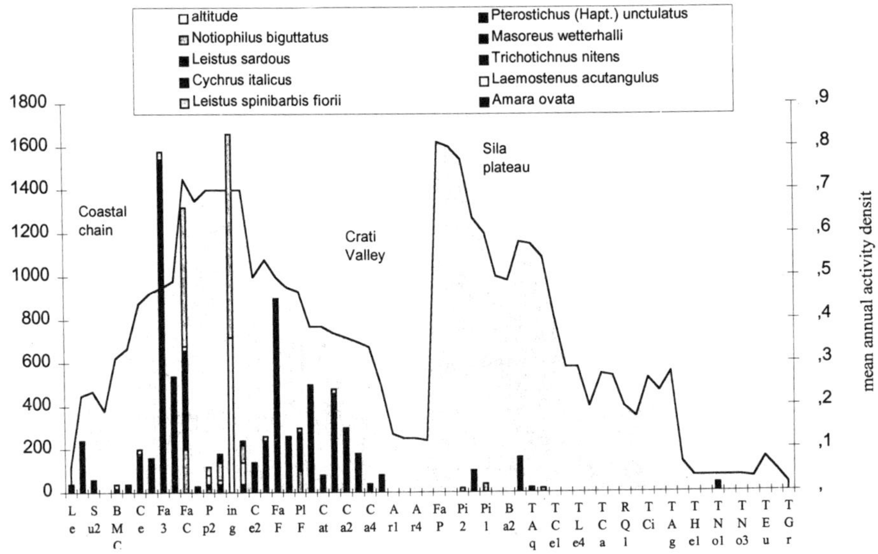

Figure 4. Strictly forest dwellers of the transect crowd on the western side of the transect, that's the most humid and fog-rich.

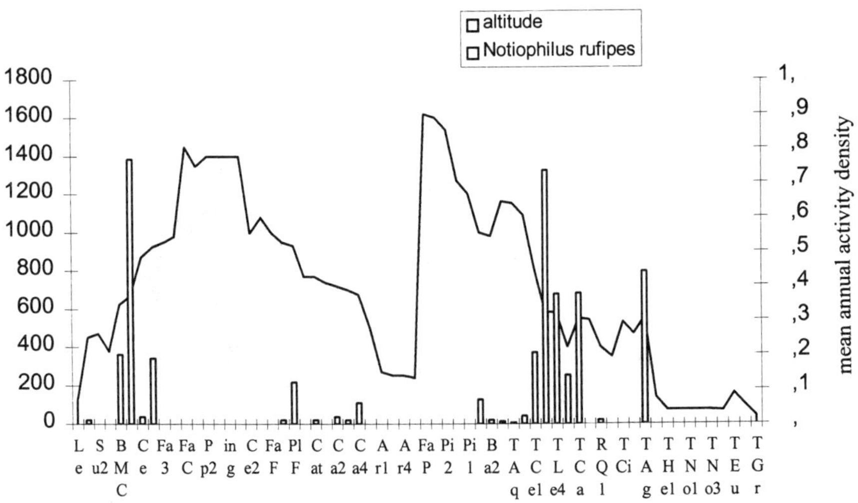

Figure 5. The abundance distribution of *Notiophilus rufipes*, an indicator of thermophilous deciduous oak forests, along the Paola-Calopezzati transect.

from the same population, with a significance level of 0.05.

In clearings and clear-cuts a species number always higher of the neighbouring forest is found, as typically in short-lived, instable "successional" situations, but in anthropozoogenous pastures on mountain bedrocks the species number doesn't surpass that of mountain forests, especially on calcareous (karstic) soils. In these pastures the genus *Cymindis* is of particular importance, *Calathus* and *Syntomus* are here very common. The maximum of α-diversity in the above sense is measured in abandoned croplands, as well as in some true croplands of the ionian side (see TGr, a durum wheat field on sodium chloride rich clays). Abandoned fields are inhabited by many opportunistic carabids with high dispersal power (that means: well developed metathoracic alae, see also the contribution of den Boer in this book) and often more or less tied to some herbaceous plants, as *Daucus carota*, *Plantago* sp., some crucifers and Graminaceae. All the warm open lands bear many species belonging to polyphagous carabids like *Harpalus*, *Amara*, or strictly seed eaters: *Ophonus*, *Ditomus*, *Carterus*, together with many true predators like: *Pterostichus*, *Poecilus*, *Calathus*, *Siagona*, *Brachinus*.

Species diversity values higher than in forests are often found also in riparian habitats, on the riverbank habitats of Mediterranean "fiumara" beds. The azonal river ecotopes are inhabited by a completely different group of habitat specialists, there are at least three ecological groups that show strong habitat partitioning inside the river ecotope: the ripicolous one, strictly bound to the water borders on fine or coarse sediments (e.g.: *Bembidion*, *Nebria*, *Omophron*, etc.), the xerophilous group, living far from the water in open habitats (many *Amara* species, *Apristus* sp., etc.) and the riparian forest dwellers (*Asaphidion*, *Agonum*, etc.).

The study of the transect reveals that in describing diversity of a country at least four or five stand categories should be dealt with separately: **climax ecosystems** versus **non-climax** ones; **zonal** versus **azonal** ecosystems (that is not the same of the precedent, see chapter 2); **anthropogenous** versus less **anthropogenous or primeval** (secondary habitats versus primary ones).

Changes in large geographical transitions can be followed at best in climax or zonal habitats. In the transect studied, especially the West-East decrease of forest dwellers "historical indicators" is easily explained by climate differences (more oceanic conditions of the western coastal chains). Variation due to bedrock changes will be easier discussed in the following.

4.2 *The landscape dimension.*

The eastern part of the transect offered us the opportunity for a more exhaustive landscape study, because the Medimont Programme was focused on the influence of man's activities on the biodiversity of a Pilot zone of about 1200 square kilometres, the "Sila Greca Pilot Zone". Our investigation was concentrated on a greater river basin, the Trionto-basin, that in the lower part of the valley behaves as a typical "fiumara", with a rain-dependent water regime marked by heavy floods in winter and low water level in summer. Around the Trionto river, on the coastal side of Sila Greca, other two important ecotopes can be identified, well different each from the other in substrate features and erodibility of the soil: i) the "lower clay-hill" and ii) the "medium height hill". The first is placed at very low altitude and completely deforested, the substrates being mostly Pliocene or Miocene clays, often sodium-chloride rich. This ecotope is occupied by pastures, *Eucalyptus*-plantations and durum-wheat cultivations, the climax remnants are given only by scattered Oleo-Lentiscetum stripes. The second ecotope is marked by very hard rocky substrates, steep slopes and poorly erodible soils. It begins at somewhat higher altitudes, the percentage of forest cover lies around 30-40%, cropland and pastures are of minor importance.

4.2.1 *The validity of the model.*

The application possibilities given by the model have been tested on a first lot of samples from

Sila Greca, 12 among the 24 year samples gained from this area, and these results concern some thousands specimens collected (about 6,000) and 67 species of carabid beetles (some further species have been collected only by hand, mostly in the fiumara ecotope after floodings, and for this reason excluded from the analyses presented in this paper).

A first proof of the validity of the three layer model could be obtained by comparing coenotic affinities "inside" ecotopes and "between" them. Inside we should find communities (species-abundance patterns) that range from very distant to very close, and clear coenoclines indicating the existence of ecological successions (seres, series) that regulate the transition from one community to the following in time or to the neighbouring in space. For example: in the "fiumara ecotope" the community of bare gravels of river banks, mainly colonised by *Nebria psammodes*, relates mainly to the *Amara fusca-Apristus subaeneus* grouping, the first stage of the colonization of gravel terraces, and this stage transforms slowly itself into the *Asaphidion curtum*- and finally in the *Asaphidion rossii* grouping, which inhabits the narrow Alnetum or Salicetum fringes originarily bordering the Trionto fiumara.

The fiumara is typically an azonal ecotope, because its “climax” (better: equilibrium-) ecosystems don’t depend on the general climate of the area, the animals living there are mostly adapted to conditions of high ecosystem instability, as it is demonstrated by the high numbers and percentages of carabid species able to fly (macropterous or dimorphic). The populations of these beetles are normally short-lived, anyway they behave opportunistic, and many elements are adapted to the regular fiumara floods, escaping them by short **intraecotope migrations**, for instance from the borders of the running waters to the borders of the river or onto the dams.

On the other side, we expected low affinities between an ecotope and the neighbouring ones, because hardly the ecological successions are able to cover and relate different ecotopes into an unique climaxcomplex (or "geosygmetum"), in other words the borders of ecotopes are mostly "**ecotones**" and not "**ecoclines**". Ecoclines prevail inside the ecotopes, but the boundaries between ecotopes are often sharpened by man's activities to counteract erosion or for other territory management policies. What we found (see table 1) are Sörensen's QS values from 8,0 to 60,9 inside single ecotopes, and no more than 28,6 between ecotopes, this means that the fauna is so less homogeneously distributed in the landscape, that a description through the ecotope level is certainly "useful", and perhaps, necessary, at least for Carabids. In other words, inside an ecotope we found both low as well high affinity values, accordingly to the hypothesis that communities “inside” are dynamically interrelated. In each ecotope a “core” of higher values is recognizable, and we marked arbitrarily this core starting from the value 50% (darkened cells). Lower values (below 35%) are found both inside ecotopes as well as between stands of different ecotopes, and in certain cases the nature of soil plays an interesting role in create links between different ecotopes, as we can see in column Tci, where the sandy soil of the *Cistus* garigue relates more to the sand banks of the “fiumara” than to the other forest stands of the medium height hills, which develop on rocky soils. Another point should be underlined: the high values core is found mostly in the less disturbed stands, forests, thickets or tree plantations, and this agrees with the idea of the successional relationships regulating evolution of communities towards “climax” conditions or at least more stable aspects.

On the other hand, if we compare the “eta” diversities of the three ecotopes, very low affinities are found (see lowest group of cells in table 1). Between the species set of the river ecotope and the neighbouring clay hills no more than 16,7 of Sörensen value has been calculated, that means that the two community complexes are clearly separate each from the other, as the substrate condition already suggested. The same holds true for the two hill ecotopes, (22,2), the carabid fauna of which show fundamental differences in general adaptations (see next chapter).

Now, the validity of such a general model should not be based only on carabid beetles, a group that is clearly tied to soil features and microclimate, and which is known to be linked to geomorphic conditions.

In table 2 the alpha, eta and gamma diversity of phanerogamous plants, birds, micromammals and butterflies are represented versus Carabid beetle ones. Each group shows a peculiar pattern

Table 1. - Soerensen similarity index between carabid groupings of the three ecotopes studied, and between ecotopes (eta-diversities) as well (lower section of the table).

	THe2	THe1	TNo1	TNo2	TNo3	TP1	TOl1	TEu	TCi	TLe2	RQ1	TCa	Stand initials
	-	-	-	-	-	E	-	O	S-SE	N-O	O-NO	N	Slope direction
	74	74	74	74	74	100	68	160	530	350	400	550	Altitude
	0	0	0	0	0	16	0	10	15	40	20	20	Slope inclination (°)
	30	40	100	85	100	85	20	50	90	90	85	60	% of vegetation covering
THe2		42,9	38,9	18,2	20,5	25,	12,5	10,3	17,6	8,	6,7	12,9	
THe1			80,	43,2	55,8	22,7	33,3	9,3	42,1	13,8	17,6	28,6	
TNo1				45,2	59,4	21,1	20,	10,8	43,7	26,1	21,4	27,6	
TNo2					58,8	28,6	14,8	23,6	41,4	20,	16,	30,8	
TNo3						34,1	30,3	15,	45,7	30,1	19,4	31,2	
TP1							47,1	58,5	27,8	22,2	12,5	24,2	
TOl1								30,3	35,7	31,6	16,7	32,	
TEu									22,9	15,4	6,5	12,5	
TCi										38,1	30,1	37,	
TLe2											47,1	55,6	
RQ1												60,9	
	2,46	2,41	12,02	7,94	10,39	0.57	2.01	2,69	1,52	1,36	3,57	1,84	total annual AD
	19	23	17	14	20	21	13	20	15	6	11	12	species number
	THe2	THe1	TNo1	TNo2	TNo3	TP1	TOl1	TEu	TCi	TLe2	RQ1	TCa	stand initials
	17	19	14	11	16	17	9	16	10	2	5	3	macropterous or dimorphic
	89	83	82	79	80	81	69	80	67	33	45	25	% of species able to fly
	1	1	1	2	1	3	1	2	2	2	3	4	endemic species (II)
	5	4	6	14	5	14	8	10	13	33	27	33	% of endemic spp.
Habitat	**Helichrysum shrubs**					**pasture**			**Cistus**				
			Oleander thickets				**olive growth**			**Holly oak**			
								Eucalyptus pl.			**Pub. oak**		
												Chestnut	
Ecotope	**F l u m a r a**					**lower clay hills**			**medium height hills**				
species n.			47				31			23			
QS							16,7			28,6			
										22,2			

Table 2 - Diversity (species) in the three ecotpes studied, more taxa. The Chestnut stand of table 1 (TCa) could not be included, because bird and plant data are lacking for this station. In the column of "lower clay hill" the losses of diversity in absolute number of species and percentage are reported for each taxon. The darkened column at right contains the species found only in the medium height hills, a number used in section 4.2.3 for the calculation of diversity losses in "lower clay hills".

ANIMAL BIODIVERSITY IN THE TRIONTO LANDSCAPE FROM LOWLAND TO 500 m a.s.l.													only in m.h.h.
ECOTOPE	FIUMARA ECOTOPE					LOWER CLAY HILLS			MEDIUM HEIGHT HILLS				
stand number(transect)	34	35	36	37	38	33	32	39	40	41	42	43	
initials	THel	THe2	TNol	TNo2	TNo3	TOl1	TP1	TEu	TLel	TLe2	TCi	RQ1	
slope exposure	-	-	-	-	-	-	E	W	SE	N-W	S-SE	W-NW	
altitude	74	74	74	74	74	68	100	160	150	350	530	400	
Vegetation													
"alpha" diversity		101	148			12	97	80	45	75	19	70	
"eta" diversity			201				178(34.6)				149		**94**
"gamma" diversity							***383***						
stand initials or ecosystem		THe	TNo			TOl	TP	TEu	TLel	TLe2	TCi	RQ1	
n. samplings/biotope type		6	5			3	5	6	3	5	4	5	
average species n./sampl.		34	42			9	31	42	29	27	11	46	
Carabid beetles													
"alpha" diversity	23	12	17	14	19	13	21	17	2	6	15	11	
"eta" diversity			46				30 (31.8)				22		**14**
"gamma" diversity							***67***						
Micromammals													
"alpha" diversity		4	0	4		5	3	0	5	7	5	5	
"eta" diversity			4				5 (37.5)				8		**3**
"gamma" diversity							***8***						
stand initials		THel	THe2	TNol		TOl1	TP	TEu	TLel	TLe2	TCi	RQ1	
Birds													
"alpha" diversity		8	6	12		22	18	19		19		24	
"eta" diversity			15				33 (31.5)				31		**15**
"gamma" diversity							***51***						
Butterflies													
"alpha" diversity	15	6	9	5	5	20	15	11	6	11	12	20	
"eta diversity"			19				27 (28.9)				34		**11**
"gamma diversity"							***42***						

in the three ecotopes, and especially plants show practically the same behaviour of carabids, with a maximum of diversity in the fiumara. All the other animal groups are more or less linked to the plants, (e.g.: to the vegetation pattern or physiognomy, or to specific plants as food or shelter, or to preys tied to certain plants or ecosystem features, etc.), higher vertebrates like birds and micromammals show anyway higher diversity in hill biotopes, carabids and plants, on the contrary, behave more or less in the same manner, with a maximum of eta-values in the fiumara.

4.2.2 - A*daptation of carabid populations to the ecotope conditions*

Some of the basic adaptations of Coleoptera Carabidae have been dealt with in the paper of den Boer in this book. The most relevant life-history traits of the ground beetles seem to be the dispersal power and the age of first reproduction. The first feature is easily to be evaluated by the status of the metathoracic alae, that varies from fully "macropterous" to "pteridimorphic" to "brachypterous", a status that is often found in stable habitats like forests. Entire faunas or communities can be rapidly checked on their dispersal power, using the morphological marker of hindwings, as it has been done firstly by Lindroth (1949) in his pioneer study. The use of hindwings in carabid populations varies highly from one species to the other, and relates strongly to the population turnover in the landscape (den Boer, 1977), brachypterous species tend to become endemic and sedentary especially in forests (Brandmayr, 1983).

Figure 6 gives us a strong evidence of how basic adaptations of carabids react to the geomorphic ecotope characteristics. In the fiumara good flyers and non-endemic elements largely prevail, owing to the highly unstable conditions of the water regime of this mediterranean kind of river. The same condition is found in Northernitalian rivers, as well in Centraleuropean ones. Also the lower height hills seem fairly unstable, but we should remember that clayey soils of the mediterranean area undergo heavy solifluction each year, partly as a result of strong deforestation by man. Only on the rocky hills, where forests are still conserved, the number of sedentary species raises highly (flyers drop to less than 50%), and endemic taxa (distribution area limited to the Appennines, or to southern Italy) raise to more than 20% in the community.

At the eta-diversity level, the brachypterous species versus total of species hitherto ascertained in the ecotopes has been found to be: 6/46 in the fiumara, 5/30 in the lower clay hill, 10/22 in the medium hills, thus decreasing more or less than in the diagram of figure 6: 13, 17 and 45% respectively.

The other "primary" life-history trait in Carabids seems to be the time (season) of reproduction. Carabid species of Europe are traditionally encased in spring- and autumn breeders (Larsson, 1939), the first ones laying their eggs in spring and with short larval development during the summer, the others laying eggs in late summer or autumn and with preimaginal development in the winter months. Thiele (1977) and Paarmann (1979) studied and explained the physiological regulation of many other rhythm types, on the whole we are now able to recognize, at least in european countries, two fundamental types, one with summer larvae that seems tied to soils with high water retention power or high watershed level, (but also to continental climate conditions), and a type with winter larvae that seems to be favourished by low water levels in the soil, but also by mediterranean climate or, generally speaking, mild winters. Summer and winter larvae differ in a very important feature: only the summer larvae develop speedly, free of dormancy, whereas winter larvae, as well as the larvae of the two-year generation type, show various kinds of obligatory interruptions of the development. In the mountains of southern Europe, till now very poorly investigated, 2 year generation is found in

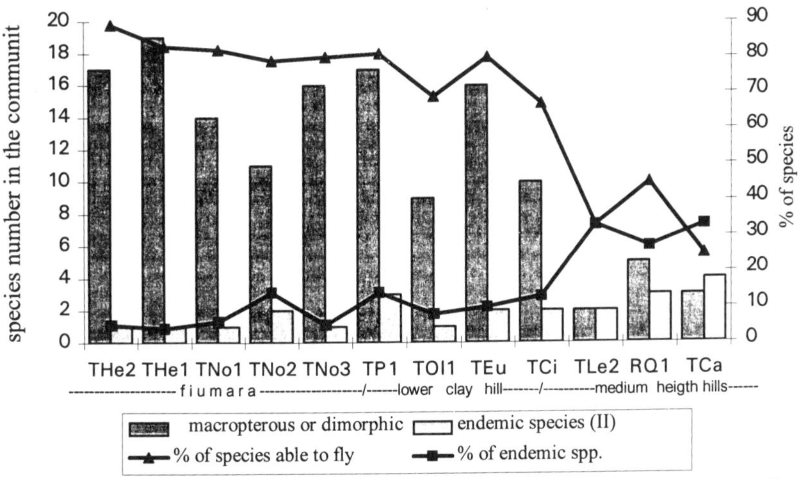

Figure 6 - Absolute species number and percentage of brachypterous and endemic taxa in each of the 12 habitats censused, ordered in the three ecotopes, ranked from left to right by increasing ecosystem stability.

many forest species (*Pterostichus*, *Abax*, *Platyderus*, etc.). Species with this kind of preimaginal development are mostly brachypterous and true forest dwellers.
Winter larvae seem to predominate in drier or forested ecotopes, in the fiumara the species bearing this reproduction type drop to a minimum of 40%. In the clayey soils of the lower hills, where the original forests are largely destroyed, the two main types account each for 50%, in the medium height hills forest species with longer lasting development (winter or two years) are of mayor importance.

4.2.3 - *Applications of the three layer model.*

The three-layer model allows accurate descriptions of the actual and potential faunal landscape of an area (the faunation of Udvardy, 1969), and comparisons between prae-anthropic (primeval) and actual faunations. We tried to do this for the litoral part of the "Sila Greca pilot zone", comparing the three main ecotopes already discussed in the previous paragraph. We should note, firstly, that at least 5 new communities or classical "zoosociological groupings" have been created by man during its progressive modification of the landscape physiognomy, and that probably the man made coenoses are far more numerous, at least some dozens.
The most striking in the faunation studied is that some ecotopes seem more vulnerable than others, perhaps because they were particularly favourable to mans activities or for they are of scarce resilience, as we found in the "**lower clay hills**", an ecotope (or physiotope in the sense of Pignatti) originally covered by sclerophyll forests and now entirely "desertified" by cropland, burnings, severe solifluction phenomena in the rainy mediterranean winter, soil degradation, overgrazing, so that now we find here true badlands, called "calanchi" in the italian language, or pastures, durum wheat crops and sparse shrubby remnants of the forest ecosystem. Being the "**medium height hills**" still rich in oak and sclerophyll forest formations, although severely degraded, we enucleated the species of Carabids found only in this ecotope, where only forests or "macchias" were sampled, as potential inhabitants of the extirpated forests of the first ecotope,bearing in mind that the fauna of a forest varies with the facies, the exposition, the water retention power of the soil, and many other biotic and abiotic factors. Of course, no open land, grassland or cropland species of this area was included in the list.
On this way, a comparison has been made between the climax ecosystem of one ecotope (medium height hills), where the climax is better conserved due to the prevalence of

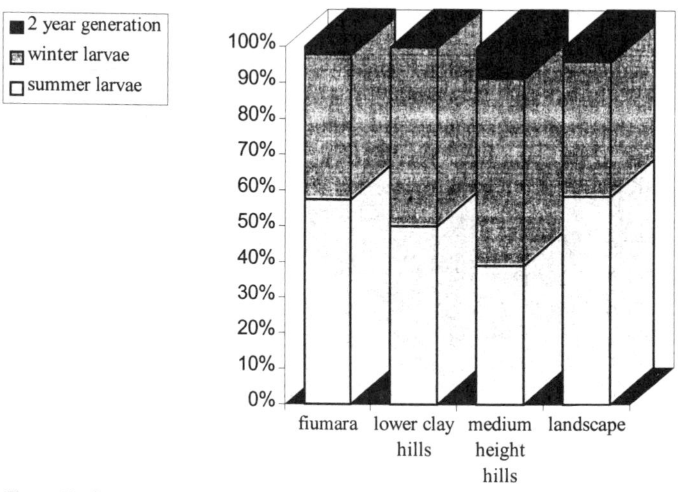

Figure 7 - Percentage of reproduction rhythms distribution in the three ecotopes studied, compared with the whole landscape. A clear minimum of winter larvae beetles is seen in the river ecotope.

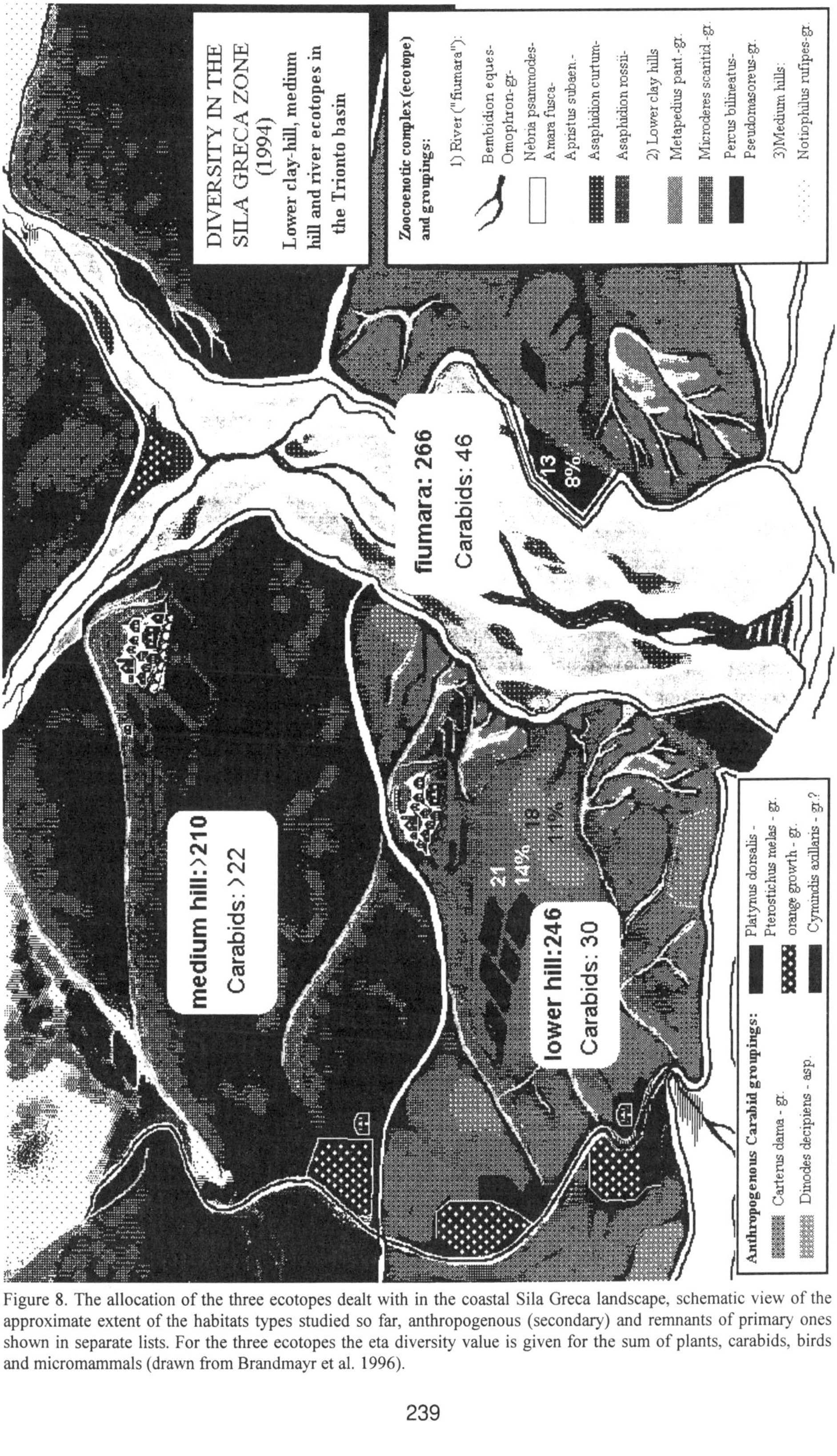

Figure 8. The allocation of the three ecotopes dealt with in the coastal Sila Greca landscape, schematic view of the approximate extent of the habitats types studied so far, anthropogenous (secondary) and remnants of primary ones shown in separate lists. For the three ecotopes the eta diversity value is given for the sum of plants, carabids, birds and micromammals (drawn from Brandmayr et al. 1996).

metamorphic "hard substrate", and a fully desertified ecotope, where no climax formations survived, although they should have been not so different from that of the previous ecotope. Carabid species numbers indicate that at least **31.8%** of the species of the "lower clay ecotope" have been lost only because climax ecosystem has been destroyed (14 species versus 30+14 = 44, see table 2 for more details).

This means that the original status of species richness in the first ecotope was of at least 30% species more.

We calculated the same numbers for plants, birds, micromammals and butterflies, keeping rigorously the same habitats (stands, both in number as well in type). The "eta" diversity losses obtained were astonishing similar: **34.6%** for plants (on a "gamma" value of 383); **37.5%** for micromammals (on a "gamma" of only 8); **31.5%** for birds, on a "gamma" of 51); **28.9%** for Rhopaloceran Lepidoptera (on a "gamma" of 42).

All these numbers are so close each other, that the idea arises, the η-diversity losses calculated on this way reveal an intrinsic property of the distribution of biota in the landscape. If these results will be confirmed by further and quantitatively better supported observations, the importance of the use of Carabid beetles or of other "key taxa" as "overall" animal diversity indicators in the landscape will surely increase.

5 SUMMARY AND CONCLUSIONS

Studies on animal communities, especially on birds, but also on other vertebrate and invertebrate taxa are since two decades increasing all over the mediterranean biome: Catling (1988), Cody (1975), Di Castri (1973), Di Castri & Vitaldi di Castri (1981), La Greca (1982) and many other authors provided for helpful information and basics of biodiversity in many kinds of mediteranean ecosystems of more continents. The present study aims to the definition of a more "compact" method that should allow rapid multitaxon surveys in the often lesser investigated mediterranean countries.

Using the newly proposed three layer model, that is based on a geomorphological concept of ecotope, not very far from the original definition, we observed, using only 12 "alpha" year samples, that ecotopes may be very different in species diversity ("eta" level), and that the carabid beetles show some typical population and especially life-history adaptations that differ from one ecotope to the neighbouring one. Moreover, the "faunula" of each ecotope seems to share very low affinity (or "β-diversity", in the new sense used today to mean diversity differences) to the next one, as the ecotopes would be separated by strong ecological boundaries, in fact resembling the ecological boundaries well known as ecotones. Also many behavioural traits of Carabidae, concerning habitat choice and intrahabitat migration, well studied especially in riverside beetles (see e.g. Thiele, 1977) agree with the above presented picture.

Some other animal taxa and plants show also similar strong differences in the "eta" values, the plants more in the river, birds, small mammals and butterflies more on the hills, but nothing is known, since today, on the possible affinities we could find between ecotopes, although we can predict that affinity values will be surely higher in good flying animals like butterflies, because of their ability to migrate from one habitat to the other also during the seasons.

The question arises if the habitat (stand) types choosen are really representative of the habitat variation existing in the ecotopes studied, otherwise the numbers discussed would reflect earlier a subjective choice of the researcher than natural phenomena. Also for this reason during the last two years our research group investigated many other crucial vegetation types in these three ecotopes. The stands now sampled are 20, the species ("gamma"-diversity) found lies around 110, (some taxa are still undetermined, what made impossible to present this new dataset), the "eta"-values of the three ecotopes increased to 72 (river), 54 (lower clay), 26 (medium hills), thus the trend already observed is confirmed.

The new “three-layer”model presented in this paper probably allows, with a combined landscape ecological and geographical approach to draw biodiversity maps on more eco-biological basis. The contemporary study of zooceonoses at both scale dimensions probably facilitates also the evaluation of diversity in climax ecosystems apparently identical but differing in the general climate.

Prof. Pietro Brandmayr is personally indebted with following co-workers and colleagues: Dr. Antonio Mingozzi: birds; Prof. Mara Cagnin, sig. Gaetano Aloise: small mammals; Prof. Giuliano Cesca, Prof. Liliana Bernardo, Dr. Carmen Gangale: phanerogamous plants; Dr. Maria C. Algieri, for help in field trips; Prof. Franco Rose, for some diaphotos.

REFERENCES

Boer, P. J., (den) 1977. Dispersal power and survival: Carabids in a cultivated countryside. *Misch. Pap. Landbouwhogesch. Wageningen*, 14:1-190.

Brandmayr, P. 1982. Lineamenti principali del paesaggio zoocenotico della pianura padano-veneta: passato e presente. *Quad. C.N.R. Str. Zooc. terr.*, 4:137-149.

Brandmayr, P. 1983. The main axes of the coenoclinal continuum from macroptery to brachyptery in carabid communities of the temperate zone. *Report 4th Symp.Carab.* '81 (1983): 147-169 Wageningen: Pudoc.

Brandmayr, P. 1990 - Le relazioni fra Zoologia ed Ecologia del Paesaggio terrestre. *Atti 53° Congr. naz. UZI, Simp. "Ricerca Zoologica ed Ambiente"*, pp. 6-7.

Brandmayr, P., Codogno, R. & Pizzolotto R. 1991. Basi ecologiche per la mappatura delle risorse naturali in Calabria: biomi ed unità ambientali minori lungo la sezione Catena Costiera - Sila Grande - Sila Greca. *S.IT.E. Atti*, 12 (1): 389-393.

Brandmayr, P. & Ghirardelli, E. 1995. *Gli animali e l'ambiente: ecologia.* Cap. 12 in: Baccetti B. *et al.*: Zoologia I, Trattato Italiano - Zanichelli, Bologna, pp. 469-603.

Brandmayr, P., L. Bernardo, M. Cagnin, T. Mingozzi & R. Pizzolotto 1996. Biodiversità multitaxon in Calabria su scala locale, paesaggistica e regionale. *4° Workshop Progetto Strategico C.N.R. "Clima Ambiente e Territorio nel Mezzogiorno"*, Lecce, 11-14/11/91, Atti(2):259-302. A cura di: A. Guerrini.

Brandmayr, P. & Pizzolotto R. 1988. Indicatori "storici" ed ecologici nella coleotterofauna terricola delle foreste dell'Appennino. *Atti XV Congr. Naz. Ital. Ent.*, l'Aquila, 1988: 589-608.

Brandmayr, P. & Pizzolotto, R. 1995. Arthropodes menacés, stabilité de la faune du sol et biodiversité dans les biomes méditerranéens de l'Italie du Sud: les communautés de Carabides. *Colloque Scient. int. Bio'mes, 6es Rencontres Ag. reg. Environm. Provence-Alpes-Cote D'Azur*, Gap, 16-18 Novembre 1994, pp. 254-262.

Cagnin, M., Pizzolotto R., Barbieri, A., Tripepi, S., Rossi, F., Ripolo, D., Aloise, P. & Brandmayr, P. 1991. Zoocenosi (Coleotteri Carabidi, Rettili, micromammiferi terricoli) in ambienti forestali e prativi della Catena Costiera di Calabria. *S.IT.E. Atti*, **12** (1): 455-459.

Cagnin, M., Venanzoni, R., Brandmayr, P., Canullo, R., Dessì-Fulgheri, F. & Pedrotti, F. 1991. Le risorse botaniche, zoologiche e paesaggistiche nello studio di impatto ambientale (V.I.A.) "Acquedotto Menta" (Aspromonte, Italia). *S.IT.E. Atti* **12**: 739-748.

Cagnin, M. & Aloise, G., 1993. La mammalofauna antica ed attuale della Calabria. *VII Congr. Ass. A. Ghigi*, Bologna, 15-16 gennaio 1993.

Caloiero, D., Niccoli, R. & Reali, C. 1990. Le precipitazioni in Calabria (1921-1980). *CNR-IRPI*, Geodata, 36, Cosenza, 53 pp..

Catling, P. C., (coordinator) 1988. Vertebrates. In: *Mediterranean-type Ecosystems, a data source book* (Specht R. L., ed.), pp. 171-194. Kluwer Academic Publ.

Cavazzani, A. (ed.), 1990. *Sistemi agricoli marginali. Rapporto a conclusione del progetto di ricerche sullo scenario Sila Graca.* Progetto finalizzato CNR-IPRA, Roma, pp. 189.

Codogno, M. & Puntillo, D. 1988. Variations of the vegetation-belt series in narrow areas of mediterranean basin (particularly in Calabria, S-Italy). *Sauteria*, 4: 141-148.

Cody, M. L. 1975. Towards a theory of continental species diversities: Bird distributions over mediterranean habitat gradients. In: Cody M. L., Diamond J. M. (eds.), *Ecology and evolution of communities* - pp. 214-257. Cambridge, Mass: Harvard Univ. Press.

Di Castri, F., 1973. Soil animals in latitudinal and topographic gradients of Mediterranean ecosystems. In: *Mediterranean-type Ecosystems. Ecological Studies* (Di Castri F. & Mooney, eds.), 7:171-190.

Di Castri, F. & Vitaldi Di Castri, V. 1981. Soil fauna of Mediterranean-climate regions. In: *Mediterranean-type Shrubland* (DI CASTRI F., GOODALL D. W. & SPECHT R. L., eds.). Amsterdam: Elsevier Scientific Publishing Company.

Fascetti, S., Bernardo, L., Cesca, G., Codogno, M. & Puntillo, D. 1988. *Carta fisionomico-strutturale della vegetazione della comunità montana della "Sila Greca"*. Orto Botanico - Università della Calabria.

Forman, R.T.T. & M. Godron 1986 - *Landscape Ecology*. New York: Wiley.

Hudson, W. E.(ed.) 1991. *Landscape Linkages and Biodiversity*. Defenders of Wildlife, Island Press, Washington, D.C.; Covelo, California

Italian Society of Biogeography, 1984. Il popolamento dell'Appennino meridionale. Dall'Appennino campano alle Serre calabre. *Biogeographia*, 10 (NS): pp. 829.

La Greca, M. (ed.) 1982. Ambienti mediterranei I. Le coste sabbiose. *Quad. str. zoocen. terrestri, 3. Coll. P. F. CNR "Promozione della qualità dell'Ambiente"*: 212 pp.

Larsson, S. G. 1939. Entwicklungstypen und Entwicklungszeiten der dänischen Carabiden. *Entom. Medd.*, 20:277-554.

Leser, H. 1976. *Landschaftsökologie*. Stuttgart: Ulmer.

Lincoln R.J., G.A. Boxhall & P.F. Clark 1993. *A dictionary of ecology, evolution and systematics*. Cambridge University Press.

Lindroth, C. H. 1949. Die fennoskandischen Carabidae. III. Allgemeiner Teil. *Göteborgs Kungl. Vetenskap. Vitterh. Samh. Handling.* S. B, 4(3):1-709.

Magistretti, M. 1962. Coleotteri Cicindelidi e Carabidi della Sila. *Mem. Mus. Civ. St. Nat. Verona*, 10:121-194.

Majer, J.D. & Greenslade, P. (Coordinators) 1988. Soil and litter invertebrates. In: *Mediterranean-type Ecosystems a data source book.* (Specht R.L., ed.), Kluwer Academic Publishers: 196-226.

Massa, B. & Mingozzi, T. 1991. Considerazioni sulla conservazione e la gestione della fauna in Italia (Uccelli e Mammiferi). In: *La gestione dei parchi e delle aree protette* (Pranzini E. & Valdrè G., eds.), Ed. delle Autonomie, 29: 82: 103.

Mingozzi, T. 1995. L'avifauna della Riserva Naturale Orientata "Valle del fiume Argentino" (Orsomarso, CS): analisi faunistico-ecologica ed indicazioni conservazionistico-gestionali. *Collana Verde, Min. Agr. Foreste*, Roma: Ministero Agricoltura e Foreste.

Mingozzi, T. & Brandmayr, P., 1991. L' évaluation cartographique des ressources faunistiques: un exemple appliqué aux ornithocénoses d'une vallée alpestre. *Rev. Ecol. Alp.*, **1**: 1-21.

Naveh, Z. & Liebermann, A.S., 1984. *Landscape Ecology. Theory and applications.* New York: Springer.

Petch J.R. & J. Kolejka, 1993 - The tradition of landscape ecology in Czechoslovakia. In: Haines-Young R., D. R. Green & S. Cousins (eds.): Landscape ecology and geographic informations systems, pp. 39-56. London, New York, Philadelphia: Taylor & Francis.

Pignatti, S., 1979. I piani di vegetazione in Italia. *Giorn. Bot. It.*, 113: 411-428.

Pignatti S., 1994 - *Ecologia del Paesaggio*. Torino: UTET.

Pizzolotto, R. & Brandmayr, P. 1990. Ground beetle coenoses in the landscape of the Nebrodi mountains, Sicily (Coleoptera, Carabidae). In: Bruno Massa (ed.): "Omaggio a G. E. Hutchinson". *Naturalista siciliano*, 4, 14(suppl.): 51-64.

Pizzolotto, R., Mingozzi, T., Cagnin, M., Brandmayr, P., Tripepi, S., Aloise, G., Barbieri, A. & Scalzo, A. 1991. Effetti della ceduazione periodica del castagneto sulle comunità di Coleotteri Carabidi, Rettili, Uccelli e micromammiferi terricoli. *S.IT.E. Atti*, **12** (1): 449-453.

Paarmann, W. 1979. Ideas about the evolution of the various annual reproduction rhythms in Carabid beetles of the different climatic zones. Misc. Pap. Landbouwhogesch. Wageningen, 18:119-132.

Thiele, H.U. 1977. Carabid Beetles in Their Environments. Zoophysiol. and Ecol., 10. Berlin, Heidelberg, New York: Springer.

Vigna-Taglianti, A. 1993. Coleoptera Archostemata, Adephaga 1 (Carabidae). Minelli, Ruffo & La Posta (eds.) *Checklist delle specie della Fauna D'Italia*, 44. Bologna: Calderini.

Whittaker, R. H. 1973. Ordination and Classification of Communities. In: Tüxen R. (ed.): *Handbook of Vegetation Science*. The Hague: V. Junk.

Population and Community Ecology for Insect Management and Conservation, Baumgärtner et al. (eds)
© 1998 Balkema, Rotterdam, ISBN 90 5410 930 0

Temporal spacing, competition and aggregation in *Aphodius*-dominated dung beetle communities

C. Palestrini, A. Rolando & E. Barbero
Dipartimento di Biologia Animale, Turin, Italy

ABSTRACT: This paper deals with temporal spacing, competition and aggregation of adult beetles in dominated communities inhabiting cattle dung pats. Data from an Italian Alpine area are re-analyzed to point out interspecific temporal spacing. Evidence for the presence or absence of competition is reviewed from north-temperate and Alpine communities. Beetles were collected from pastures at Palanfrè in the Valle Grande (Cuneo Province, north-western Italy) at about 1400 m a. s. l. from June to September in 1994. The entire material collected comprises 9000 adult beetles of 25 different species. The dendrogram produced by Cluster Analysis clearly shows that species can be classified into three groups: 1) species arriving early in the season, occurring in large numbers in June-July (*A. foetens*, *A. haemorrhoidalis*, *A. fossor*), 2) late-arriving species, concentrated in September (*A. borealis*, *A. uliginosus*, *A. sticticus*, *A. corvinus*, *A.contaminatus*), 3) species occurring throughout the period (*A. rufipes*, *A. erraticus*, *A. rufus*, *A. depressus*). *O. fracticornis* was classified as a late-arriving species, but at a rather high Euclidean distance from the other species of the same group. *Anoplotrupes stercorosus* was clustered apart from all the other species. The ordination of species obtained by Principal Component Analysis placed *A. erraticus* well apart from the other species, hence emphasizing the dominance of this species, which accounted for about half the individuals collected.

The present data together with those from other north-temperate and Alpine communities give no convincing evidence that competition is a major organizing force in *Aphodius*-dominated adult dung beetle communities. In many areas, like in the Alps, the low number of adults per dung pat would maintain intra- and interspecific coexistence at a non-competitive level. Occasional situations of local abundance (also observed in the Alps) are probably not important for competition because adults use only small amounts of dung for feeding or reproducing and, at any rate, they can emigrate to less crowded droppings. Given this picture, aggregation of adults might be interpreted as a mechanism, whose cause is still unknown, that works independently of competition. Because of the overall low relevance of competition, there is no reason to interpret dung beetle aggregation on the basis of the Aggregation model.

INTRODUCTION

Dung beetles represent a rather intriguing ecological problem since they appear to challenge the logic of interspecific coexistence.

The classical idea of resource partitioning, derived from the niche theory, predicts that ecologically similar species can coexist provided there is sufficient dissimilarity between the resources they use (MacArthur 1972, Schoener 1974, Pianka 1976). However, the dung is the only resource used by beetles when they inhabit the pats and, even though temporal and spatial interspecific segregation have been described (Holter 1982, Yasuda 1984, Yoshida & Katakura 1985, Hanski 1986, Lumaret & Stiernet 1991, Wassmer 1994, 1995), a large degree of interspecific overlap in resource use remains in most communities (Holter 1982, Giller & Doube 1994, Palestrini et al. 1995). Intra- and interspecific competition are known to occur, especially in tropical and subtropical areas where the dung is rapidly used by beetles (Giller & Doube 1989, 1994, Hanski & Cambefort 1991b).

Given this picture, it is somewhat surprising that dung beetles tend to aggregate in droppings. However, intraspecific aggregation is often higher than interspecific aggregation, and certain theoretical models have used aggregation as a basic premise to explain the coexistence of competing species. The so-called Aggregation model (Shorrock & Rosewell 1986, Ives 1988), in particular, predicts that an increase in the relative strength of intraspecific aggregation with respect to interspecific aggregation tends to facilitate coexistence. It must be underlined, however, that this model has been proposed for organisms that develop in ephemeral patches of food, which, in our case, would be the larvae of dung beetles, not the adults.

Species' dimensions and breeding biology significantly affect coexistence among dung beetles. Intra- and interspecific competition may be severe in large Scarabaeidae dung beetles which are able to rapidly remove a portion of the resources in a dropping for their exclusive use (Giller & Doube 1989, Hanski & Cambefort 1991b, Kohlmann 1991). These beetles reproduce by tunnelling and burying dung beneath the dung pat (tunnelers, paracoprids) or by pushing fragments of dung to a previously dug burrow far from the dung pat (rollers, telecoprids) (Halffter & Matthews 1966, Halffter & Edmonds 1982, Zunino & Palestrini 1986, Zunino 1991). Conversely, competition may or may not be strong in small Aphodiidae dung beetles which usually reproduce within the dung (dwellers, endocoprids) (Hanski & Cambefort 1991b).

Communities of the two groups approximately correspond to different types of dynamics. "Lottery dynamics", which refers to a few individuals dominating and partly monopolizing a resource patch, may be applied to large Scarabaeidae. "Variance-covariance" dynamics, which refers to assemblages showing an extensive variation in the number of individuals between habitat patches, may be applied to small Aphodiidae which cannot monopolize habitat patches (Hanski 1991a). This type of dynamics would work according to the ideas of the Aggregation model. In fact, increasing intraspecific aggregation (variance) across patches would amplify intraspecific competition relative to interspecific competition and coexistence would be facilitated; increasing interspecific aggregation (covariances) would work in the opposite direction (Hanski 1991a). Nevertheless, the dynamics of dung beetles cannot be easily divided according to these two models and they span a continuum from the lottery to variance-covariance types.

This paper deals with temporal spacing,competition and aggregation of adult beetles in *Aphodius* dominated communities inhabiting cattle dung pats. These communities are characterized by the dominance of small *Aphodius* beetles which can coexist with a few larger Geotrupidae and Scarabaeidae. Data from an Italian Alpine area which were presented in a previous paper are re-analyzed to point out interspecific temporal spacing. Evidence for the presence or absence of competition is reviewed from north-temperate and Alpine communities.

STUDY AREA AND METHODS

Beetles were collected from pastures at Palanfrè in the Grande Valley (municipality of Vernante, Cuneo Province, north-western Italy) at about 1400 m a. s. l. from June to September in 1994, at a time when the pastures were utilized by cattle (420 cows). Woods of the area are characterized by beech (*Fagus silvatica*) up to 1600 m a. s. l. and by *Pinus uncinata* at higher altitudes.

Rainfall is mainly concentrated during spring and autumn. August and July are the warmest months whilst January and February are the coldest.

Six sampling bouts were carried out from 27 June until 28 September. During each bout, 90 artificial dung pats (fresh weight about 1.4 kg, 26 cm in diameter) were made from thoroughly mixed 0-9 h-old cattle dung and distributed evenly within the experimental pasture. At intervals of 1-3-5-7-9-11 and 13 days after the distribution, twelve dung pats were sampled. Hence during each sampling bout 84 dung pats were sampled (12 pats x seven intervals) for a total of 504 dung pats during the whole sampling period (84 pats x six bouts). Beetles were extracted by flotation and tunnelers were looked for by digging to a depth of about 20 cm beneath the dung pats.

Principal Component Analysis (PCA) was performed to obtain an ordination of species according to their temporal distribution and abundances while Cluster Analysis provided a classification according to their interspecific temporal overlap.

The temporal distribution of species was analyzed in detail with the different sampling days separated. Each element of the Cluster Analysis triangular matrix (species x species) represented the temporal overlap between two species as calculated by the formula

$$C_{ih} = 1\text{-}0.5 \mid p_{ij} - p_{hj} \mid \text{(Colwell \& Futuyma, 1971)} \qquad (1)$$

where p_{ij} and p_{hj} are the proportion of individuals of species i and h collected on the same day j. The Euclidean distance and the average linkage method were used in the Cluster Analysis. Each element X_{ij} of the PCA rectangular matrix (species x sampling days) represented the percentage of abundance of the i-th species on the j-th sampling day. Normalization was obtained by subtraction of 100/number of species from all X_{ij} (Digby & Kempton, 1987). Rotation (varimax) of factors and factor scores was carried out. Factors were considered significantly correlated with sampling days when the rotated loading was higher than 0.9.

Computations for PCA and Cluster Analysis were carried out by the Systat (1985) computer package.

RESULTS

The entire material collected comprises about 9000 adult beetles of 25 different species. Apart from two species of Geotrupidae (one *Anoplotrupes* and one *Geotrupes*), three of Scarabaeidae (two *Onthophagus* and one *Euoniticellus*), the remaining twenty species belong to the family Aphodiidae (one *Euheptaulacus* and 19 *Aphodius*). Species of the genus *Aphodius* accounted for 93.4% of collected individuals, *Onthophagus* 5.7%, *Anoplotrupes* 0.6% and *Geotrupes* 0.3%. Since very few beetles were sampled in 9-13 day-old pats (Palestrini et al., submitted), only four dung ages (1,3,5,7 day-old dung pats) were considered during each sampling bout. Accordingly, 24 sampling days (six sampling bouts x four days) in total were taken into account. Eleven species with less than 40 individuals collected through the 24 days were discarded from the analysis because their presence in the dung pats was occasional. The phenologies and the total number of individuals collected for each of the fourteen remaining species are shown in Fig.1.

The dendrogram produced by Cluster Analysis (Fig.2) clearly shows that species can be classified into three groups: 1) species arriving early in the season, occurring in large numbers in June-July (*A. foetens*, *A. haemorrhoidalis*, *A. fossor*), 2) late-arriving species, concentrated in September (*A. borealis*, *A. uliginosus*, *A. sticticus*, *A. corvinus*, *A. contaminatus*), 3) species occurring throughout the period (*A. rufipes*, *A. erraticus*, *A. rufus*, *A. depressus*). *O. fracticornis* was classified as a late-arriving species, but at a rather high Euclidean distance from the other species of the same group. *Anoplotrupes stercorosus* was clustered apart from all the other species. Within each group temporal separation was usually poor and, in fact, Euclidean distances were very low.

Rotated loadings produced by PCA indicated that Factor(1) was positively correlated with the 1st and 2nd sampling days of the 1st, 2nd, 3rd and 5th sampling bouts, Factor(2) was negatively correlated with the 2nd and 3rd days of the 5th and 6th bouts and Factor(3) was positively correlated with the 3rd and 4th days of the 3rd and 4th bouts. Factor(1) explained 34.2% of total variance, Factor(2) 19.1% and Factor(3) 17.4%. The ordination of species obtained by PCA (Fig. 3) placed *A. erraticus* well apart from the other species on account of the high scores of Factor(1). The other thirteen species were plotted separately along the Factor(2) and Factor(3) gradients.

DISCUSSION

Temporal spacing

Seasonality is well documented in local assemblages of *Aphodius*, which usually show some phenological differences between species (Holter 1982, Yasuda 1984, Yoshida & Katakura 1985, Hanski 1986, Lumaret & Stiernet 1991, Palmer 1995). Conversely, even though differences between species with regard to their succession in the droppings (i.e. the occurrence in relation to dung age) have been observed, it has been concluded that species are poorly segregated along the successional axis (Holter 1982, Hanski 1991b).

A. fossor N=86

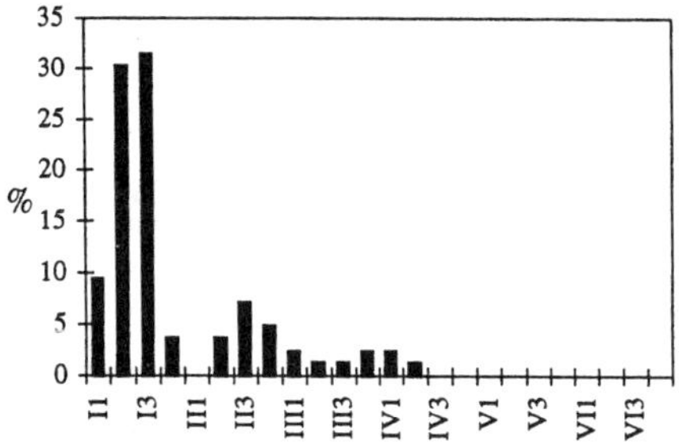

A. borealis N=512

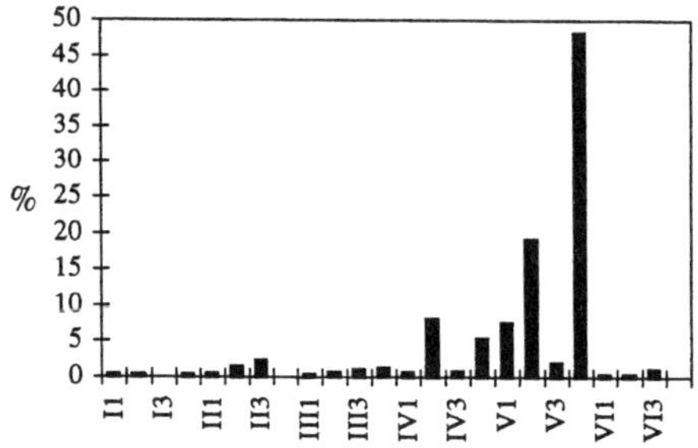

A. foetens N=164

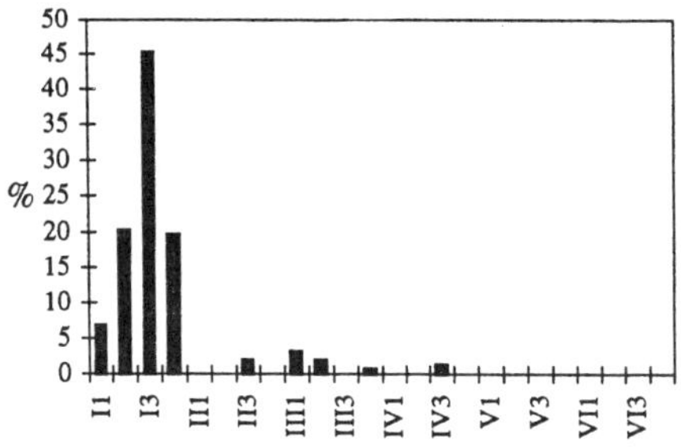

A. uliginosus N=333

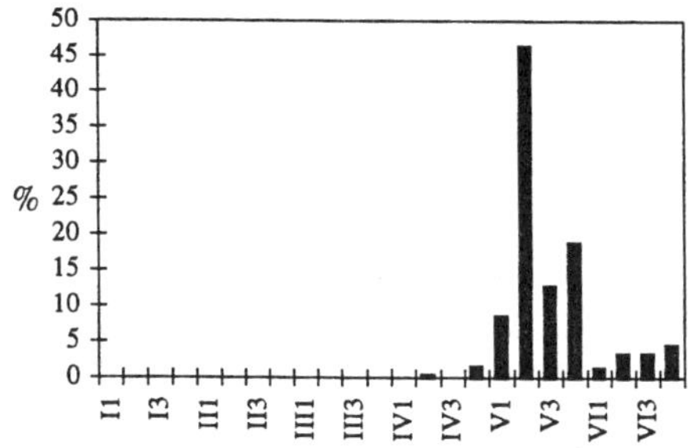

A. haemorrhoidalis N=596

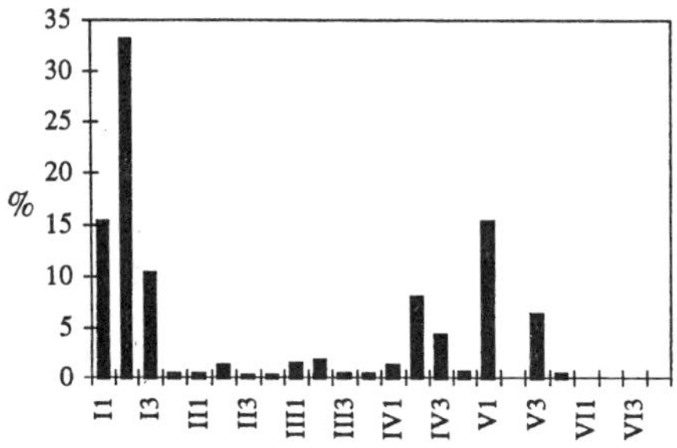

A. sticticus N=141

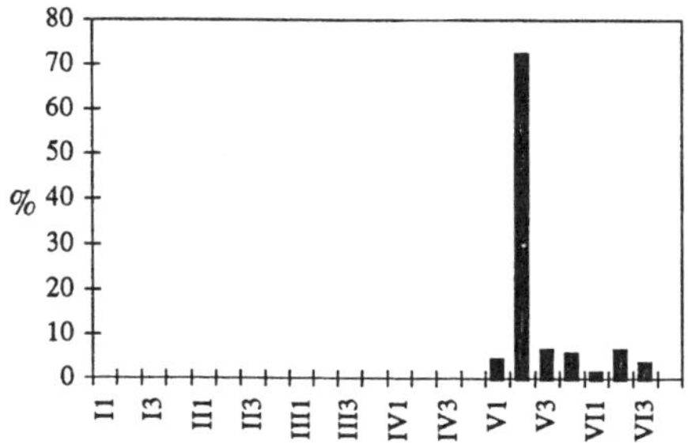

A. rufipes N=851

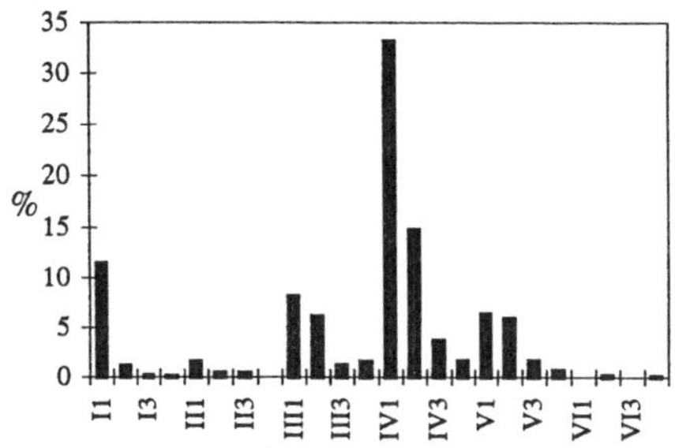

A. contaminatus N=78

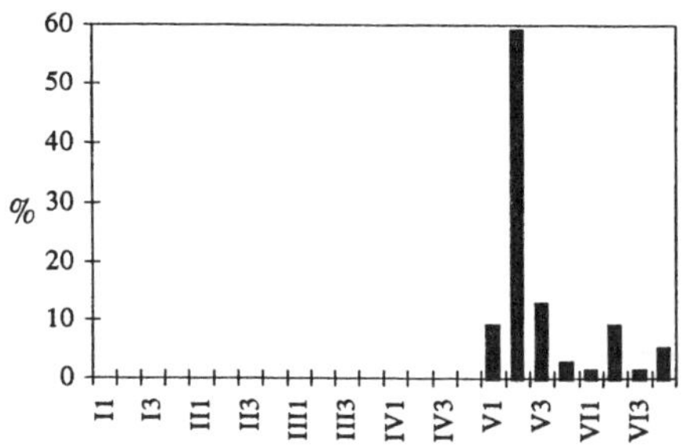

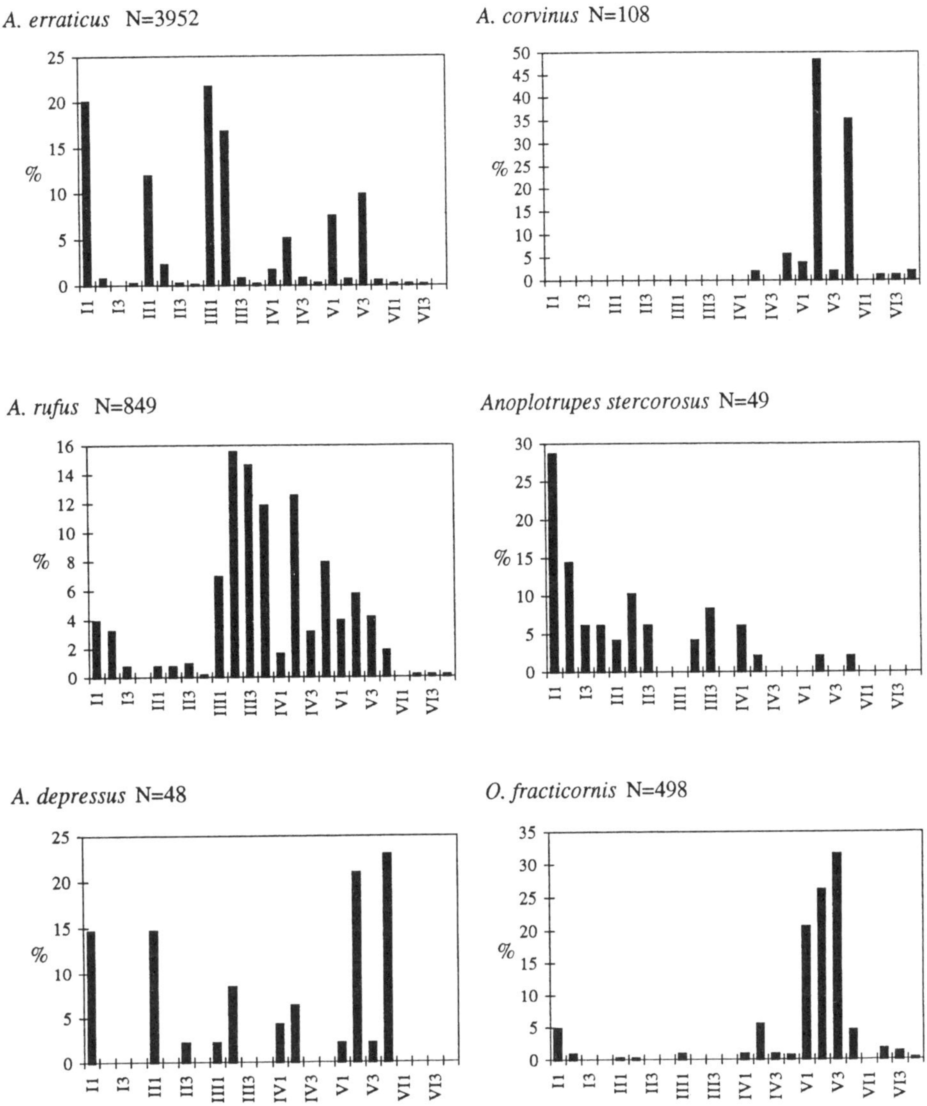

Figure 1. Percentages of occurrence of the fourteen species selected from June to September 1994 at Palanfrè. The 24 sampling days are divided according to the 6 sampling bouts (I-VI); during each bout dung pats were sampled at 1,3,5 and 7 days after their deposition. N= Number of individuals collected.

At Palanfrè interspecific temporal differences pointed out by Cluster Analysis mostly depended upon seasonality. The three clusters obtained reflected differences in seasonal occurrence, i.e. early-arriving species (June-July), late-arriving species (September) and species occurring throughout the period. Within each seasonal group, however, overlap was rather high and species occurred evenly in pats of equal age. It must be underlined that *A. stercorosus* was clustered apart from all other species and that *O. fracticornis* was rather well separated from the other *Aphodius* species of its cluster (late-arriving species). These two differentiations suggest that even though the temporal habits of Geotrupidae and Scarabaeidae partly overlap with those of Aphodiidae, some differences can persist. The

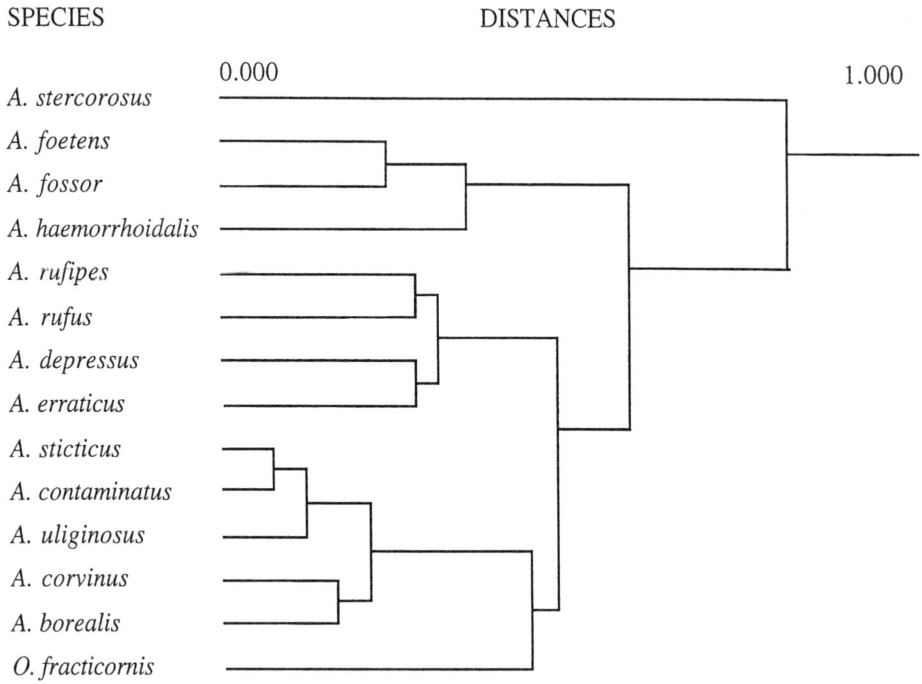

Figure 2. Dendrogram of temporal similarity in the use of the dung obtained through Cluster Analysis. Distances are Euclidean distances (average linkage method).

three most abundant species (*A. erraticus*, *A. rufus* and *A. rufipes* in decreasing order) occurred throughout the reproductive season. They were also the species with the highest mean numbers of beetles per pat (Palestrini et al., submitted). It could be suggested that temporally generalist species have an ecological advantage over the specialist ones.

Cluster Analysis was carried out with data that were independent of specific abundances. PCA, to the contrary, used data reflecting specific abundances and, accordingly, emphasized the dominance of *A. erraticus*, which accounted for half the individuals collected. This species monopolized the variance of data explained by Factor(1) which, in fact, was correlated with all the days during which *A. erraticus* was abundantly sampled. The dominance of *A. erraticus* is particularly apparent when we consider that 11 species were not taken into account because of their occasional presence in the droppings. Accordingly, even though at Palanfrè the guild of dung beetles was composed of 25 species, only one would characterize and eventually control the organization of this community (perhaps via intra- and interspecific interactions).

Competition

It has been suggested that temporal spacing may be due to interspecific competition (Hanski 1980). Palmer (1995) showed that *Aphodius* species on the Balearic Islands were more evenly distributed in terms of temporal overlap than expected by chance. He pointed out that *A. ictericus* occurs from July to August in Sweden whereas it occurs from April to June in the Balearics. Citing *A.borealis* from Finland and England (Hanski 1980), he concluded that seasonal occurrence is labile and that it could be modified also by species-species interactions. However, given the climatic differences between Sweden and the Balearic Islands, differences in the occurrence of some *Aphodius* species could well be expected. It must be emphasized that at the Balearics, because of the mild climate, the growing season spans the whole year (Palmer 1995). Hence, comparisons with areas (like Sweden) characterized by a very short breeding season are quite misleading. Moreover, it has been

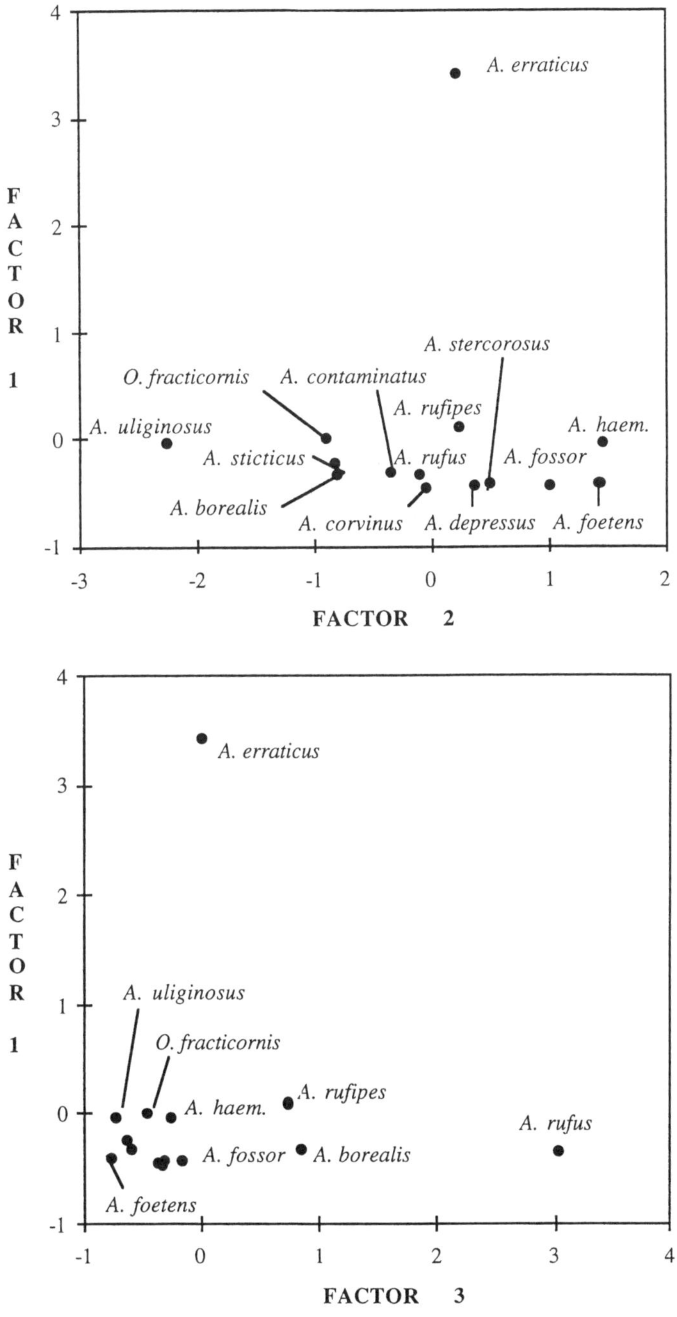

Figure 3. Principal Component Analysis. Plots of scores for the dung beetle species on the first three factors (F1 x F2 and F1 x F3)

shown that the seasonal occurrences of many *Aphodius* species remain substantially constant (apart from minor differences easily attributable to climatic factors) all over Europe, irrespective of the systematic composition of the different dung beetle communities (Wassmer 1994, Palestrini et al. 1995).

Inferring the existence of competition from seasonal distributions may be a purely academic exercise if we do not consider the availability of resources. In fact, a necessary pre-condition for the occurrence of intra- and interspecific competition is the scarcity of resources, which must be limiting. Numbers of adult beetles in *Aphodius*-dominated communities may be low in relation to resource (dung) availability. Pats may remain apparently undigested since the overall beetle biomass is much lower than the dung biomass and the amount of dung used by each beetle is rather low. It is well known that decomposers (bacteria and fungi) and detritivores (dung beetles, in our case) tend to have low population densities when temperatures are low (Remmert 1973, 1978, Begon et al. 1986), and this may be common in the Alps. At Palanfrè, for instance, the mean number of beetles per 3-5 day-old pat was less than 30 individuals (Palestrini et al., submitted), whereas in another area in the south-western Italian Alps the number was even lower, i.e. 4 individuals per dung pat (Palestrini et al. 1995). Conversely, on some occasions, several species of *Aphodius* are so numerous in cattle pats that they disintegrate the pat in a matter of hours or days (Hanski 1991b).

Adult dung beetles may directly compete for dung either because it is used as food or because it is used for their reproduction. With regard to the dung as food, it is usually assumed that dung beetles occurring in cattle pats feed on the dung. However, this may not be true. A few species are known to be saprophagous at the adult stage, a few others at the larval stage. At Palanfrè, for instance, adults of some species were also caught by pitfall traps baited with progressively putrefying meat. In particular, *Anoplotrupes stercorosus* (notwithstanding its scientific name suggesting a coprophilous habit) was sampled more frequently in these traps than in dung pats. With regard to the dung as nest material, it must be considered that *Aphodius* species do not build nests, even though exceptions have been described (Zunino & Barbero 1990, Zunino et al. 1994, Barbero & Palestrini 1995, Palestrini et al. 1997). Accordingly, eventual direct competition among adults for dung may be diminished both by the different feeding habits and by the scarce use of dung as nest material.

Species' dimensions may affect coexistence as well, with larger species dominant over the smaller ones (Hanski & Cambefort 1991b). To check this hypothesis, some manipulative tests were carried out at Palanfrè. Dung pats were covered with metallic nets whose mesh was chosen in order to prevent larger beetles from reaching the dung and, conversely, some dung pats were manipulated by the addition, immediately after the deposition of the dung, of 1-5 pairs of large or medium-size beetles. The results obtained suggested that small beetles were not excluded by larger ones, probably because they belonged to different ecological groups (Palestrini et al., submitted).

Intra- and interspecific competition in larvae have been poorly studied. Yet, it is likely that interactions are more severe among larvae than among adults. Second instar *Aphodius* larvae require 2 to 4 hundred times their volume of dung for optimal development (Landin 1961). Hence, competition for dung and space may be commonplace in larvae.

Aggregation

Intraspecific aggregation is well known in dung beetle communities (Hanski & Cambefort 1991a). Even though a negative association between species, or none at all, has been observed (Hanski 1986, Kohlmann 1990), the usual pattern of interspecific aggregation is some positive association (Hanski & Cambefort 1991a).

The results obtained at Palanfrè suggested that beetles of the same species tended to aggregate in certain pats and simultaneously occurring species aggregated in the same pats, that is, there was a clear tendency both for intra- and interspecific aggregation (Palestrini et al., submitted). Even though many hypotheses have been advanced, there is not as yet consensus about the cause of aggregation. It has been suggested that the ultimate cause of aggregation is mate-finding (Holter 1982) or, alternatively, that individuals might aggregate because the very presence of beetles within a dung pat would modify its chemical status and improve its quality (Palestrini et al., submitted).

Whatever the answer, life history could be usefully considered to understand aggregation in adults. The need of reproducing is likely an important factor controlling movements in dung beetles. In this regard, sexes could display different behaviours. Emigration from droppings seems to be density-dependent, with beetles moving out from the most crowded droppings (Landin 1961). It is possible that emigrating individuals are mostly mature females which have already copulated. Females are expected to lay eggs in non-crowded pats to assure survival of their offspring and, in fact, the rate of oviposition of certain species is known to decrease with the increasing density of beetles (Holter 1979). Also, the distribution of eggs in cattle pats was significantly less aggregated than the distribution of adults (Holter 1979). Therefore, we cannot exclude that aggregation is due to the behaviour of males and females prior to copulation, while mature females before laying may act against aggregation.

Competition and aggregation act in opposite ways; the former promotes interspecific segregation while the latter opposes it. Yet, paradoxically, they both occur in dung beetles. The Aggregation model seems to suggest a solution to the dilemma by predicting that interspecific coexistence may be facilitated when intraspecific aggregation is higher than interspecific aggregation (Shorrock & Rosewell 1986, Ives 1988). However, the model has not yet been satisfactorily tested for dung beetles, especially since studies have exclusively considered adults. On the contrary, the Aggregation model of coexistence has been formulated with organisms that develop in ephemeral patches of food (i.e. larvae of dung beetles, carrion flies and *Drosophila*) (Sevenster 1996). Indeed, if eggs of Aphodiidae in cattle pats are significantly less aggregated than adults, as suggested by Holter (1979) for *A.rufipes*, the Aggregation model could not properly be applied to dung beetles.

To conclude, there is no convincing evidence that competition is a major organizing force in *Aphodius*-dominated adult dung beetle communities. In many areas, like in the Alps, the low number of adults per dung pat would maintain intra- and interspecific coexistence at a non-competitive level. Occasional situations of local abundance (also occasionally observed in the Alps) are probably not important for competition because adults use only small amounts of dung for feeding or reproducing and, at any rate, they can emigrate to less crowded droppings. Given this picture, aggregation of adults might be interpreted as a mechanism, whose cause is still unknown, that works independently of competition. Because of the overall low relevance of competition, there is no reason to interpret dung beetle aggregation on the basis of the Aggregation model. Indeed, since this model has been proposed for organisms developing within resource patches, it is perhaps not appropriate to apply it to adult dung beetles.

REFERENCES

Barbero, E. & C. Palestrini 1995. *Aphodius (Coprimorphus) scrutator* (Herbst, 1789): descrizione della larva e note di biologia riproduttiva (Coleoptera, Scarabaeoidae, Aphodiinae). *Fragm. Ent.*, 26:341-352.

Begon, M, J. L. Harper & C. R. Townsend 1986. *Ecology. Individuals, populations and communities*. Oxford: Blackwell Scientific Publications.

Colwell, R.K. & D. J. Futuyma 1971. On the measurement of niche breadth and overlap. *Ecology* 5:567-576.

Digby, P. G. N. & R. A. Kempton 1987. *Multivariate analysis of ecological communities*. London: Chapman and Hall.

Giller, P. S. & B. M. Doube 1989. Experimental analysis of inter- and intraspecific competition in dung beetle communities. *J. Anim. Ecol.* 58:129-142.

Giller, P. S. & B. M. Doube 1994. Spatial and temporal co-occurrence of competitors in Southern African beetle communities. *J. Anim. Ecol.* 63:629-643.

Halffter, G. & E. G. Matthews 1966. The Natural History of Dung Beetles of the subfamily Scarabaeinae (Coleoptera, Scarabaeidae). *Folia Ent. Mex.* 12-14:1-312.

Halffter, G. & W. D Edmonds,. 1982. *The nesting behavior of dung beetles (Scarabaeinae): an ecological and evolutive approach.* Instituto de Ecología, México, D. F.

Hanski, I. 1980. Patterns of beetle succession in droppings. *Ann. Zool. Fennici* 17:17-25.

Hanski, I. 1981. Coexistence of competitors in patchy environment with and without predation. *Oikos* 37:306-312.

Hanski, I. 1986. Individual behaviour, population dynamics and community structure of *Aphodius* (Scarabaeidae) in Europe. *Acta Oecol., Oecol. Gen.* 7:171-187.

Hanski, I. 1990. North Temperate Dung Beetles. In: Hanski, I. & Cambefort, Y. (eds), *Dung Beetle Ecology*. New Jersey: Princeton University Press, pp. 75-96

Hanski, I. 1991b. The dung insect community. In: Hanski, I. & Cambefort, Y. (eds), *Dung Beetle Ecology*. New Jersey: Princeton University Press, pp. 5-21.
Hanski, I. & Y. Cambefort 1991a. Spatial processes. In: Hanski, I. & Cambefort, Y. (eds), *Dung Beetle Ecology*. New Jersey: Princeton University Press, pp.283-304.
Hanski, I. & Y. Cambefort 1991b. Competition in dung beetles. In: Hanski, I. & Cambefort, Y. (eds), *Dung Beetle Ecology*. New Jersey: Princeton University Press, pp.305-329.
Holter, P. 1979. Abundance and reproductive strategy of the dung beetle *Aphodius rufipes* (L.) (Scarabaeidae). *Ecol. Entomol.* 4:317-326.
Holter, P. 1982. Resource utilization and local coexistence in a guild of scarabaeid dung beetles (*Aphodius* spp). *Oikos* 39:213-227.
Kohlmann, B. 1991. Dung beetles in subtropical North America. In: Hanski, I. & Cambefort, Y. (eds), *Dung Beetle Ecology*. New Jersey: Princeton University Press, pp. 116-132.
Ives, A. R. 1988. Aggregation and the coexistence of competitors. *Ann. Zool. Fennici* 25:75-88.
Landin, B. O. 1961. Ecological studies on dung beetles (Col. Scarabaeidae). *Opusc. Entomol. Suppl.* 19:1-228.
Lumaret, J.-P. & N. Stierniet 1991. South temperate dung beetles. In: Hanski, I. & Cambefort, Y. (eds), *Dung Beetle Ecology*. New Jersey: Princeton University Press, pp. 242-254.
MacArthur, R.H. 1972. *Geographical ecology*. New York: Harper & Row.
Palestrini, C., A. Rolando & E. Barbero 1995. Analysis of temporal segregation in a dung-inhabiting beetle community at a low-altitude area of the Italian Alps. *Boll. Zool.* 62:257-265
Palestrini C., E. Barbero & A. Rolando (submitted). Intra- and interspecific aggregation among dung beetles (Coleoptera, Scarabaeoidea) in an Alpine pasture.
Palestrini C., L. Borghesio & E. Barbero 1997. A new pattern of nesting behaviour in the genus *Aphodius* (Coleoptera, Scarabeoidea). (Abstract). XXV Intern. Ethol. Congr. Vienna 20-27.VIII.1997.
Palmer, M. 1995. Testing for seasonal displacement in a dung beetle guild. *Ecography* 18:173-177.
Pianka, E. 1976. Competition and niche theory. In: May, R. M. (ed.), *Theoretical Ecology, Principles and Applications*. Oxford: Blackwell. pp. 114-141.
Remmert, H. 1973. Uber die Bedeutung warmblütiger Pflanzenfresser für den Energiefluss in terrestrischen Oekosystemen. *J. Orn.* 114:227-249.
Remmert, H. 1978. *Oekologie. Ein Lehrbuch*. Berlin: Springer-Verlag.
Schoener, T. W. 1974. Resource partitioning in ecological communities. *Science* 185:27-39.
Sevenster, J. 1996. Aggregation and coexistence. I. Theory and analysis. *J. Anim. Ecol.* 65: 297-307.
Shorrocks, B. & J. Rosewell 1986. Guild size in drosophilids: a simulation model. *J. Anim. Ecol.* 55:527-542.
Wassmer, T. 1994. Seasonality of coprophagous beetles in the Kaiserstuhl area near Freiburg (SW-Germany) including the winter months. *Acta Oecol.* 15:607-631.
Wassmer, T. 1995. Selection of the spatial habitat of Coprophagous beetles in the Kaiserstuhl area near Freiburg (SW-Germany). *Acta Oecol.* 16:461-478.
Yasuda, H. 1984. Seasonal changes in the number and species of Scarabaeid dung beetles in the Middle Part of Japan. *Jap. J. Appl. Ent. Zool.* 28: 217-222.
Yoshida, N. & H. Katakura 1985. Life cycle of *Aphodius* dung beetles (Scarabaeidae, Coleoptera) in Sapporo, northern Japan. *Japan. Environ. Sci., Hokkaido Univ.* 8:209-229.
Zunino, M. 1991. Food Relocation Behaviour: a multivalent strategy of Coleoptera. In: Zunino, M., Bellés, X. & M. Blas (eds.), *Advances in Coleopterology*. European Association of Coleopterology, Barcelona, pp. 297-314.
Zunino, M. & E. Barbero 1990. Food relocation and the reproductive biology of *Aphodius fossor* (L.) (Coleoptera, Scarabaeidae:Aphodiinae). *Ethol,. Ecol., Evol.* 2:334.
Zunino M., L. Canino & E. Coletta 1994. Feeding and nesting behaviour of *Aphodius* (Colobopterus) *erraticus* (L.) (Coleoptera Scarabaeoidae) Aphodiinae). *Ethol., Ecol., Evol.*, 6:451.
Zunino, M. & Palestrini C. 1986. El comportamiento telefágico de *Trypocopris pyrenaeus* (Charp.) adulto (Col., Scar.: Geotrupidae). *Graellsia* 42:205-216.

Population and Community Ecology for Insect Management and Conservation, Baumgärtner et al. (eds)
© 1998 Balkema, Rotterdam, ISBN 90 5410 930 0

Author index